UNDERSTANDING DATA

Principles & Practice of Statistics

DAVID GRIFFITHS

W. DOUGLAS STIRLING

K. LAURENCE WELDON

JOHN WILEY & SONS

Brisbane • New York • Chichester • Weinheim • Singapore • Toronto

First published 1998 by
JACARANDA WILEY LTD
33 Park Road, Milton, Qld 4064

Offices also in Sydney and Melbourne

Typeset in 10.5/12 pt New Baskerville

National Library of Australia
Cataloguing-in-Publication data

Griffiths, David.
Understanding data: principles and practice of statistics

Includes index.
ISBN 0 471 33734 X.

1. Statistics. I. Stirling, W. Douglas. II. Weldon, K. L. (Kenneth Laurence), 1942–. III. Title.

519.5

Cover photograph: Horizon Photo Library

Printed in Singapore

10 9 8 7 6 5 4 3 2 1

To our families

CONTENTS

LIST OF DATA SETS

Graphs only (no numerical data)

PREFACE

To the instructor

Scope and approach

This book is intended for an introductory course in the application of statistical methods. The book is suitable for a one semester course, but could be extended to a two semester course if more emphasis were given to data collection projects and supplementary computing exercises.

Statistics concerns the use of data to obtain information about real-life situations and problems. There are various statistical tools that help extract useful information from data. Techniques for summarising and displaying data are among the most important statistical tools, so we concentrate on looking at, and summarising, the various common data structures throughout chapters 2 to 5.

However, data are only of any use if they help answer questions about real life. The context of the data — their source and method of collection and the objective of the analysis — is of critical importance, as is the audience to which the information will later be presented. Chapters 6 and 7 explain the crucial role of these topics.

Finally, in most practical situations, we have to allow for uncontrolled variation — randomness. Chapters 8 and 9 introduce the strategies of inference.

Our belief is that if students think hard about information in data, and are given the tools to examine the data, they will have the basis for making proper use of data in their working lives. The extensive use of real data sets in the explanation of concepts helps to demonstrate to students that the concepts have practical relevance.

The mathematics requirements of the book are minimal, and we demonstrate most statistical concepts rather than prove theorems about them. This approach provides most students with a deeper understanding of the properties of statistical methods and their use in practical problems than do traditional mathematical explanations.

This book contains little mention of probability. Probability modelling is a very useful basis for the description of situations involving uncertainty, and therefore underlies most advanced statistical techniques. However, students must be a little more comfortable with mathematics to appreciate probability models, so we prefer to describe probability models and their use in the analysis of data in a separate book. It is important even for mathematically capable students to be immersed in data-based studies as portrayed here, so they do not come to believe that the probability models are more important than the information in the data modelled! Moreover, the role of models will be better understood if students are familiar with typical data-based questions.

Computer software

The ready availability of sophisticated statistical software has drastically changed what must be learned in statistics. Repetitious exercises involving evaluation of formulae and drawing of graphical displays by hand are no longer necessary since students will use computers to produce them. It is more important that the student learns when to use such formulae and what the resulting values imply about real-life questions, rather than memorising the actual formulae themselves. We therefore *strongly* recommend that students be allowed, and even encouraged, to rely on computers to evaluate statistical summaries and produce statistical graphics. If students do not need to concentrate on details of implementation, they are more likely to spend time thinking about what their analysis is telling them about the actual question of interest.

Because of the wide range of statistical software that might be available to students, we have not included specific instruction about computer use. The availability of computer facilities is not essential for use of this book, but computer-based exercises are a highly recommended supplement. The Macintosh-based exercises in *Statistical Exercises Using Models'n'Data*[1] complements the approach taken in this book, but other software such as MINITAB can also be effectively used.

Order and selection of topics

In a statistical study, the logical sequence of activities is:

- setting the research objective
- study design
- data collection
- data analysis
- inference
- data presentation and summary.

We have chosen a different ordering for our chapter sequence:

- data analysis
- data presentation and summary
- data collection
- setting the research objective
- study design
- inference.

1. W. D. Stirling, *Statistical Exercises Using Models'n'Data*, John Wiley & Sons, Brisbane, 1995.

This ordering was chosen for reasons of pedagogy — study design and data collection are much easier to understand after the student has an appreciation of the types of data that commonly arise and how useful information may be extracted from them. Our approach also presents topics in order of increasing conceptual difficulty, and is effective in maintaining interest. All the steps of data-based studies are presented in our book and the natural sequence for applications will be apparent by the end of the course.

It is, however, possible to vary the order of the chapters, and even to omit some chapters. The relationships between the chapters are shown in the following diagram.

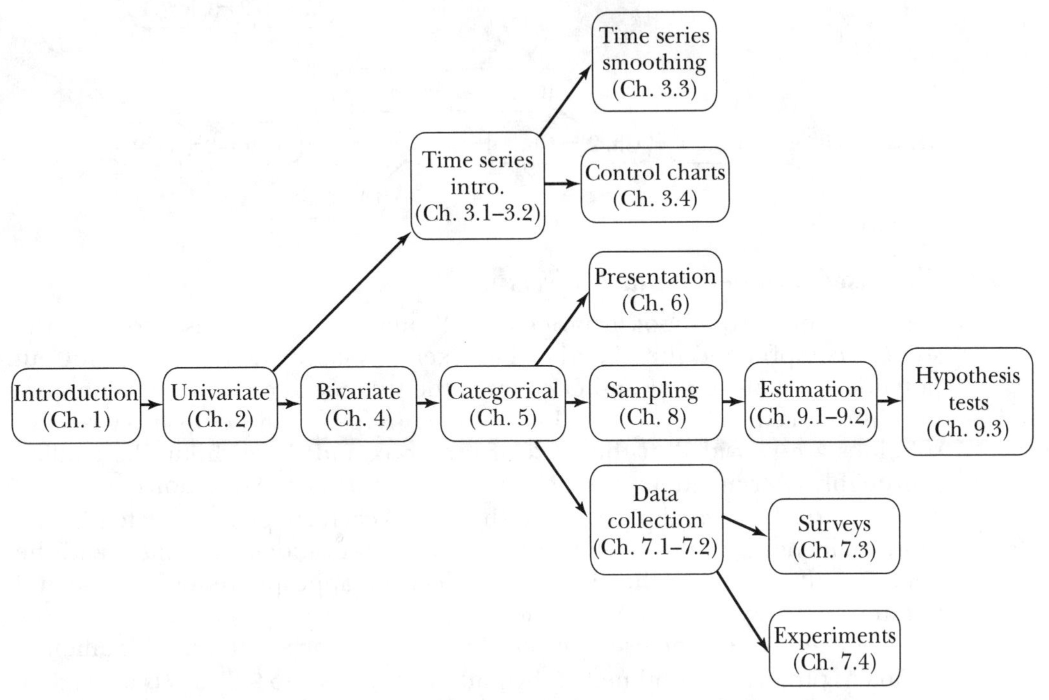

Some possible selections are indicated below.

General

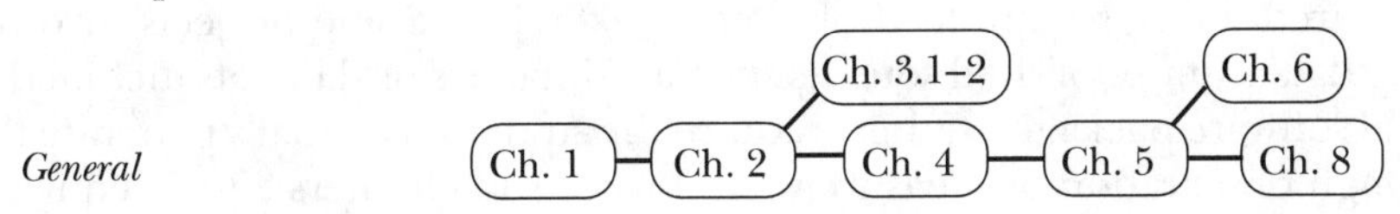

Business

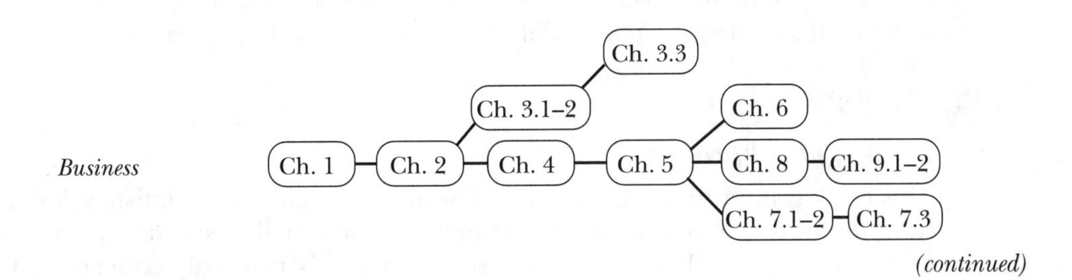

(continued)

(continued from previous page)

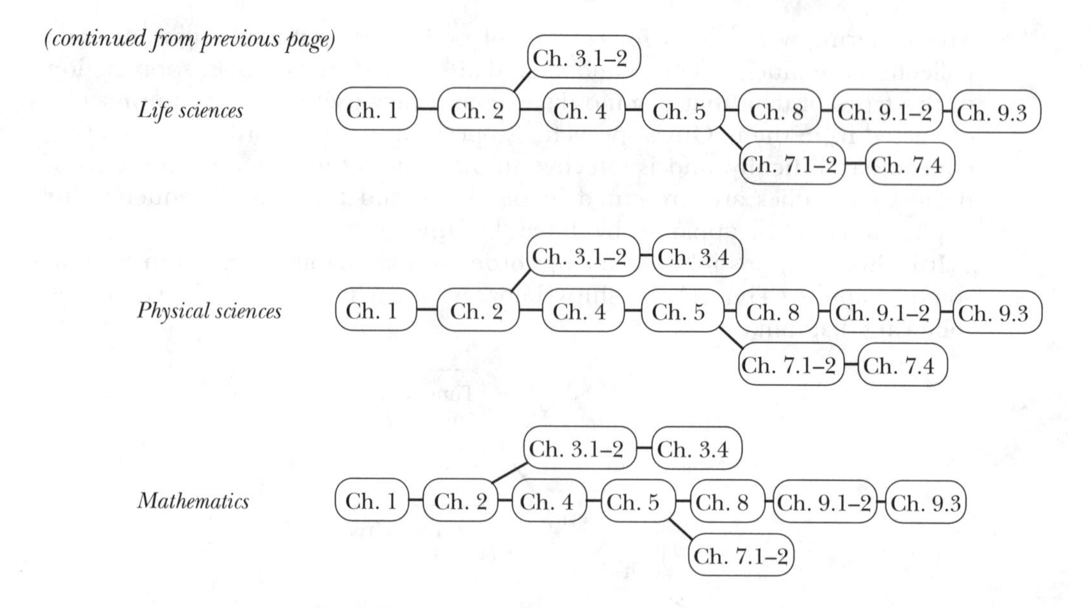

Exercises, problems and projects

Every section of the book from chapter 2 onwards has exercises to help the student comprehend the material. The exercises clarify definitions, point out potential misconceptions and encourage the student to try out the techniques just introduced. The role of the exercises is primarily to instruct — full solutions are provided at the back of the book and these should be studied thoroughly after an attempt has been made to answer the questions.

There are also several problems at the end of each chapter. These tend to be more challenging than the exercises, requiring the student to understand the concepts introduced in the chapter, to select the appropriate tools to use and to interpret the results in novel contexts. Answers are not provided for the problems so that they can be used for term assignments or as tutorial instruments.

Finally, projects are outlined at the end of chapters 2 to 9. Projects are assignments that require the student to consider the use of statistical techniques in less structured settings than do problems or exercises. Some projects involve gathering data from published sources such as Almanacs or data sets included in statistical software packages. Others require the student to actually generate the data through observation or measurement. Projects in chapters 2 to 5 require only a very simple design and report format, since these topics are covered in more detail in chapters 6 and 7. The aim of these projects is to involve the student in the full range of activities associated with data-based research, but in a context of general knowledge such that all students will be able to participate.

To the student

Scope and approach

This book is intended as an introduction to the discipline of statistics. You may have the impression that statistics is mathematics, or that statistics is the study of numerical facts but, as you will see, statistics is primarily concerned with techniques for using data to understand real-life problems. The mathematics requirements in this book are minimal, and there are very few formulae that need to be learned. Nor will you have to memorise numerical facts. You will need to become familiar with a few basic statistical concepts, but these can be explained and understood without mathematics. The most important concepts are related to words with which you will already be familiar, such as *distributions, mean, variation, association, dependence, sample* and *test,* but these words have special meanings for statistical use, and these meanings take time to absorb.

The best way to understand these concepts, and how to use them to extract meaning from data, is to engage in some supervised experience with data. The ability to make use of facts about the context of data while the data are being analysed is important. This is why memorising the principles is not enough — you really need practice in applying them. We hope you enjoy the variety of data sets. (See 'List of Data Sets', page vii.)

Practice in applying the principles of statistics is obtained by doing exercises and problems. Exercises appear in each section from chapter 2 onwards and will assist you in learning the ideas. Answers are provided and you should study them thoroughly after making an attempt yourself. The problems in each chapter are a bit more challenging and will stretch your understanding. The projects at the end of chapters 2 to 9 are perhaps even more useful; they provide valuable experience in data collection, presentation and reporting.

Relationship to computers

Modern statistics uses computers. Good statistical software spares you from the drudgery of arithmetic and drawing graphs. It encourages you to be creative in your work since trial-and-error is feasible. As a student, it can even help you to learn by encouraging numerical experimentation. Although this book does not require the use of a computer, if you want your course to be useful you will have to gain some experience with statistical software. Programs such as *Models'n'Data* are valuable supplements to a course at this level, since they make numerical experimentation and graphical display very straightforward — *Models'n'Data,* in particular, is designed as a teaching tool. Another popular statistical software package is MINITAB, but this requires more direction if you have not used statistical software before. Your instructor may have other suggestions to introduce you to statistical computing.

Concepts, not facts

You will discover that memorisation of the *facts* in this textbook is not very useful; you really need to *understand the concepts* to use statistical methods properly. It is not enough to memorise formulae — there is no substitute for practice in using these methods in real (or realistic) settings. The good part is that, like riding a bicycle, once you learn to use the concepts, you won't forget them. The facts will fade but the concepts will stay. But, like riding a bicycle, the way to learn is with fearless attempts and practice — not just learning the rules. We hope this book gives you that kind of practical experience, with only a few falls along the way!

Acknowledgements

We would like to recognise the support given to us by our families during the writing of this book. We also thank the staff at Jacaranda Wiley for their tireless work and continued support, encouragement and assistance.

The authors and publisher would like to thank copyright holders, organisations and individuals for their assistance and permission to reproduce material in this book. Every effort has been made to trace the ownership of copyright material. Holders of information that will enable the publisher to rectify any error or omission in subsequent editions will be welcome to contact the Permissions Department at Jacaranda Wiley. In these instances, the publisher will be happy to pay the usual permissions fee to the copyright holder.

CHAPTER 1

Why statistics?

Data! Data! Data! I can't make bricks without clay.

Arthur Conan Doyle (1859–1930)
(Sherlock Holmes in *The Adventure of the Copper Beeches*)

1.1 SOME OPPORTUNITIES FOR STATISTICS

Plastics and the Top 20

In the 1968 movie *The Graduate*, Dustin Hoffman plays the role of a college graduate who returns to his home town, having excelled academically and on the sporting field. (He nevertheless might, at least initially, be described by many of today's young adults as a nerd.) His proud parents throw a welcome home party for young Dustin. At that party, he is waylaid by a business associate of his father. He harangues Dustin, saying:

'Son, I have one word to say to you: Plastics'.

Unfortunately, the talkative industrialist has more than one word to say. At the end of his diatribe, any chances of his succeeding in luring Hoffman into a career in the plastics industry have vanished.

In November 1984, a special issue, 'Science '84', of *Science* magazine listed the consensus views of 20 leading scientists as to the top 20 scientific advances of the twentieth century. One of the chosen top 20 was plastics. This may seem surprising, but a few moments of reflection might convince you that the choice was entirely appropriate. What were the other choices? Among those advances listed, some are fairly obvious:

- antibiotics
- computers
- lasers
- television

... others are somewhat contentious:

- pesticides
- IQ tests, ...

Also on the list was *statistics*. That statistics is a twentieth century advance may surprise many. Before this century, much of statistics was devoted to what was sometimes called political arithmetic; that is, the collection, aggregation, analysis and presentation of numerical information concerning the state of the nation. Such socioeconomic data remain important but statistics has grown enormously as a discipline.

There was no organised body of statistical theory before the twentieth century. It was people such as Karl Pearson, William S. Gossett and especially R. A. Fisher who put together the foundations of modern statistics. Fisher might reasonably be called the father of modern statistics. After an extraordinary, distinguished career in England in government scientific research and academia, he spent his last years working with CSIRO in Adelaide where he died in 1962.

Other key advances included the advent of the modern computer which has changed the face of statistics in the last thirty or so years (so much so that some of the very elementary methods you will meet in chapter 2 had not been developed previously). A parallel development with statistical theory and statistical computing power was the recognition of the wealth of applications — indeed, the universality — of statistics. That statistics deserves its place on the list should be of little doubt to any well-educated person. H. G. Wells was one of the best known writers who speculated on what the future would be like as a result of scientific advances. He said that 'Statistical thinking will one day be as necessary for efficient citizenship as the ability to read and write.'

Sydney 2000

In September 1993, Juan Antonio Samaranch announced that Sydney would be the venue for the 2000 Olympic Games. When, in the year 2000, should they be held? There are many determining factors but, just as it may have been important in winning the Olympic Games for Sydney, climate is one of particular importance. What, then, are the climatic factors that would lead to the choice of September?

Temperature and rainfall are important determinants of climate suitability. The temperature at the time of the Olympics should be neither too hot nor too cold. Sydney's spring climate pretty much guarantees this, but what is the evidence?

The evidence is based on the historical record and the assumption of long-term stability of the climate. The data and the assessments of those data are entirely statistical. Rainfall is somewhat more problematical than temperature because of its much greater inherent variability. But, again, the issue must be addressed with statistics. The recent historical record indicates that in Sydney, on average, there is less rain in September than in any other month. But averages do not tell us everything. How many rainy days can we expect in September 2000? And on how many of those days will there be thunderstorms? We cannot predict the future with any certainty. But we can use statistics to help our predictions.

Will Bill Clinton die in office?

There is only one way of telling whether the president of the United States will die in office. Wait and see. But there are other ways of shedding light on the question; all rely on collection and interpretation of statistical information. We could start with his age (Clinton was born on 19/8/46). We also could take account of the likelihood of his winning a second term, the security arrangements in place to minimise the chances of success of any assassination attempt, his medical history and that of his family, and so on.

Rather than attempting to answer the question directly, let us pose an easier one. What is the historical evidence on age at death of presidents of the United States, all of whom (to date) are, like Clinton, white males? We can use this and other statistical information to help assess the likelihood of Clinton dying in office.

It's not cricket

Cricket is popular in England and most of its more recent former colonies, especially in the Indian subcontinent (India, Pakistan and Sri Lanka), the West Indies and the Southern Hemisphere (Australia, New Zealand, South Africa and Zimbabwe). Second only, perhaps, to baseball, cricket is a source of a vast amount of statistical information of enormous interest to a substantial but small minority of the population; in the rest of the population, cricket statistics invoke a range of responses — hatred (long-suffering families, presumably), dislike, disinterest and mild interest.

We anticipate that many of the readers of this book would be in the last three categories, but some will have a great passion for cricket and, perhaps, cricket statistics. (To find out, statisticians might design and conduct a survey; see chapter 7.)

Most of that statistical information relates to the accumulated records of individuals and teams, but it is not on these data that we focus now. As with many other sports, the game of cricket has undergone substantial change in recent years. The sedate 'gentlemen's game' of three to five days, even then often completed without a result, still exists but with a more modern image. It now has a companion which, in an exciting (to some) one-day format, virtually guarantees a winner and provides action and entertainment which attracts large crowds and television audiences. It has borrowed and modified the colourful uniforms of baseball and the pizzazz, even the musical accompaniment, of basketball.

... and, to keep the viewer informed and interested, television has enlisted the support of statistics. Even the statistically uninformed viewer is given guidance to the current state of the game, the likely outcome and, in the event of a tight match, the tension leading to the finish. This is provided mainly through a variety of graphical presentations.

Flirting with fashion

How does a retail clothing store decide on quantities to stock for its new range of clothing? Past sales, marketing initiatives and a good buyer to choose designers and styles (and, hence, largely determine costs) are all relevant to the decision, but none of these can take the variability out of forecasting sales.

Suppose we restrict ourselves to a much simpler question. For a particular style of summer dress, say, what mix of sizes should the clothing store order? It is important to know to target market (for example, women aged 18–28) and how dress sizes vary among the women in that target population. It is equally important to know whether and how that store's customers differ from others in the population. It is well known, for example, that the cheaper and larger retail chains need to supply a wider range, especially in the larger sizes, than many of the smaller, up-market boutiques.

A simpler but similar problem arises for large organisations which supply a standard range of uniform clothing to their employees. For such organisations, information collected from a random sample of their employees (see chapter 7) enables them to order the right mix of uniform sizes. It may be cheaper to have hand-made clothing for a small subset of employees who require special sizes rather than to cater to their needs as part of the standard order.

The university lottery

In many countries, student performance in an external examination at the end of secondary education determines entry into the majority of university courses. In some disciplines at some universities, interviews and auditions are also essential parts of the selection process.

In Australia, the examinations (and hence entry to university) are State-based. However, there are strong similarities in the systems among some States and reasonable procedures for allowing transfer of results between States and, hence, gaining access to interstate universities.

We will look further at the State of New South Wales. There, most students sitting the final Higher School Certificate (HSC) examination take a suitable mix of subjects to receive an overall single mark which is the sum of the scaled marks in their best ten units. (Over the years, different constraints have been introduced. Most recently, for example, one unit of English must be included in the ten units which are used to determine the scaled total.) There is a great deal of contention as to the appropriateness of a process which converts marks in a range of subjects into a single total. Is that one number a fair measure of student performance for gaining access to preferred university courses? Is that one number relevant to students who are *not* seeking access to universities? What is the purpose of the scaling process? Does it achieve that purpose? How is it achieved?

Partly because the scaling process is poorly understood by most students, teachers, parents, community members, education bureaucrats, politicians and university academics, grave concerns are often expressed about that process and the parity between different subjects.

Combining several numbers which measure different things into a single number can lead to some difficulty of interpretation, be quite misleading or even plain dumb! The average of the numbers shown in the drawing below is just over 2040, but this statistic is of no interest or use.[1]

The scaling process for HSC examinations is a complicated statistical procedure which weights subjects according to the 'quality' of the candidature of those subjects. The aim is to achieve better parity between subjects.

1. Drawing by Dana Fradon; © 1976. *The New Yorker Magazine*, Inc.

Before the overall scaling process is implemented to obtain a single total, the results of the individual subjects are separately scaled (another statistical procedure!). The total of these subject-specific scaled marks is transformed (another statistical procedure!) to obtain the student's final overall grade — the TER, or Tertiary Entrance Rank. The transformation process involves ranking the students from first to last (another statistical procedure!) and then grouping those ranks into percentile bands (another statistical procedure!). Students in the top band receive a TER of 100; those in the second band receive a TER of 99.95; in the third band the TER is 99.90, and so on. Thus, each band contains (approximately) 0.05% of the students who qualified to obtain a single overall mark.

In conjunction with interviews (for some courses) and prerequisites (for example, in English and Mathematics), the TER determines entry into university courses. Each course has a cut-off or minimum TER. Applicants whose TER equals or exceeds the cut-off and who meet any further prerequisites gain entry to the course. To ensure that all this happens, the statistical procedures which are needed to carry out all these processes are computerised. This is a massive exercise which needs to, and does, work smoothly.

There are other statistical procedures used in association with the Higher School Certificate. For example, quality control of the marking process (see chapter 3) is necessary to ensure that different markers assess equally or as nearly equally as is humanly possible. (Because of the size of the candidature, different markers are needed to mark all the papers in any examination, and even to mark an individual question. Imagine one person marking, say, 20 000 papers!)

With the rider that many people believe that there is no sense in trying to obtain such parity and that the concept of a single overall mark is inappropriate and unfair, it is generally accepted that the scaling process substantially meets these parity aims. There is, however, vigorous and healthy debate as to whether the relative scalings of different subjects over-reward the candidates of the more difficult subjects. (The relativities between different mathematics subjects, different English subjects, different languages, science and humanities and so on are all matters of contention.)

Why is there so much contention? It is a simplification, but we will summarise this by saying that the statistical model (see chapter 9) which underlies the scaling process is, at best, a good approximation. The model is poorly understood and often misrepresented by both advocates and opponents of the scaling process. A detailed discussion of the model is beyond the scope of this book so, tantalisingly, we must leave it there.

We are the champions!

In many team sports, the annual competition consists of two parts: the regular season and the finals series or play-offs. In the regular season, each team plays the same number of games. At the end of the regular season, teams are ranked according to their success in all games played. A chosen number of top teams then play in a finals series to determine the champion team. In the finals series, teams are progressively eliminated until just two teams are left to play for the championship.

What does this have to do with statistics?

Before answering this question, let us reflect on the variety of ways in which a competition may be organised. In some sports, more than one competition is contested independently during the season. Some are knockout competitions in which teams are eliminated in each round of the competition. Others are based on a regular season. The results during the regular season are used to rank the teams; the top ranked team is the one which wins the most matches during the regular season (or is allotted the most points in a system which allocated different numbers of points for wins, draws and losses). In competitions with no finals series, the champion team is the top ranked team from the regular season. In other competitions, the top few teams compete in a knockout finals series to choose the champion team. The format for the finals series can be the same as that of a knockout competition but, often, a more convoluted system is adopted.

We will examine further a competition that consists of a regular season followed by a finals series. Statistics can be used to separately investigate the two parts of the competition.

The regular season

If each team plays each other team the same number of times during the regular season the competition's draw is balanced and fair.[2] Balance is a principle of statistical design (see chapter 7) whose application ensures that no team is disadvantaged compared to others.

In some sports, the number of teams in the competition is such that if teams play each other twice, the season would be too long. If they play each other twice, the competition is balanced for home-and-away effects, an impossible outcome if they play each other only once. With teams playing each other just once, the competition could achieve a lesser degree of balance if each team played the same number of home matches. Since a season in which teams play each other just once might be too short, in some competitions each team plays every other team more than once but less than twice; there is then a different statistical design or balance problem.

Suppose, for example, there are 16 teams, but the regular season consists of 22 rounds. Each team will play all 15 other teams once and then play 7 of the teams a second time. It is desirable to balance the competition draw to ensure that the 7 teams that your team plays twice are of comparable average strength to the 7 teams played twice by any other team. This notion is familiar to sports administrators but, in general, the statistical design principles which allow fine tuning of the draw to achieve the best possible balance are not.

---

2. Even in such a draw, some interesting statistical design issues can arise. For example, if your team follows the competition 'tough' team through the draw and therefore each week plays opponents who played the tough team in the previous week, the opponents might be weakened through injury, thus advantaging your team. Statisticians are able to use the principles of experimental design to eliminate such undesirable residual effects.

Finals series

We will now turn to the finals series. Fashions and finances have dictated many changes in the structure of such series over the years. Typically, we have seen increases in the number of teams involved in the finals series. We have also seen a variety of procedures introduced with the intent of giving some advantage to teams which finished nearer the top during the regular season.

The simplest possible arrangement for a finals series is for the two top teams to play one match. This could be played at the home ground of one of the two teams or at a neutral venue. In some sports, this single match is replaced by a best of three, or best of seven (or whatever) series of matches. In addition to the financial incentive of extra matches, the notion of chance (or probability, see chapter 8) underlies the motivation for a series of matches between the two top teams. There are enough uncertainties that the 'luck of the bounce' or some other circumstance might prevent the 'best' team from winning a single match. But how many matches should be played to ensure that the best team wins the championship? Is three sufficient, or seven? Or do we need as many as 101 matches? (And why play an odd number of matches?)

To answer such questions, we need to introduce probabilistic models (see chapter 8) for the number of matches won by a particular team. Probability allows us to pose and answer a variety of questions relating to such series of games. For example, if, in a seven match finals series, your team loses three of the first four matches, what is the chance of them coming from behind to win the series? Such questions seem to fascinate some American baseball fans, some Australians (who, according to myth, will bet on which of two flies will first land on a wall) and even some statisticians, although they generally prefer to address more meaty and important issues.

Turning to finals series involving more than two teams, a variety of systems has been used. To evaluate the relative merits of the play-off systems, we need to understand principles of experimental design and have a facility for computing probabilities of sequences of outcomes. Statistics is not only relevant, but essential, to the proper management of sporting competitions.

1.2 AN OVERVIEW OF STATISTICAL STUDIES

As we have seen through example, statistics fulfils many roles in our society. Statistics is the study of variation in data: the implications for the collection, analysis and synthesis of data and for the interpretation and communication of information contained in data which exhibit variation.

The various stages of a statistical project are:

- **design** — stating objectives, planning strategy, and data collection
- **analysis** — exploratory data analysis and modelling
- **synthesis** — drawing conclusions and presenting results.

Design

Every well-designed study begins with a clear statement of the objective of the study, and this objective must guide decisions about what data to collect and how to collect them — the design of the study. There are good and bad ways of collecting data! Any conclusions we make are only as good as the data collected.

An important consideration in the design of a statistical study is the question: 'On what measurable characteristics are we interested in obtaining information?' We may, for example, be interested in obtaining information on how people will vote at an election and, perhaps, to relate this to such things as education and socioeconomic status. The design of the study will determine from which individuals information is obtained. The set of data available to the statistical analyst consists of the values of one or more variables of interest, collected from each of these individuals. Often, we are particularly interested in comparing data from two or more groups.

Analysis

There are two stages in data analysis:

- exploratory data analysis (organising, describing, summarising)
- modelling, estimating and hypothesis testing (drawing inferences).

Exploratory data analysis (EDA) should always be the first step in analysis. It has always been part of the practice of competent statisticians but has received renewed emphasis in recent years. EDA includes what has traditionally often been regarded as the 'easy' part of the subject — descriptive statistics. However, EDA is a crucial part of statistical practice: it reveals data structure and assists the statistician to choose sound methods of statistical inference. Exploratory data analysis uses numerical summaries and graphical representation to give a quick overview and to highlight features of the data of which the statistician should take particular note. Features such as unusual values and errors or simply the pattern of the data may indicate that the usual and intended mode of analysis is inappropriate.

In order to go beyond data exploration, it is useful, and often necessary, to consider that the data collected *might* have been different — if the data were to be collected again, different values might be recorded. This *randomness* is described by a statistical *model. Statistics* is used to infer characteristics of the model (and, hence, the unknown nature of 'the world'), based on observed data.

Estimation is statistically guided, educated guessing of an unknown quantity (a parameter). Estimates are necessarily almost always wrong; that is, they do not usually give the exact 'correct' answer. But they should represent the best possible assessment, given the available data and how they were collected. We use statistics to find the 'best' estimates and to indicate the degree of uncertainty associated with such estimates. Another form of statistical inference that is based on the same theoretical framework as estimation is hypothesis testing. In this, sample data are assessed for evidence of their consistency with a particular theory or 'model'.

Synthesis

In the analysis section, we pull the data apart and examine it carefully, but we have to bring the findings of this analysis to bear on the original objective. This requires synthesis and interpretation of our findings, and the expression of those findings in summary form via graphs, tables and words. Whenever statistics are used to make sense of numbers, it is vital to ensure that this sense is effectively presented and communicated to others.

In the remaining chapters of this book we describe how the collection, analysis and presentation of data should be approached so that they help us to answer questions of practical importance.

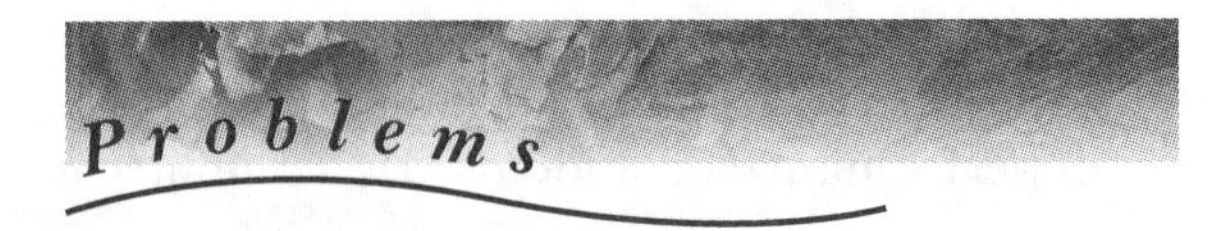

The problems in this chapter are different from those in the rest of the book. Most of the problems that you will meet later are based fairly directly on the theory and methods introduced in the preceding chapters. The problems which follow are based on the contexts introduced earlier in this chapter, and ask you to use your general knowledge to see how far you can go in answering the statistical questions. Their purpose is to orient your thinking so you will understand what a statistical problem is. It is not expected that you will be able to answer these problems now as well as you will be able to answer them after studying the entire text. Do not be discouraged if you feel a bit 'at sea' with them. If you spend some time thinking about the questions, you will be well prepared for the theory which follows in successive chapters.

1. **Plastics**
   (a) How do you think the editor of *Science* magazine determined the consensus views that resulted in the published list of 'Top 20' scientific advances of the twentieth century?
   (b) How do you think the computer has changed 'the face of statistics'?

2. **Sydney 2000**
   (a) Sydney's average rainfall in September is the smallest of any month. What other feature of the rainfall would be of interest to the organisers of the Olympic Games in the year 2000?
   (b) What features of temperature would the organisers want to know if the data were provided in a monthly summary?

3. **Bill Clinton**
   (a) If we look at the age of death of all US presidents prior to President Clinton, would this help us to predict the age at death of President Clinton?
   (b) In estimating Bill Clinton's age at death, how would you use the knowledge that three presidents of the United States (Lincoln, McKinley and Kennedy) were assassinated?

4. **Cricket**
   (a) How would you determine the proportion of your class who have an interest in cricket statistics?
   (b) During televised sports events, do broadcasters supply the past performance statistics for the competitors? Why or why not?

5. **Fashion**
   (a) If you were buying inventory for a particular type of shoe for sale in a shoe store in your area, what size mix would you choose in the order? Why?
   (b) When the inventories of a certain model of shoe get low, would you expect more shoes left over in the middle sizes or in the extreme sizes? Give your reasoning.

6. **The university lottery**
   (a) At your educational institution, can you think of an index in common use for measuring the overall performance of students? Describe how it is computed.
   (b) Is the index you described in part (a) subject to the same controversies as the index used for evaluating the academic potential of high school graduates in New South Wales? Explain.

7. **Champions**
   (a) Suppose a sports league consists of 10 teams and during the regular season each team plays each other team twice. From each game, a given team will record either a loss (0 points), a tie (1 point) or a win (2 points). Thus, the 90 games result in a total number of points for the entire league of 180 points, with an 'average team' receiving 18 points. The possible range of points is 0 to 36. If the teams are of identical quality in the sense that each game is equally likely to be won by either team, how much variation in the teams' points for the regular season would you expect?
   (b) In view of the hypothetical league suggested in part (a), comment on the custom of using a finals series among the top teams from the regular season in order to sort out the real champion.

8. **Overview**
   (a) Section 1.2 of this chapter describes a process through which data-based studies progress: objective or research question, study design, data collection, data analysis, modelling, data synthesis and presentation of results. Suppose you wanted to determine which of two brands of beer was preferred by 10 guests at your house party. Using the headings in the process outlined in the overview, describe how you would do this study. (Remember to keep it simple — you don't want to spoil your party!)
   (b) What do you think is meant by 'the scientific method'?

CHAPTER 2

Exploring univariate data

Example is always more efficacious than precept.

Samuel Johnson (1709–1784)

2.1 DATA — MORE THAN NUMBERS

In chapter 1, we saw how numbers can hold information that is important to us all. In this and subsequent chapters, we will examine how to extract information from these numbers.

What materials and tools do we need? The materials we need are data. The best tools to work these materials depend upon the structure of the material. There are many types of data structures. The tools we use work the data by presenting them in numerical and graphical forms which allow us to discover information that may be hidden, or at least not readily noticed, in the raw data. They also allow us to present the data to others and to summarise effectively.

Many data sets have been collected with a specific purpose in mind, perhaps to answer questions such as those raised in chapter 1. Quite commonly, however, we may simply be using curiosity to drive the exploration of available data. This is precisely the motivation we use with some of the examples in this chapter and those which follow.

Sometimes we have a specific question in mind, but do not have data that will allow us to directly (or perhaps adequately) address the question. Some related data may still give us insight, perhaps providing a partial answer to the question and helping us to pose further questions. And just occasionally, we are lucky, or we have made our own luck. There is a very specific question we wish to ask and our data allow us to answer that question with a reasonable degree of certainty.

The simplest data structure contains a single measurement (*observation*) on a single characteristic (*variable*) of each of several 'items' (individuals, plants, towns, components, days, ...). A data set of this type is called a *univariate* (or one-variable) data set. For example, the record of students' scores on a final exam is a univariate data set; however, if the record also contained scores on assignments and tests for each student, the data set would be *multivariate* (or many variables) and would be more complicated to analyse. In this chapter, we will consider only univariate data; multivariate data will be examined in chapters 4 and 5.

Sometimes a data set contains measurements made at several successive times; when the ordering of the data is important, the data are called a *time series.* The measurement of your weight each morning during a diet regime would be a univariate time series. When the ordering of the data is unimportant (or we choose to ignore it), the data are called a *batch* of values. This chapter will consider only univariate batches; chapter 3 will examine univariate time series.

In chapter 1, we addressed the question of whether the president of the United States, Bill Clinton, will die in office. One of the suggested ways of answering this question was to compare historical evidence of age at death of presidents of the United States with the current age of the incumbent.[1]

1. Bill Clinton was born on 19 August 1946.

The following table gives the ages at death of the first 36 presidents of the United States.[2] The list of dead presidents is a list of the whole population of interest. Each member of the population can be identified by name. This is typical of, but not necessary for, data sets which represent whole populations.

Age at death of US presidents

Washington	67	Harrison	68	Johnson	66	Roosevelt	60	Eisenhower	78
Adams	90	Tyler	71	Grant	63	Taft	72	Kennedy	46
Jefferson	83	Polk	53	Hayes	70	Wilson	67	Johnson	64
Madison	85	Taylor	65	Garfield	49	Harding	57	Nixon	81
Monroe	73	Fillmore	74	Arthur	56	Coolidge	60		
Adams	80	Pierce	64	Cleveland	71	Hoover	90		
Jackson	78	Buchanan	77	Harrison	67	Roosevelt	63		
Van Buren	79	Lincoln	56	McKinley	58	Truman	88		

Comparable data on the sixteen Australian prime ministers who have died is given in the following table.

Name	Year of birth	Year of death	Age at death	Year of appointment
Barton	1849	1920	71	1901
Deakin	1856	1913	57	1903
Watson	1867	1941	74	1904
Reid	1845	1918	73	1904
Fisher	1862	1928	66	1908
Cook	1860	1947	86	1913
Hughes	1862	1952	90	1915
Bruce	1883	1967	84	1923
Scullin	1876	1953	76	1929
Lyons	1889	1939	49	1932
Page	1880	1961	81	1939
Menzies	1894	1978	83	1939
Fadden	1895	1973	78	1941
Curtin	1885	1945	60	1941
Forde	1890	1983	93	1945
Chifley	1895	1951	65	1945
Holt	1908	1967	59	1966
McEwen	1900	1980	80	1967
McMahon	1908	1988	80	1971

2. R. Famighetti (ed.), *The World Almanac and Book of Facts 1995*, Funk and Wagnalls, Mahwah, New Jersey, 1995, p. 634. Reprinted with permission.

For now, we will not pursue the comparison of the ages at which the leaders of Australia and the United States have died.[3] However, it is worth noting that there are other differences of interest between the two populations. Both lists include all leaders who have died, but there have been many more presidents than prime ministers; the United States gained its independence from Britain in 1789, whereas the Australian States were British colonies until 1901.

In chapter 1, we saw how weather records might have assisted the organisers to choose September as the month in which to hold the Olympic Games in the year 2000. Here are some records from the University of Wollongong Climate Station.

Maximum daily temperatures (°C) in July 1994

July 1	14.9	July 9	20.0	July 17	14.4	July 25	18.3
July 2	16.9	July 10	18.2	July 18	17.4	July 26	16.3
July 3	18.2	July 11	19.8	July 19	17.3	July 27	15.8
July 4	22.3	July 12	15.6	July 20	18.1	July 28	18.3
July 5	21.9	July 13	16.7	July 21	17.4	July 29	20.8
July 6	21.5	July 14	17.0	July 22	13.1	July 30	20.0
July 7	21.2	July 15	18.6	July 23	13.6	July 31	13.8
July 8	18.0	July 16	17.2	July 24	15.0		

Daily rainfall (mm) in January 1993

Jan. 1	–	Jan. 9	–	Jan 17	–	Jan 25	0.8
Jan. 2	–	Jan. 10	–	Jan 18	–	Jan 26	0.8
Jan. 3	–	Jan. 11	0.6	Jan 19	1.2	Jan 27	2.4
Jan. 4	11.6	Jan. 12	–	Jan 20	–	Jan 28	2.8
Jan. 5	–	Jan 13	2.2	Jan 21	–	Jan 29	–
Jan. 6	–	Jan 14	–	Jan 22	17.2	Jan 30	–
Jan. 7	46.4	Jan 15	–	Jan 23	19.0	Jan 31	–
Jan. 8	2.0	Jan 16	–	Jan 24	2.6		

3. In making comparisons, ancillary data such as those provided in the table for Australian prime ministers would be relevant.

Number of days in month with rain, 1977–1990

	Jan.	Feb.	Mar.	Apr.	May	June	July	Aug.	Sept.	Oct.	Nov.	Dec.
1977	11	13	12	5	11	8	2	5	15	6	8	8
1978	18	9	17	10	7	14	8	7	12	10	15	19
1979	9	10	20	9	15	9	6	4	8	12	13	7
1980	14	12	8	7	15	12	7	4	3	10	11	12
1981	15	15	5	8	14	7	6	7	6	12	17	13
1982	11	10	19	5	2	11	12	3	9	11	3	12
1983	9	12	9	12	13	8	8	12	9	16	8	18
1984	15	20	11	10	8	6	15	6	10	8	15	10
1985	9	10	8	19	13	6	7	10	12	18	19	16
1986	15	10	11	12	11	3	9	9	10	11	18	9
1987	14	11	10	7	9	11	11	16	4	17	8	16
1988	11	15	14	18	12	10	6	10	10	3	14	19
1989	21	11	22	23	22	20	12	9	7	6	13	15
1990	14	21	22	19	17	10	12	9	13	13	8	13

Each of these three data sets consists of a batch of meteorological data, but there are subtle differences between the types of information they record. Each data set provides a single figure for each period (day or month) which summarises a continuous process. The three data sets give examples of a maximum (temperature), an aggregate (rainfall) and a count (number of rainy days); and the nature of the variable being measured affects the subsequent analysis of the data.

Another important distinction is between data sets that are, in some sense, 'complete', such as the US presidents' ages at death, and data that are 'representative' of a more general process or population. The following data set is a sample which comes from exploration and testing done in preparation for the design of improvements to runways and taxiways at the Chicago O'Hare International Airport.[4] The values are unconfined compressive strengths (tsf) from a sample of soils and can be considered to be 'representative' of the characteristics of the soils in the area.

4. R. D. Holtz and R. J. Krizek, 'Statistical evaluation of soils test data', in *Statistics and Probability in Civil Engineering*, P. Lumb (ed.), Hong Kong University Press, Hong Kong, 1972, p. 266.

Unconfined compressive strength (tsf) of soil samples from an airport runway									
7.7	1.0	2.0	4.6	6.7	4.0	5.0	1.0	4.0	5.0
1.0	3.5	0.5	3.5	3.0	6.0	3.1	3.5	4.0	5.3
0.8	1.3	2.2	4.5	4.5	1.5	4.2	3.5	0.2	0.4
1.0	2.1	1.0	1.5	0.5	4.0	3.0	3.4	2.1	6.6
1.5	1.1	1.7	2.4	3.0	3.3	5.0	3.0	4.0	2.2
6.0	2.5	4.5	1.5	2.0	1.0	6.0	2.2	6.4	2.9
2.0	3.0	2.0	2.0	5.0	6.0	1.8	3.4	1.1	6.7
3.5	3.0	3.0	3.5	2.4	4.3	1.5	3.5	5.5	
6.1	1.7	4.0	2.0	2.0	4.4	3.1	3.9	2.1	
1.5	1.0	2.5	2.0	3.0	3.5	1.6	3.5	6.2	
2.0	4.0	6.4	3.6	3.8	4.0	2.5	1.0	4.3	
2.0	0.8	2.8	2.1	5.0	3.0	3.5	3.0	6.0	
2.5	0.5	3.0	2.0	4.0	5.0	6.6	1.3	1.5	
3.0	5.0	1.5	2.5	6.0	3.0	5.6	4.9	6.3	
1.9	4.5	4.0	3.8	6.0	5.0	1.0	0.2	6.6	

It is not always obvious what measurements should be collected and analysed to give the most helpful information about a question of interest. For example, the second column of the table on page 19 describes emissions of carbon dioxide by various jet aircraft in each takeoff/landing cycle.[5] As with the presidents and prime ministers data sets, the whole of the population of interest is represented here. (There are, of course, other jet aircraft but they may have been judged by the investigators to be of little significance in commercial aviation.) Although it is of interest to examine the CO_2 emissions from the various aircraft, it is perhaps of more interest, in view of the different sizes of the aircraft, to consider standardised CO_2 emissions (emissions divided by the weights of the aircraft). The final column of the table shows emissions *per 1000 kg of aeroplane weight.* Choosing the most meaningful measurement(s) to examine from each individual in the population requires careful consideration.

In many of the data sets in this section, we know more about each individual than a single numerical measurement. In the CO_2 emissions data, we have additional information about the weight of each aircraft, and we also know the name of each aeroplane. Similarly, we know the names of the US presidents, which provides us with further information about them, such as which were assassinated. In extracting information from data sets, we must keep in mind the labels attached to each individual — these may contain pertinent information.

5. B. G. Woodmansey and J. G. Patterson, 'New methodology for modeling annual aircraft emissions at airports', *Journal of Transportation Engineering*, 120, 1994, p. 345. Reproduced by permission of the American Society of Civil Engineers.

In contrast, we have no further data about the individuals (soil samples) in the O'Hare soil example beyond their unconfined compressive strengths, and there is no further structure to the data. The example which follows the CO_2 emissions table below, and involves a group of male students at the University of Hong Kong, has similar characteristics.

Aircraft	CO_2 emissions (kg per takeoff/ landing cycle)	Weight (kg × 1000)	CO_2 emissions per 1000 kg weight
B747-400	10 822.72	394.00	27.469
B747-200	11 673.04	351.50	33.209
MD-11	8 115.23	274.00	29.618
DC10-30	7 313.15	251.70	29.055
L1011-200	8 283.17	195.00	42.478
A300	5 633.66	165.00	34.143
DC8-63	6 246.50	161.00	38.798
A310	4 880.16	150.00	32.534
B707-320B	6 246.50	148.30	42.121
B767-300	5 351.02	137.00	39.059
B757-200	4 614.96	108.80	42.417
B727-200	4 866.75	88.30	55.116
A320	2 898.05	73.50	39.429
B737-300	2 758.79	63.00	43.790
B737-100	3 244.50	52.30	62.036
DC9-50	3 272.23	48.90	66.917
BAe-146	1 855.08	40.50	45.804
BAC 111-400	2 516.73	39.40	63.876
Fokker 28	2 034.24	29.40	69.192
Dassault Falcon 20	1 117.43	12.80	87.299
Gates Learjet 36	536.79	8.10	66.270
Gates Learjet 35	536.79	7.70	69.713
Gates Learjet 24D	1 221.07	6.10	200.175
Cessna Citation	462.08	5.20	88.862

The following table describes the 'maximum voluntary isometric strength' (MVIS) of a group of 41 male students at the University of Hong Kong.[6] Each student was asked to exert maximum upward force on a horizontal bar which was close to floor level, with his feet 400 mm away from the bar. The force was averaged over a five second period and is recorded in kilograms. The individuals are identified (with codes '1', '2', and so on) but their identities do not provide us with relevant information and are not of any importance to this study.

6. W. A. Evans, 'The relationship between isometric strength of Cantonese males and the US NIOSH guide for manual lifting', *Applied Ergonomics*, 21, 1990, p. 139.

Subject	MVIS (kg)	Subject	MVIS (kg)	Subject	MVIS (kg)
1	33	15	26	29	20
2	16	16	10	30	13
3	35	17	12	31	25
4	33	18	20	32	26
5	47	19	31	33	41
6	40	20	12	34	14
7	18	21	19	35	20
8	54	22	36	36	22
9	18	23	23	37	19
10	44	24	22	38	18
11	21	25	20	39	23
12	29	26	15	40	26
13	12	27	15	41	19
14	12	28	16		

There are many questions we might ask about data such as those in the examples discussed so far in this section.

- Do all the subjects have about the same strength, as measured here?
- Is there any evidence of there being two or more classes of individuals, at least with respect to strength?
- Are there any individuals who distinguish themselves as either exceptionally weak or exceptionally strong?

These questions are thinly disguised applications of the following generic questions:

- How variable are the numbers?
- Are there clusters?
- Are there any outiers?

The questions we ask are directed at finding patterns which represent and describe the overall variation in the set of numbers and at displaying these so that any important features are highlighted.

Our questions, and the graphical and numerical tools that will be described in this chapter to help answer them, become more interesting when we wish to compare several batches of data. In particular, we would then also ask:

- Do the batches differ?
- If so, how do they differ?

Organising data to be more informative

A data set is usually presented to us in the form of a table, as was done with the examples given earlier. Throughout this chapter, techniques are presented to highlight patterns and other features of such data.

The following data are typical of what we might obtain in sampling a variety of populations. They could arise from measurements of the percentage content (by weight) of some group of compounds (for example, fats) in samples of a particular food or they could be obtained by measuring the percentage of damaged tissue from a group of patients after a medical procedure. Our artificial data set consists of 20 values, and each value will be assumed to be the percentage (by weight) of contaminant in a sample from the output from an industrial process. There are many other scenarios in which we might collect similar samples of values.

Contamination of samples from an industrial process (%)

6.1	5.2	7.9	2.3	3.4
1.4	5.3	7.1	3.2	2.8
5.1	6.9	6.1	3.4	5.2
5.5	2.0	1.3	4.9	6.4

Understanding the nature of the measurement, which is a percentage, tells us that all values (not just the 20 sample values we have obtained) should lie between 0 and 100. A scan of the data enables us to make some general statements about the values. In a small data set like this, it is easy to notice that all the values lie between 0 and 10%. (In a much larger batch of data, even this assessment would be difficult.) Without effort or experience, it is not easy to scan such a table and obtain further useful information. The data need to be organised and summarised in useful ways to make them more informative.

The initial table lacks structure. We need to scan the whole table even to detect the maximum contamination. One useful way to give the numbers more structure is to list them in order of magnitude.

Ordered data: contaminations (%)

1.3	1.4	2.0	2.3	2.8	3.2	3.4	3.4	4.9	5.1
5.2	5.2	5.3	5.5	6.1	6.1	6.4	6.9	7.1	7.9

A quick look at the list of ordered values now reveals the smallest (1.3%) and largest (7.9%) values. Further scanning of the list reveals other features, such as clusters of values around 3% and 6% and a gap between 3.4% and 4.9%.

However, such information is better presented graphically. It is difficult to spot patterns in lists or tables of numbers since the digits must be individually read. In chapter 6, we discuss methods to improve tabular displays of data, but well-presented graphs generally reveal information more rapidly than tables do.

EXERCISE 2.1

Consider the data sets introduced so far in this chapter:

(a) US presidents
(b) Australian prime ministers
(c) Wollongong temperatures, July 1994
(d) Wollongong rainfall, January 1993
(e) Wollongong days with rain by month, 1977–1990
(f) soil compressive strength
(g) aircraft pollution
(h) isometric strength

1. Which of the data sets are univariate data sets (univariate batches or univariate time series)?
2. Sorting a batch of data results in a tabular display that is easier to absorb than the unsorted data, but sorting does destroy the original ordering. Which of the data sets are presented in an order that is important for understanding the data?
3. Which data sets include data that are not numerical?
4. Sometimes a useful first step in analysing a data set is to split it into several subgroups and then summarise and compare these subgroups. For which of the above data sets would this approach be helpful?
5. What is the most basic tabular reorganisation that helps the viewer absorb a batch of data? When is this reorganisation not a good idea?

2.2 DOT PLOTS

A *dot plot* represents data as a set of dots or crosses along a line.

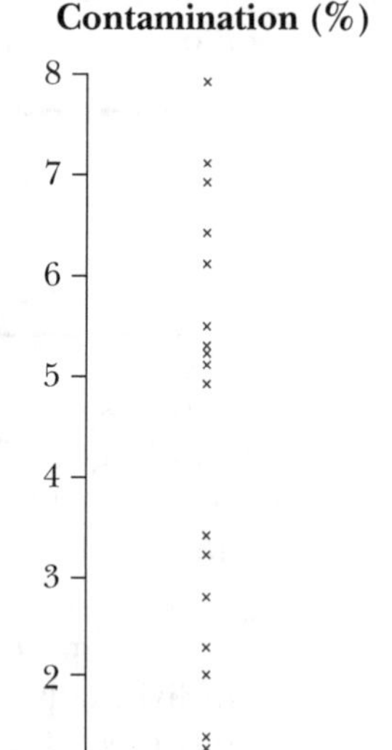

The smallest and largest values, the clusters and the gap are immediately apparent when a dot plot (right) is used to display the contamination data. The graphical display does involve a small loss of detail (for example, it is difficult to tell whether the largest value is 7.9 or 8.0).

The dot plot also provides some information about *location* (where the data are 'centred'), and about *spread* and how the density of values changes over the range of the data.

In this data set, there are three repeated values (3.4, 5.2, 6.1) which cannot be distinguished on the dot plot. The plot also needs to be examined with care to correctly count the number of points in 'dense' bands (for example, the five plotted values in the interval 4.9–5.5).

In general, if there are too many data values the dot plot gets cluttered with dots or crosses drawn over each other. One way around this is to *jitter* the points in the dot plot. Jittering is done by displacing the points 'arbitrarily' or 'at random' in a horizontal direction (or in a vertical direction if the dot plot is horizontal).

Here are three different jittered dot plots for the contamination data. The first two use different 'random' horizontal offsets. The horizontal and vertical axes have been swapped in the third dot plot so that the contamination percentages are displayed along the horizontal axis and the jittering is done vertically. In all cases, the information revealed to the viewer about the data is essentially the same.

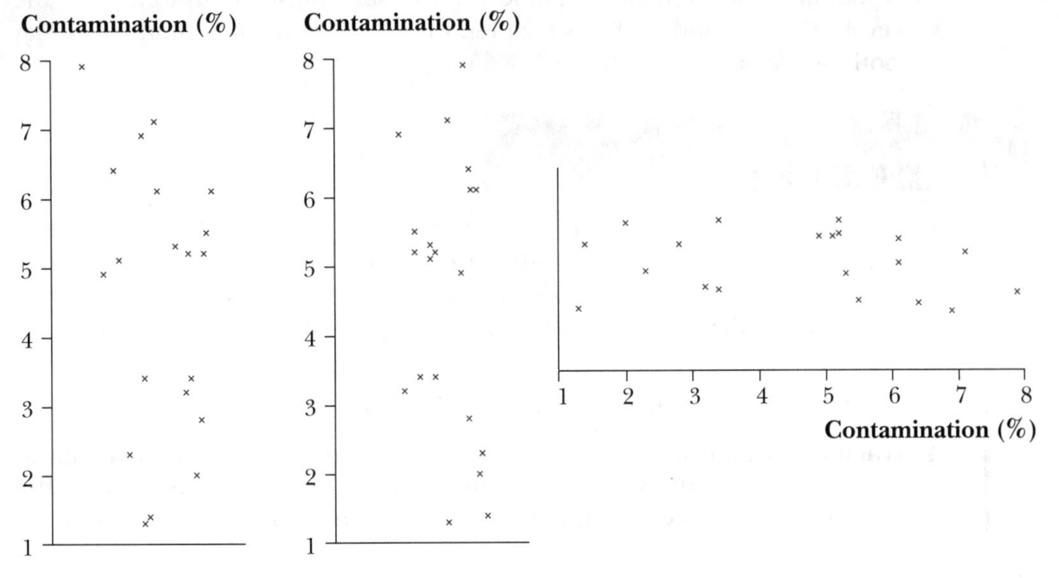

Jittering is rarely done without a computer. In drawing a dot plot by hand, it is easier to stack repeat points (and points which would otherwise be too close to easily distinguish from points already plotted), as shown on the horizontal dot plot below.

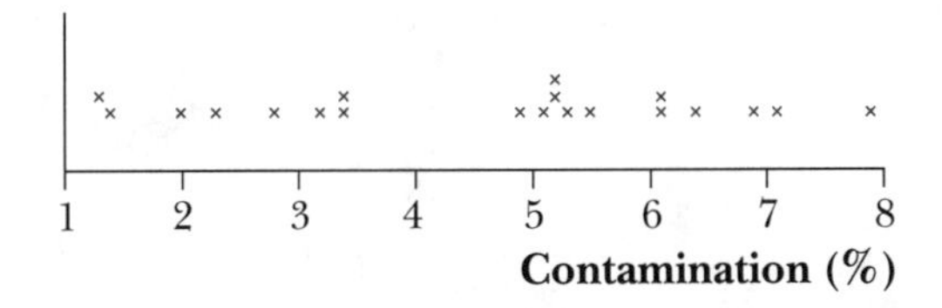

Computer-drawn dot plots provide as much information about the values in a batch as the plotting resolution of the screen or printer will allow. Some computer software displays dot plots in 'character' displays, where the maximum resolution is a character width. For example, the dot plot below shows the contamination data in this form.

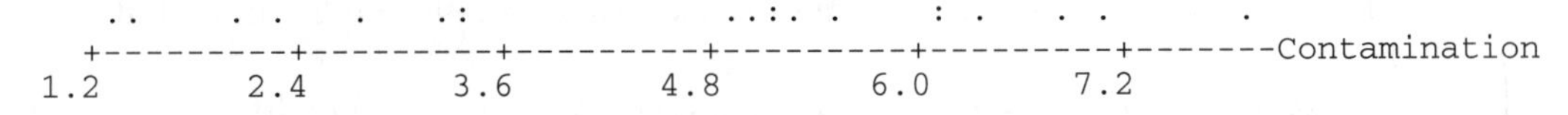

Although the horizontal resolution of this display is limited (since each horizontal character position contains a range of 0.12 values), there can be no overlap between adjacent categories and the dots line up in neat vertical stacks. The loss of information resulting from this grouping should not affect interpretation of the data. Stacked dot plots can be drawn with finer horizontal grouping of values when the display is not character-based, limited only by the resolution of the screen or printer. Many users find the randomness of jittering disconcerting and find it easier to interpret these grouped stacked dot plots.

Dot plots are almost essential as an initial display of a batch of data; they often show clearly features of a data set that are masked by other summaries. However, an important role of the graphical display should be to summarise, and dot plots show too much detail in large data sets. This visual 'noise' tends to distract the viewer. In the remainder of this chapter, we will examine data summaries that 'smooth over' some of this noise.

EXERCISE 2.2

1. The contamination data have been displayed as a sorted table and as a dot plot. Which do you find more effective in conveying the pattern of values?
2. What features of dot plots are easily described in words?
3. Examine the batch of data below. Does any pattern emerge? If not, try sorting or a dot plot.

   169, –55, 193, 137, 73, 197, 185, 199

4. When a dot plot of continuous data is displayed on a computer screen, it can fail to represent exactly the numerical values in the data because of the resolution of the screen. Is this a serious problem for the dot plot technique?

2.3 STEM-AND-LEAF PLOTS

The diagram below shows a stacked dot plot of a large, artificial data set.

```
                         :   :
                  :. .   :: .:
           .     .::.: .::: ::  .
    .  .   :.    :::::.:::::::. :
    :  :   :::: :::::::::::::::.:....:     .
:   :::::.:::::.::::::::::::::::::::::.::: :  .       .
---+---------+---------+---------+---------+---------+---
21.0      22.0      23.0      24.0      25.0      26.0
```

In large data sets, stacked dot plots show the distribution of values in a batch through the *heights* of the stacks; these represent the *density* of values. Regions of the axis with low stacks (such as between 25 and 26 above) have a low density of values, whereas regions with high stacks (such as between 23 and 24 above) have a high density of data values.

A *stem-and-leaf* plot is a related data display that also uses height to represent the density of values in different ranges. However, rather than representing each value with a dot or cross, further information about the value is retained by using

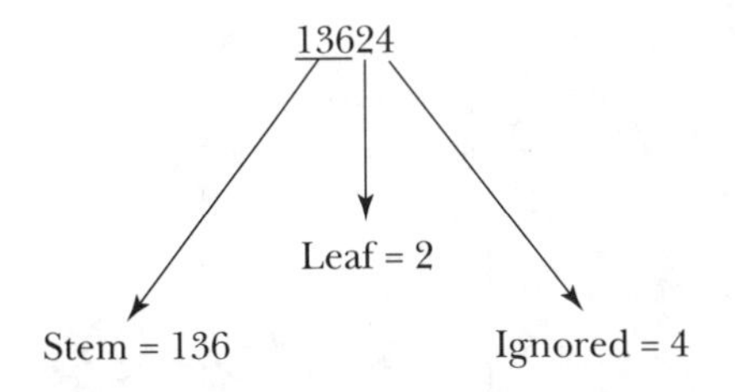

a digit. Stem-and-leaf plots are easy to draw by hand. They clearly show up clusters and outliers, as well as other features of the data.

A stem and leaf plot is a graphical display in which the data values are shown numerically. Two parts, a stem and a leaf, are extracted from each value. More precise guidelines about which parts of the values should be used as stems and leaves will be given later, but there should typically be between 5 and 20 distinct stems in the data set.

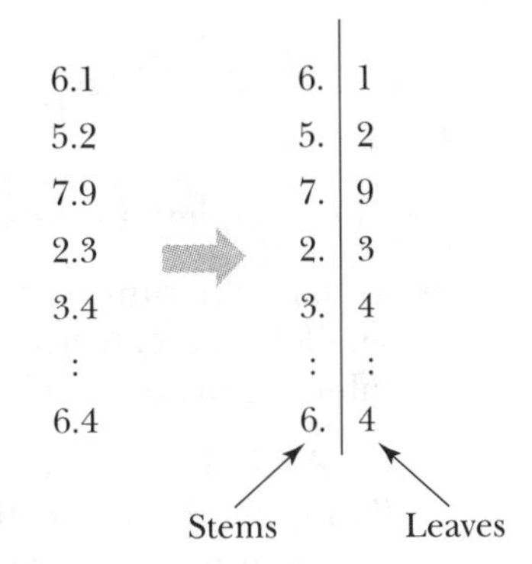

To start a stem-and-leaf plot, decide on the stem and leaf units. For the contamination data, all values consist of a single digit before the decimal point and another after. We will therefore define the first digit of each value (the 'units' digit) to be the stem and the second digit (the 'tenths' digit) to be the leaf, as shown on the right.

We now separate the numbers into their stem and leaf parts. We first list in order all distinct stems in the data, plus any intermediate stem values that are not present in the data.[7] The first data value from the batch, 6.1%, is split into its stem digit '6' and its leaf digit '1' and the leaf digit is written to the right of the corresponding stem. The remaining values are similarly split into stem and leaf digits and added to the display in the order in which they appear in the raw data (page 21).

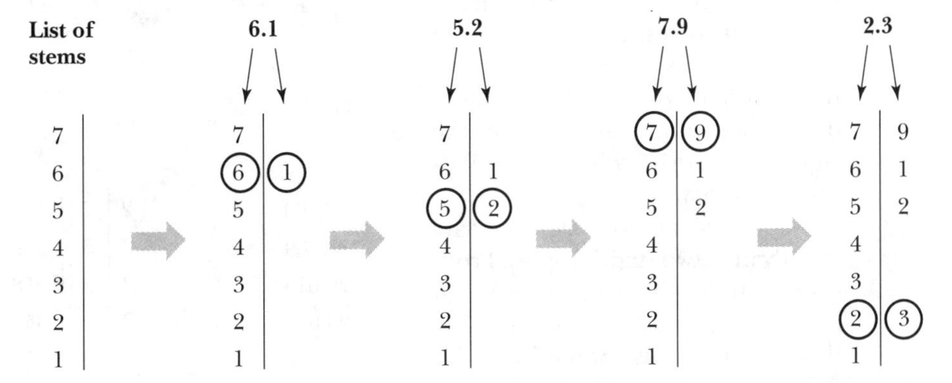

When a value is encountered whose stem has already been used, its leaf digit is stacked to the right of the earlier leaves on that stem. Finally, the leaves on each stem are usually sorted into increasing order, and the units for the stems and leaves are written on the display, as shown in the following diagram.

7. Some authors list the stems with the smallest stem at the bottom; others list them with the smallest stem at the top. When reading a stem-and-leaf plot, carefully note the ordering used.

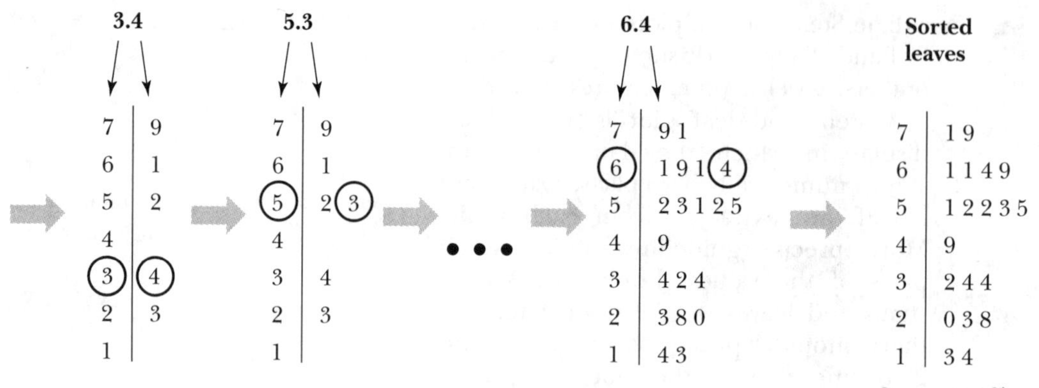

The completed stem-and-leaf plot:

- displays the numbers in increasing order of magnitude. (For these data, the smallest value, 1.3%, can be read directly from the plot, as can all other values in order up to the largest, 7.9%.)
- displays the numbers in a format that tells us about 'distribution'. (The *length* of the leaf stack against each stem shows the number of leaves and hence represents the relative *density* of data values for that stem.)

If a stem-and-leaf plot is drawn by hand, care must be taken to space the stems and leaves evenly so that a correct impression of shape is given. It is particularly important to resist the temptation to space leaves of '1' more closely than other digits! The hand-drawn stem-and-leaf plot on the right is poorly aligned and is unsatisfactory.

1 | 34
2 | 038
3 | 244
4 | 9
5 | 12235
6 | 1149
7 | 19

stem unit: 1%
leaf unit: 0.1%

Stem-and-leaf plots preserve data exactly in a 'two-digit' display. From each leaf in the display, we can reconstruct the exact data value that it represents. Sometimes the data values have more than two significant digits, so some of the digits must be omitted so the plot can be drawn without too many distinct stems. For example, the aircraft emissions data (CO_2 emissions per 100 kg weight) could be split as shown on the right.

Data	Stems	Leaves	Ignore
27.469	2	7	.469
33.209	3	3	.209
29.618	2	9	.618
29.055	2	9	.055
42.478	4	2	.478
:	:	:	:
69.713	6	9	.713
200.175	20	0	.175
88.862	8	8	.862

Some information in the data is lost by ignoring some of the digits. However, these omitted digits rarely contain useful information in summaries; they are 'noise' that does not help us to understand the 'distribution' of the data.[8]

Note that *truncation* (sometimes called 'cutting') of the values, as described above, is better than *rounding* the numbers to the nearest whole number for two reasons.[9]

- The leaves are displayed against the same stem as they would have been if two-digit leaves had been used (that is, if the leaves in the aircraft emissions data were taken to be the units *and* tenths digits of the values, '74', '32', '96', ..., '88').
- Truncation is easier — at least if the stem-and-leaf plot is being drawn by hand rather than by computer!

The resulting stem-and-leaf plot for the aircraft CO_2 emissions is shown on the right. Note that stems in the middle of the display should not be omitted, even if there happen to be no leaves stacked against them. To interpret the shape of the display, it is important that the stem values be equally spaced with no missing values.

Stem	Leaves
20	0
19	
18	
17	
16	
15	
14	
13	
12	
11	
10	
9	
8	7 8
7	
6	2 3 6 6 9 9
5	5
4	2 2 2 3 5
3	2 3 4 8 9 9
2	7 9 9

Stem unit: 10
Leaf unit: 1

In this data set, an outlier is apparent (the Gates Learjet 24D, for which the stem-and-leaf plot shows the truncated value to be 200). The stem-and-leaf plot also indicates that there are several clusters of planes, grouped around 27–45, 62–69 and (perhaps) 87–88.

Adjusting the number of stems

A stem-and-leaf plot most effectively provides information about clusters, unusual values and other aspects of the shape of the distribution of values in a data set through the 'canopy' shape made by the ends of the rows of leaves. With the correct number of stems, the shape of the stem-and-leaf plot (and hence the shape of the distribution of the data) is clearly shown by this canopy.

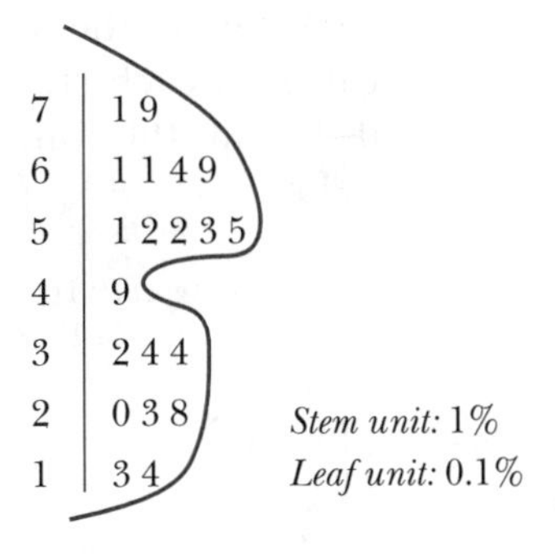

8. See chapter 6 for a longer discussion of the retention of only two or three 'varying' — or information-carrying — digits in displays of data.
9. The loss of information is the same for either rounding or truncating. The rounded number 27 represents a number in the interval (26.5, 27.5); the truncated number 27 represents a number in the interval (27.0, 28.0).

Sometimes, the initially chosen stem-and-leaf plot does not provide an effective display because it has too few or too many stems. With too few stems, information regarding shape tends to be 'smoothed out' and lost. With too many stems, and few leaves on any of them, a much noisier picture results. For example, two stem-and-leaf plots of the isometric strength data are shown, one below and one on the right.

Stem	Leaves
5	4
4	0 1 4 7
3	1 3 3 5 6
2	0 0 0 0 1 2 2 2 3 3 5 6 6 6 9
1	0 2 2 2 2 3 4 5 5 6 6 8 8 8 9 9 9

Stem unit: 10
Leaf unit: 1

Stem	Leaves
54	0
53	
52	
51	
50	
49	
48	
47	0
46	
45	
44	0
43	
42	
41	0
40	0
39	
38	
37	
36	0
35	0
34	
33	0 0
32	
31	0
30	
29	0
28	
27	
26	0 0 0
25	0
24	
23	0 0
22	0 0 0
21	0
20	0 0 0 0
19	0 0 0
18	0 0 0
17	
16	0 0
15	0 0
14	0
13	0
12	0 0 0 0
11	
10	0

Stem unit: 1
Leaf unit: 0.1

With stem unit 10, most of the data are stacked on stems '1' and '2', so the stem-and-leaf plot shows little shape information within the interval 10–29. On the other hand, when the stem unit is 1, each leaf is '0' since the raw data were recorded in whole numbers. The canopy is too jagged to allow shape information to be clearly conveyed.

Stem-and-leaf plots with stem units of 10, 100, 1, 0.1 and any power of 10 do not give us enough flexibility in choosing the number of *bins* in which to stack the leaves. To produce a stem-and-leaf plot with an intermediate number of bins, we need to modify the plot by *splitting* each bin in the standard stem-and-leaf plot. We can split each stem into two bins, one for leaves 0–4 and the other for leaves 5–9, as shown below.

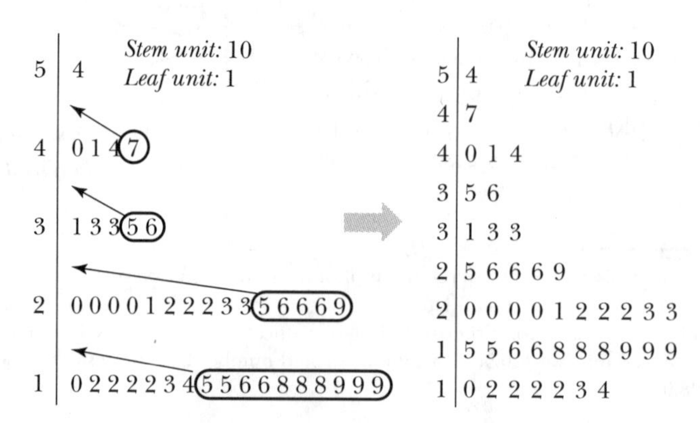

In the plot at the bottom of the opposite page, stems are repeated twice. Each stem can also be repeated five times, but other numbers of repeats are not allowed (since we require each bin to be 'home' for the same number of distinct leaves and 2 and 5 are the only non-trivial factors of 10). When a stem is repeated five times, the bins contain the leaves 0–1, 2–3, 4–5, 6–7 and 8–9. For the isometric strength data, this would give the plot shown on the right.

For these data, however, the 9-bin stem-and-leaf plot with stems repeated twice provides the best representation.

In general, it will not always be clear as to whether the initial display or an expanded or contracted version is best, but it will be clear that one of two or three alternative good plots is best and that further contraction (or expansion) beyond the most compact (or extended) of the two or three good stem-and-leaf plots will render the display uninformative and ineffective.

Stem	Leaves
5	4
5	
5	
4	
4	7
4	4
4	
4	0 1
3	
3	6
3	5
3	3 3
3	1
2	9
2	6 6 6
2	5
2	2 2 2 3 3
2	0 0 0 0 1
1	8 8 8 9 9 9
1	6 6
1	4 5 5
1	2 2 2 2 3
1	0

Stem unit: 10
Leaf unit: 1

Experimenting with stem-and-leaf plots with different numbers of bins is both a good guide and a good teacher. It is good practice to let the data speak for themselves.

Some common errors

In a stem-and-leaf plot, the stem digits followed by the leaf digit represent the data value. (Knowledge of the stem unit and leaf unit will determine whether you have to multiply the number by some power of 10.) It is wrong to represent stems and leaves in any other way. The representation:

$$6 \mid 1$$

for stem '6' and leaf '1' means that the number represented is 61 or 0.61 or 61000 or .... (If the numbers were truncated to produce the plot, it might mean 61.*xyz* or 0.61*xyz* or 61*xyz* or ..., where *xyz* are the truncated digits.)

For this reason, the stem unit of a stem-and-leaf plot must always be a power of 10 (..., 0.01, 0.1, 1, 10, 100, ...). The leaf unit is usually one-tenth of the stem unit.[10] Given that the stem unit is '1' and the leaf unit is '0.1', we therefore know that the number being represented above is 6.1 (or 6.1*xyz* in the case of truncation).

10. The only exception is when two-digit leaves are used, in which case the leaf unit is one-hundredth of the stem unit.

Two common mistakes with stem-and-leaf plots are redundant zeros in stems and repeated digits in both the stem and the leaf. For example:

60 | 1 must not be used to represent 61.

15 | 5 must not be used to represent 15.

Negative values

Some data sets include negative values. How are these values represented on a stem-and-leaf plot?

The following table describes consumption of plastic resins in Australia in 1980 and 1989.[11]

	Quantity ('000 tonnes)		
Market	**1980**	**1989**	**Change (%)**
Packaging	125	257	106
Building	90	112	24
Plumbing	70	117	67
Furniture and bedding	56	80	43
Material handling	50	65	30
Transportation	34	56	65
Electrical	48	52	8
Houseware	41	38	–7
Agricultural	24	36	50
Appliances	22	25	13
Footwear and clothing	10	10	0
Marine	12	9	–25
Other	117	149	27

We will consider the final column of the table, the percentage change in plastic resin consumption between 1980 and 1989. A stem-and-leaf plot of these data is shown on the right.

The lowest number displayed in the stem-and-leaf plot is –25, represented by a stem of '–2' and a leaf of '5'. Similarly the value –7% is split into a stem of '–0' and a leaf of '7'. Note that there must be stems for both '0' and '–0'. While 0.0 and –0.0 represent the same number, it is a convention to display a zero data value with a leaf of '0' against the '0' stem.

Stem	Leaf
10	6
9	
8	
7	
6	5 7
5	0
4	3
3	0
2	4 7
1	3
0	0 8
–0	7
–1	
–2	5

Stem unit: 10%
Leaf unit: 1%

11. Industry Commission, *Recycling, Volume 2: Recycling of Products*, AGPS, Canberra, 1991, p. 82. Permission also from Mitek Pty Ltd.

A general rule regarding truncation, which applies to both negative and positive values, is as follows:

- The leaf is the digit of the value in the 'leaf unit' position.
- The stem consists of all digits to the left.
- All digits to the right of the leaf are ignored. Thus, both positive and negative numbers are truncated towards zero.

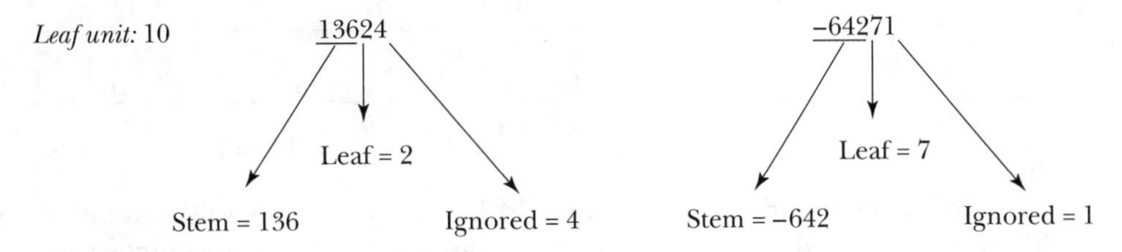

This discussion about stem-and-leaf plots has had a dual purpose. Firstly, you have been introduced to a back-of-the-envelope method of constructing a frequency distribution. More importantly, however, you have been introduced to the effect of grouping data on the resulting frequency distribution. This grouping phenomenon is an important feature of data analysis practice and statistical theory.

EXERCISE 2.3

1. (a) Construct a stem-and-leaf plot of the presidents data set. Use stems of 4, 5, 6, 7, 8, 9.
   (b) How would you expand this plot to provide a little more detail?
2. In section 2.3, we showed a stem-and-leaf plot for the aircraft emissions data after it had been standardised by the weight of the aircraft (page 27). Construct a stem-and-leaf plot for the raw, unstandardised CO_2 data. Which of the two stem-and-leaf plots is more easily interpreted?
3. In what circumstances might more stems than data points be needed?
4. Is it feasible to draw a stem-and-leaf plot of a data set consisting of 10 000 values? With a computer?

2.4 HISTOGRAMS

Stem-and-leaf plots show data stacked in classes (bins), with each class represented by a stem. For large data sets (more than, say, 100 values), the stacks of leaves can become too long for display. Also, retention of the values of the leaves is less informative because there is too much detail to take in quickly. An alternative graphical display should be used for large data sets.

The shape of a data set can be represented in a similar way to a stem-and-leaf plot by stacking up rectangles, instead of leaf digits, above each class. For the contamination data, the display would be as shown in the following diagram.

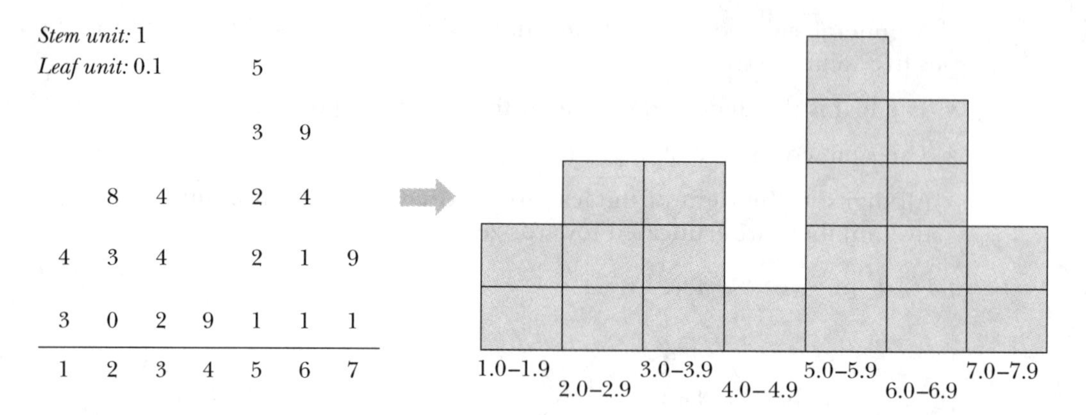

Unlike stem-and-leaf plots, this display can be successfully extended to arbitrarily large data sets by reducing the height of the rectangle representing each individual.

The display is usually drawn above a horizontal axis with a carefully marked scale, with the rectangles stacked above the range of values from the data that would fall in that class. The horizontal lines separating the rectangles are usually omitted. The resulting graphical display is called a *histogram.* (To contrast this with stem-and-leaf plots, some books refer to stem-and-leaf plots as *number histograms.*)

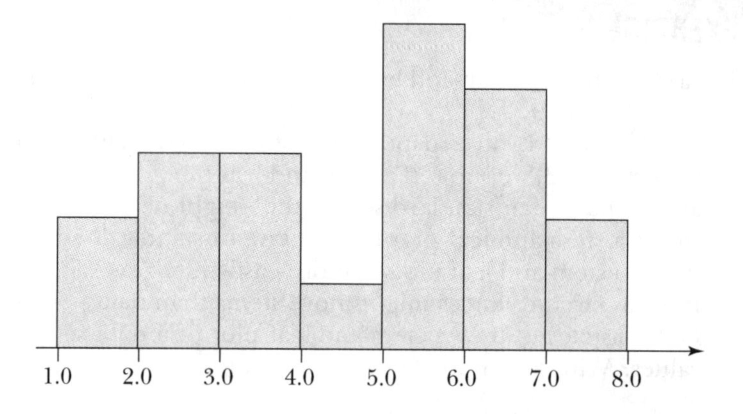

The height of the rectangle above each class is proportional to the number of data values that fall in that bin.

Although you would normally use a computer to draw histograms, it is instructive to consider how you might draw one by hand. The data set from section 2.1 describing the ages at death of the first 36 presidents of the United States will be used to illustrate the calculations involved.

The minimum and maximum ages at death are 46 (Kennedy) and 90 (Adams and Hoover) respectively, so the data cover 45 years (including both extremes). We will use a class width of 10 years for our initial histogram, giving 6 classes. Starting the initial class at 40, this leads to the first two columns in the following table.

Data values	Class	Frequency
40–49	$40 \le x < 50$	2
50–59	$50 \le x < 60$	5
60–69	$60 \le x < 70$	12
70–79	$70 \le x < 80$	10
80–89	$80 \le x < 90$	5
90–99	$90 \le x < 100$	2

A reported age such as '63' would represent an actual exact age of between 63 and 64 years (since ages are always reported in truncated form), so data values of 60–69 correspond to a class of 'exact' ages between 60 and 70. If the data values had been rounded rather than truncated, the classes would have been '$39.5 \le x < 49.5$' and so on. Note that each data value must fall quite clearly into a single class.

Next, the data set is scanned to count the number of values in each class, resulting in the third column of the table. (The table is called a *frequency table.*) The counts become the heights of the rectangles that are drawn above the section of the axis specified by the corresponding class, as shown in the diagram below.

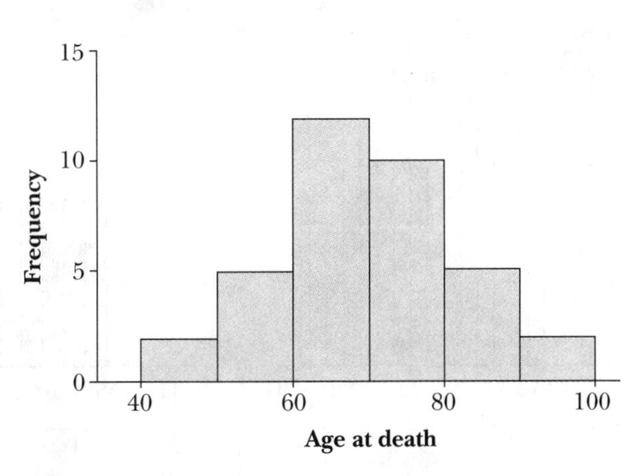

In addition to their effective display of large data sets, histograms have various further advantages over stem-and-leaf plots:

- The classes are not restricted to being of width 1.0, 2.0, 5.0 and so on. A histogram can be drawn with classes of width such as 3.0 or 4.0, giving more flexibility.
- The starting point for each class may be chosen more flexibly than for a stem-and-leaf plot.

The three histograms at the top of page 34, for the US presidents data, illustrate various choices of histogram classes. The first two both use classes of width 5 years, but use different starting values for the classes. The third histogram uses classes of width 4 years. A stem-and-leaf plot could be drawn with bins corresponding to the classes of the first histogram, but not for either of the other two.

Further histograms of the same data, using different class widths and starting points, are also shown on page 34.

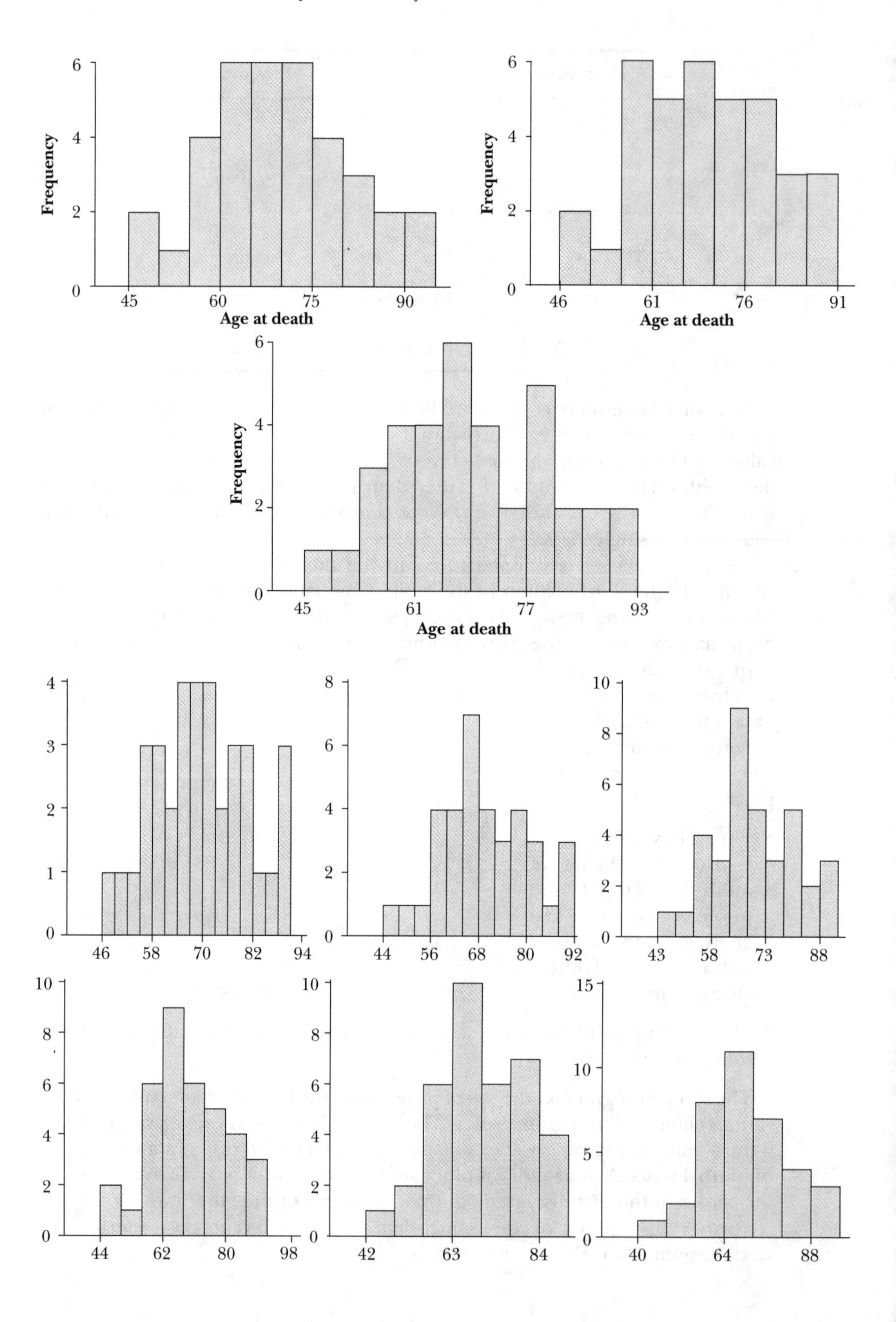

Frequency
Age at death

While all of the histograms on page 34 indicate a distribution with higher frequencies towards the middle, several suggest bimodal or multimodal distributions. Some indicate much flatter peaks than others, yet all represent the same data. This suggests a clear warning:

> Do not over-interpret the shape of histograms (or stem-and-leaf plots).

This warning is especially important when the batch of values is not the whole of the population of interest. In our contamination example, we are less interested in the values from the particular sample of items than in describing the characteristics of contamination levels in the whole underlying process. Collecting a fresh sample from the output of the process and measuring the contaminations from these items would result in a different histogram (or other graphical summary) so, again, you should not over-interpret features in graphical summaries.

All classes do not need to be the same width in a histogram, but the method must be modified slightly. Care needs to be taken if the class widths are not equal, so that the visual impact is correct. If two adjacent classes of equal width are combined, it is essential that the height of the combined rectangle becomes the *average* of the two previous rectangles, not their sum. Each data value is represented in the histogram by a small box. If the width of a class is changed, the area of the box should remain the same, so its height should be decreased, as illustrated below.

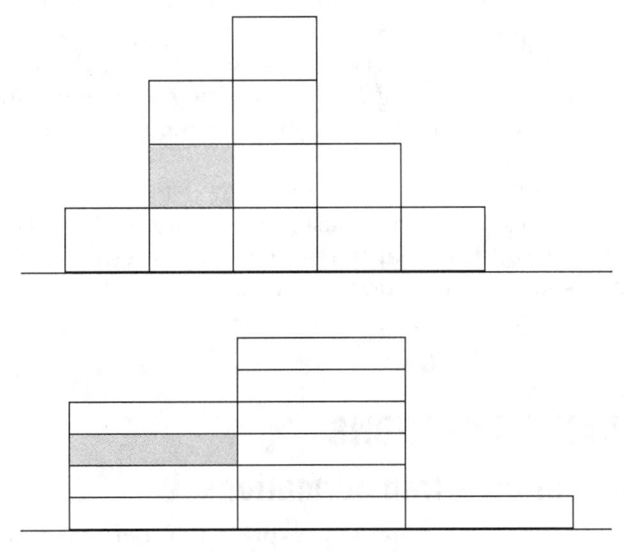

Drawing the histogram correctly ensures that the proportion of values (often called *relative frequency*) in a particular class is represented by area in correct proportion to other classes.

> Area $\propto$ Relative frequency.

This relationship between relative frequency and area is illustrated in the diagram below.

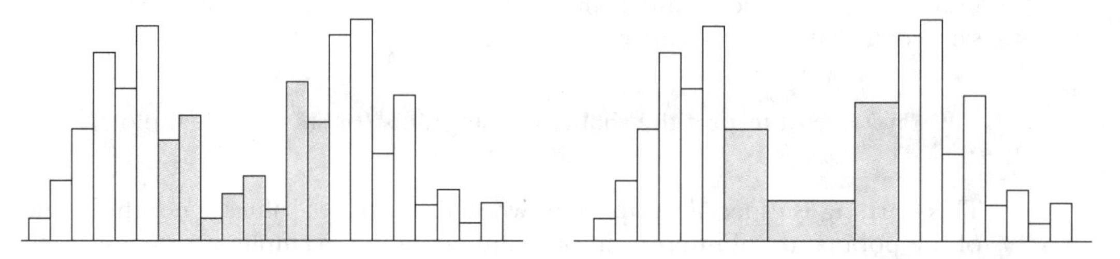

When classes are combined, their total area remains the same. Since all original classes had the same width in this example, the height of the resulting rectangle is therefore the average of the heights of the rectangles being combined.

EXERCISE 2.4

1. Examine the histogram of the presidents data and the stem-and-leaf plot of these data (see exercise 2.3, question 1). One of these graphical techniques is usually more useful for a first look at the data, and the other is usually more useful for a summary presentation of the data. By thinking about the particular case of the presidents data, can you guess which is which?
2. (a) Draw histograms of the rain data for each quarter (January–March, April–June, July–September and October–December) in a way that allows easy comparison of these quarters. Use the same 'bins' for each histogram and align the four histograms one underneath the other.
   (b) Comment on the extent to which the seasonal rain pattern is revealed.
3. The weight variable for the aircraft data is already ordered. What classes would you choose for a histogram of these data? (Keep all weight classes the same width.) What frequencies would result? Comment on the shape of the resulting distribution.

2.5 TRANSFORMATIONS

Linear and nonlinear transformations

Sometimes it is convenient to express numbers on a different scale. This is called *transformation* or *re-expression*.

Rescaling of physical measurements such as length and weight from imperial to metric measurements are examples of linear transformations. Linear transformations change a value x into a value $a + bx$ and preserve the *shape* of the data's distribution. A single dot plot of the data can be drawn with dual scales. Two examples of linear transformations follow.

1. *Distances recorded in miles and kilometres* (1 kilometre ≈ $\frac{5}{8}$ mile)

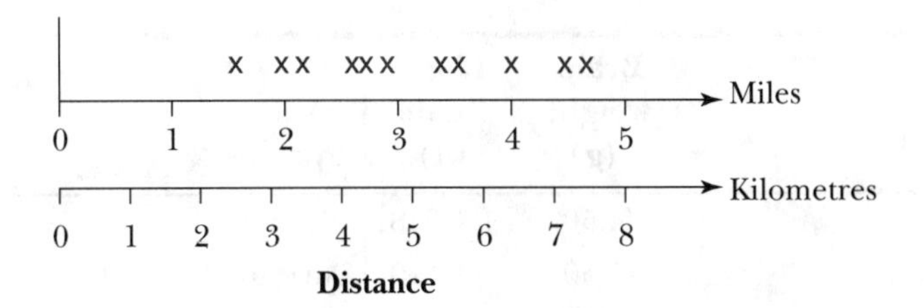

2. *Temperatures recorded in degrees Celsius and Fahrenheit* (Fahrenheit = 32 + 1.8 × Celsius)

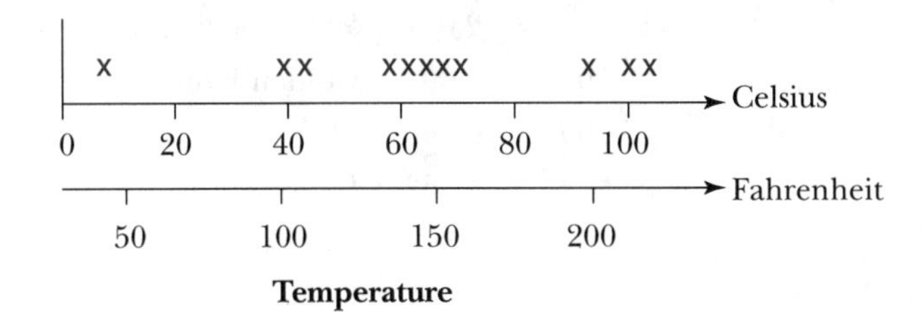

You can see from the diagrams above that linear transformations do not help us to understand the distribution of the data. Nonlinear transformations, however, can reveal extra information about the distribution of values in a batch.

An example of a nonlinear transformation is the Richter scale, which transforms the measured intensity of earthquakes to a logarithmic scale. A second example of a nonlinear transformation is the piano keyboard, which is a logarithmic re-expression of the physical scale (frequency in cycles per second) on which the pitch of sound is measured.

Nonlinear transformation preserves neither relative nor absolute distances between data values. Nonlinear transformation therefore changes the shape of the distribution of a set of data. Good re-expression may:

- spread out dense clusters of points at one tail of a display, thereby revealing clusters, outliers or other features of the distribution's shape

- condense the points at a tail of a display where values are widely spaced, making unusual observations or 'outliers' look less unusual and more like 'normal' data values.

Simultaneously with meeting the above aims, nonlinear transformations:

- reduce asymmetry (or *skewness*); that is, they make the data more nearly symmetric about the 'centre'. Symmetry simplifies the measurement of variability, since you need not describe the amount of variability in each direction separately.

- make commonly used numerical summaries (see section 2.9) more representative of the data.

To illustrate the effect of nonlinear transformations, consider the following table, which gives the average brain weights of 62 species of mammals.[12]

Species	Brain weight (g)	Log_{10} (brain wt)	Species	Brain weight (g)	Log_{10} (brain wt)
Arctic fox	44.50	1.648	Human	1320	3.121
Owl monkey	15.50	1.190	African elephant	5712	3.757
Mountain beaver	8.100	0.908	Water opossum	3.900	0.591
Cow	423.0	2.626	Rhesus monkey	179.0	2.253
Gray wolf	119.5	2.077	Kangaroo	56.00	1.748
Goat	115.0	2.061	Yellow-bellied marmot	17.00	1.230
Roe deer	98.20	1.992	Golden hamster	1.000	0.000
Guinea pig	5.500	0.740	Mouse	0.400	−0.398
Vervet	58.00	1.763	Little brown bat	0.250	−0.602
Chinchilla	6.400	0.806	Slow loris	12.50	1.097
Ground squirrel	4.000	0.602	Okapi	490.0	2.690
Arctic ground squirrel	5.700	0.756	Rabbit	12.10	1.083
African giant pouched rat	6.600	0.820	Sheep	175.0	2.243
Lesser short-tailed shrew	0.140	−0.854	Jaguar	157.0	2.196
Star-nosed mole	1.000	0.000	Chimpanzee	440.0	2.643
Nine-banded armadillo	10.80	1.033	Baboon	179.5	2.254
Tree hydrax	12.30	1.090	Desert hedgehog	2.400	0.380
N. American opossum	6.300	0.799	Giant armadillo	81.00	1.908
Asian elephant	4603	3.663	Rock hyrax (*P. habess.*)	21.00	1.322
Big brown bat	0.300	−0.523	Raccoon	39.20	1.593
Donkey	419.0	2.622	Rat	1.900	0.279
Horse	655.0	2.816	E. American mole	1.200	0.079
European hedgehog	3.500	0.544	Mole rat	3.000	0.477
Patas monkey	115.0	2.061	Musk shrew	0.330	−0.481
Cat	25.60	1.408	Pig	180.0	2.255
Galago	5.000	0.699	Echidna	25.00	1.398
Genet	17.50	1.243	Brazilian tapir	169.0	2.228
Giraffe	680.0	2.833	Tenrec	2.600	0.415
Gorilla	406.0	2.609	Phalanger	11.40	1.057
Gray seal	325.0	2.512	Tree shrew	2.500	0.398
Rock hyrax (*H. brucei*)	12.30	1.090	Red fox	50.40	1.702

12. S. Weisberg, *Applied Linear Regression*, John Wiley and Sons, New York, 1980, p. 128. Reprinted by permission of John Wiley and Sons Inc.

A dot plot of the brain weights is given below (left). The weights are highly skew, with a long tail towards the higher values (or two outliers?). The logarithms of the brain weights are also shown in the table and their dot plot is shown on the right below.

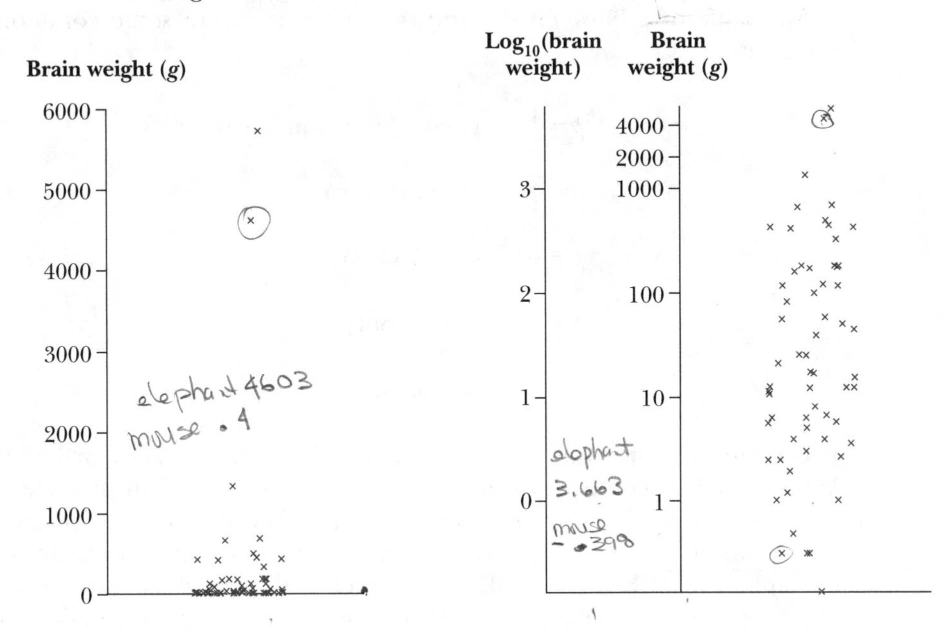

The axis for the transformed data can be labelled either with log-weights which are regularly spaced on the axis (that is, the distance between '0' and '1' is the same as the distance between '1' and '2') or with the original weights, in which case the labels are not evenly spaced (that is, the distance between '10' and '100' is much greater than the distance between '1010' and '1100' even though the difference between them — 90 — is the same).

With nonlinear transformations, the axes are often labelled with the original units, not the transformed ones, since they are usually easier to interpret. In the example above, it is more informative to label the axes with weights rather than log-weights (or both labels can be used, as done here).

The family of power transformations

Many different mathematical functions may be used to transform data. Some particular functions are useful for specialised types of data.[13] However, one particular family of simple transformations has proved useful in a wide range of situations.

---

13. For example, the logistic function $\ln \frac{y}{1-y}$ has proved to be a useful re-expression when the raw data are proportions or concentrations that are constrained to lie between 0 and 1.

The family of power transformations may be used as a general purpose tool for re-expressing data to meet the aims described in the previous section. A power transformation simply changes a data value x by raising it to some power p, giving the transformed value x^p.

Any value of p between $-\infty$ and ∞ can be used, but some commonly used values are:

$p = 1$ (that is, not transformed)

$p = \frac{1}{2}$ (square root)

$p = -1$ (reciprocal)

$p = \frac{1}{3}$ (cube root)

$p = 2$ (square).

It is relatively uncommon to use a value of p greater than 2 or less than -2. Values of p between 0 and 1 are generally the most useful in practice.

A value of $p = 0$ does not produce a useful transformation (because $x^0 = 1$ for all finite $x \neq 0$). However, for very small positive or negative p (for example, $p = 0.01$ or $p = -0.02$), the effect of the transformation is to change the shape of the data in essentially the same way as a logarithmic transformation.

The following table shows an extended list of power transformations, including the logarithmic transformation, and ordered by the value of p.

p	
2	(square)
1	(that is, not transformed)
$\frac{1}{2}$	(square root)
$\frac{1}{3}$	(cube root)
0	(equivalent to logarithm)
$-\frac{1}{3}$	(reciprocal cube root)
$-\frac{1}{2}$	(reciprocal square root)
-1	(reciprocal)
-2	(reciprocal square)

Values of p less than 1 increase the spread in the lower tail of data values and decrease the spread in the upper tail; they are therefore appropriate if the stem-and-leaf plot shows that the data distribution has a long upper tail and a high density of values at its lower end. The values in the upper tail are brought closer to the body of values, and the values in the lower tail are spread out, showing more detail of the distributional shape. The lower the value of p, the stronger the effect of pulling in the upper tail and spreading out the lower tail.

Values of p greater than 1 have the opposite effect, increasing the spread in the upper tail and decreasing it in the lower tail. Few data sets require such transformations.

Values of p less than 0 have the effect of reversing the order of data values in a batch. For example, $5 < 10$, but $5^{-1} > 10^{-1}$ (that is, $\frac{1}{5} > \frac{1}{10}$). To retain the original order, it is therefore common to transform values to $-x^p$ rather than x^p when p is negative.

The table below gives the 1987 populations of all countries in Africa.[14]

Country	Population ('000s)	Country	Population ('000s)	Country	Population ('000s)
Algeria	23 060	Ghana	13 599	Nigeria	106 736
Angola	8 756	Guinea	6 470	Rwanda	6 454
Benin	4 315	Guinea-Bissau	924	Sao Tome	114
Botswana	1 146	Ivory Coast	11 069	Senegal	6 969
Burkina Faso	8 330	Kenya	22 097	Seychelles	66
Burundi	4 978	Lesotho	1 629	Sierra Leone	3 845
Cameroon	10 927	Liberia	2 327	Somalia	5 712
Cape Verde	343	Libya	4 057	South Africa	33 285
Central African Rep.	2 727	Madagascar	10 894	Sudan	23 214
Chad	5 273	Malawi	7 629	Swaziland	713
Comoros	424	Mali	7 768	Tanzania	23 884
Congo	2 020	Mauritania	1 858	Togo	3 254
Djibouti	362	Mauritius	1 042	Tunisia	7 481
Egypt	50 954	Morocco	22 968	Uganda	15 655
Equatorial Guinea	390	Mozambique	14 591	Zaire	32 655
Ethiopia	44 788	Namibia	1 218	Zambia	7 196
Gabon	1 047	Niger	6 798	Zimbabwe	9 001
Gambia	797				

14. S. Moroney (ed.), *Handbooks to the Modern World: Africa*, Facts on File Publications, New York, 1989, pp. 626–7.

A jittered dot plot of the untransformed populations ($p = 1$) shows a single outlier (Nigeria) and a highly skewed distribution. Power transformations with $p < 1$ are therefore also shown below.

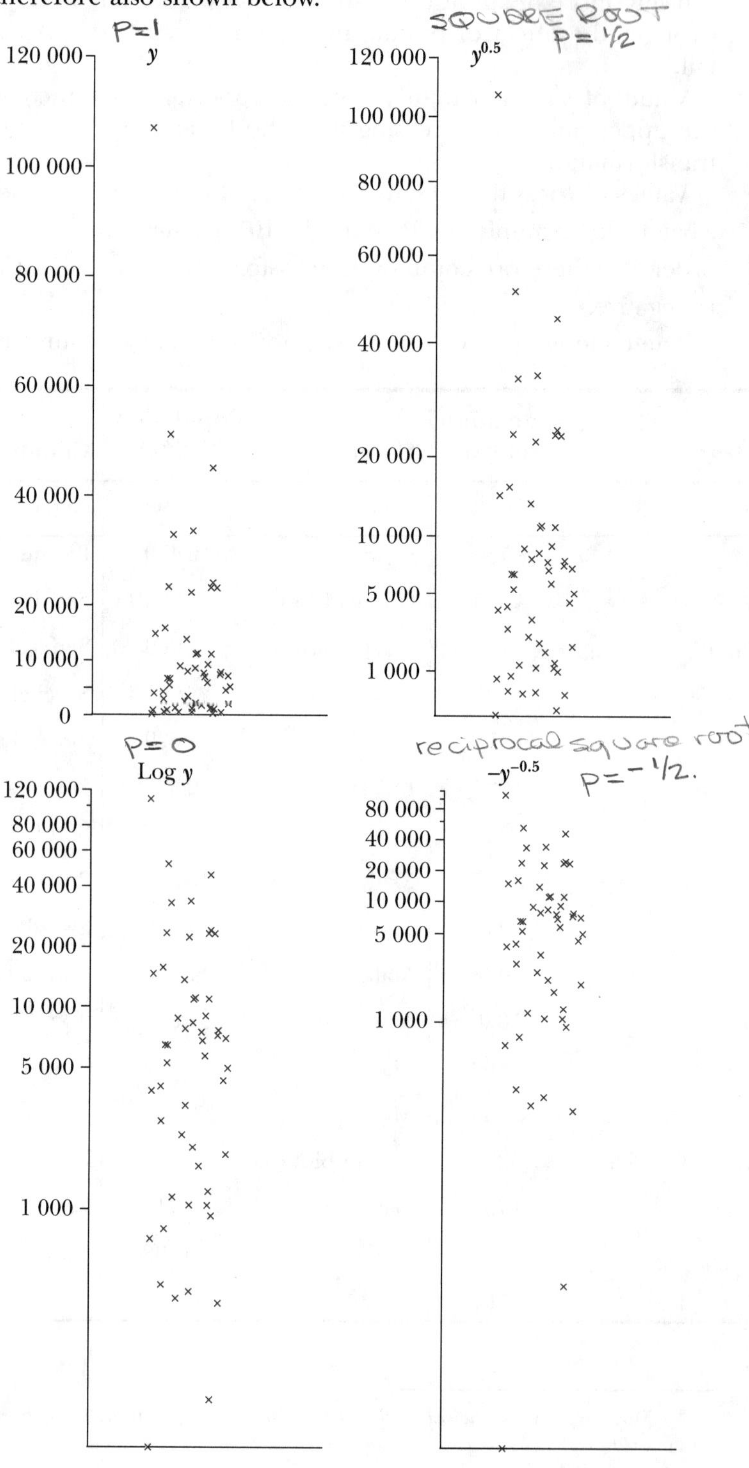

Decreasing the value of p reduces the visual effect of the 'outlier' and spreads out the lower values. A logarithmic transformation ($p = 0$) results in a reasonably symmetric distribution, from which patterns in both tails of the distribution can be seen. Further decreasing the power ($p = -0.5$) makes the lowest country (the Seychelles) appear to be an outlier, and the distribution looks skewed the other way. Note that, for comparison with the other plots, we have plotted '$-y^{-0.5}$' on the axis, rather than '$y^{-0.5}$'. If we had not done so, the least populous country (Seychelles) would have appeared at the top of the plot rather than at the bottom.

Many data sets are inherently non-negative, but for those data sets which contain negative values (and sometimes even for data without negative values), power transformations are sometimes modified by adding a suitable constant before applying the transformation. Without such modification, power transformations *cannot be used* with negative values.[15]

EXERCISE 2.5

1. Why does the transformed brain weight dot plot appear to have two vertical scales? Explain the position of 'elephant' on both scales and then do the same for 'mouse'. Is the graphical position determined by the actual or transformed data?
2. Approximately how large would a brain weight have to be to appear as an outlier on the logarithmic scale? And how small?
3. How many orders of magnitude are spanned by the brain weight data?[16]
4. Why is taking the logarithm of the brain weight data helpful before summarising the data?
5. Consider the following seven data sets.
   (i) 1, 10, 100, 1000, 10 000
   (ii) 1, 9, 25, 49, 81, 121
   (iii) 3, 24, 81, 192, 375, 648
   (iv) 1, 1.4, 1.7, 3.0, 3.2, 3.3, 3.4, 3.6, 4.4, 4.5, 4.6
   (v) 0, 9, 100, 1000, 10 000
   (vi) −2, 0, 3, 4, 5, 8, 10
   (vii) 0.028, 0.04, 0.06, 0.11, 0.25, 1.0

   For each data set, select a transformation from the list below to make it symmetric.
   (a) square
   (b) the identity transformation! That is, don't transform at all.
   (c) square root
   (d) cube root
   (e) log base 10

15. If a data set includes any zeros, logarithmic transformations and transformations with negative powers will not work either.
16. Orders of magnitude is a term used to express the relative size of numbers in a rough way, by indicating how many powers of 10 are required to reach one number from another number. For example, 1435 is about two orders of magnitude larger than 12.

(f) raise to power –0.5; that is, either invert then take square root or take square root then invert

(g) add 1 and then try log to the base 10 — no power transformation will do the job

6. Draw the dot plot for each data set in question 5 and also for the transformed data. Observe the effect of the transformation. (If you don't have a computer to do this, you may wish to think about each example a little, and then check the answers.)
7. The logarithm plot worked well at making the African country population distribution more symmetric. If you only had data for the largest 26 countries, would this same transformation result in a fairly symmetric distribution?

2.6 DISPLAYING DISCRETE DATA

Batches of numerical measurements can be classified into two major types.

1. *Discrete data.* These consist of measurements of a variable which, by its nature, may take values only from a (countable) subset of the real line. Most commonly, discrete data are *counts* of objects or events, and the possible values are the non-negative integers or a subset of them. For example, a batch of values may consist of numbers of road deaths, recorded daily in a city over one month. The numbers of daily deaths are integers, so this is a discrete data set.
2. *Continuous data.* These consist of measurements which could potentially take any value within a section of the real line. For example, temperature varies continuously and is measured on a continuous scale.

All the graphical displays of data described in earlier sections of this chapter can be used for continuous data. Most of them are also appropriate for discrete data, *provided there is a wide range of values in the data.*

The following table gives scores from 106 volunteers on a 'motivation scale'. The subjects were presented with 37 situations and could choose one of two possible actions in response to each situation. One of these satisfied short-term gains and the other was a more morally 'right' action. The score for each subject was the number of morally right actions chosen (a count between 0 and 37).[17]

13	12	11	19	24	2	13
17	15	2	17	15	7	15
13	27	4	16	13	9	5
8	19	4	17	12	5	28
7	23	13	13	6	21	20
10	6	10	7	17	18	19
10	2	13	9	27	17	14
21	9	19	12	3	18	11
18	11	25	11	10	12	14
17	5	14	30	7	15	4
19	18	11	19	1	13	8
15	20	4	4	14	13	10
15	24	14	11	22	15	7
23	15	12	18	16	6	23
12	14	23	18	10	25	18
						24

17. R. J. Larsen and M. L. Marx, *Statistics*, Prentice-Hall, Englewood Cliffs, New Jersey, 1990, p. 90.

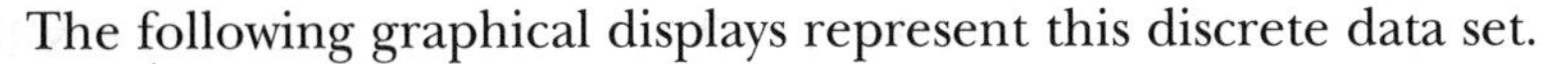

The following graphical displays represent this discrete data set.

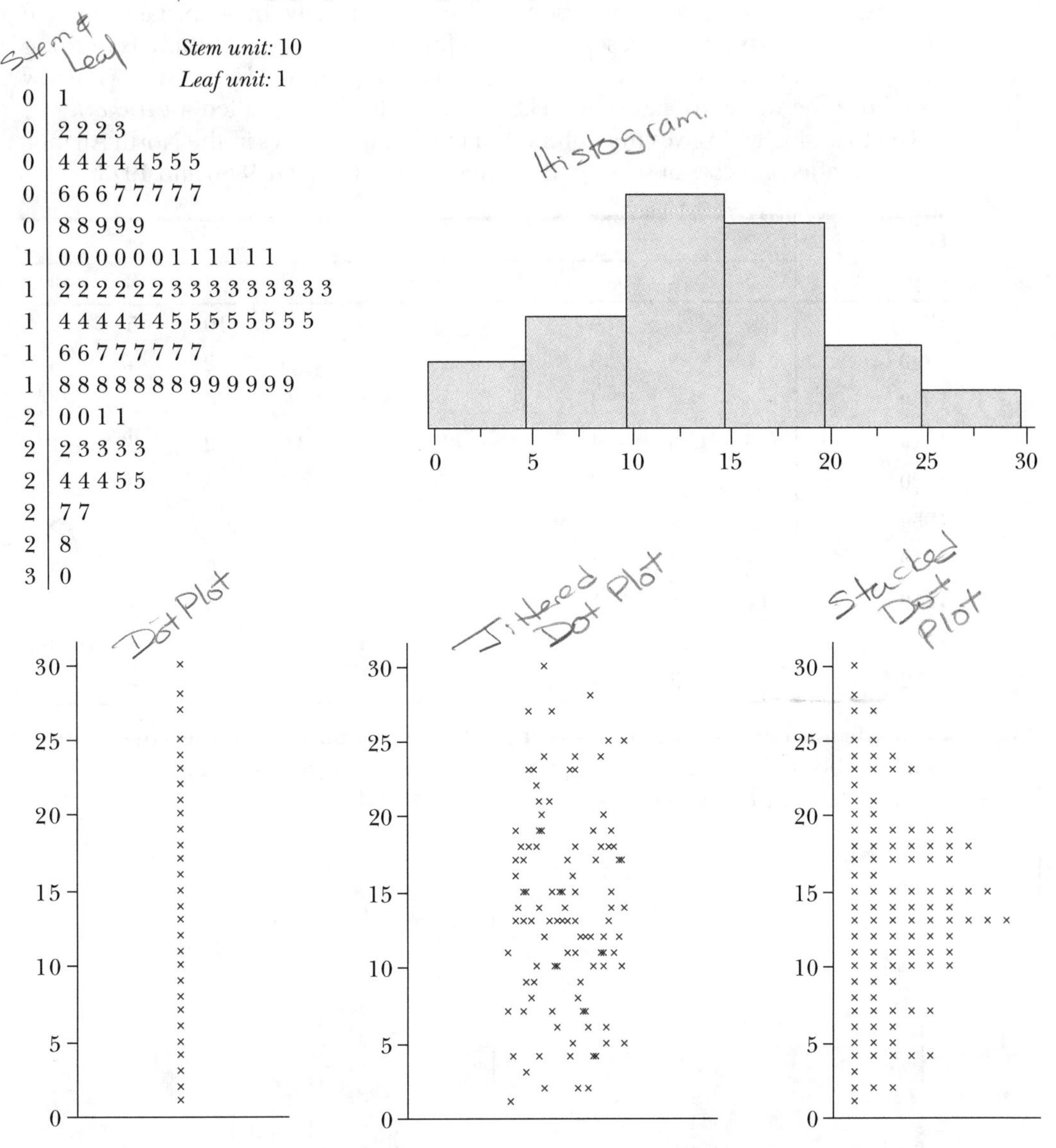

Note that:

- the histogram classes are '–0.5 to 4.5', '4.5 to 9.5', and so on, to ensure that data values do not occur on the boundary of two classes
- the unjittered dot plot does not adequately display the data values since some crosses represent single values and others represent multiple values. The jittered and stacked dot plots work better but there are still potential problems of overlapping crosses in the jittered dot plot.

If there are few distinct values in a discrete data set, the above displays are less effective. Stem-and-leaf plots would have 1's as the stem unit, and all leaves would be '0'. Dot plots would have too many superimposed crosses in their unjittered versions and jittering is not the best way to show density. In such data sets, the best histogram would have a separate class for each value. While there is nothing wrong with this, the discrete nature of the data is highlighted by drawing a narrow bar above the value instead of the class's rectangle. This is called a *bar chart*.

The following table gives the numbers of tropical hurricanes in the North Atlantic each year affecting coastal states of the United States between 1886 and 1973.[18]

Decade beginning	**Year**									
	0	**1**	**2**	**3**	**4**	**5**	**6**	**7**	**8**	**9**
1880	–	–	–	–	–	–	9	10	5	5
1890	1	8	5	10	5	2	6	2	4	5
1900	3	3	3	8	2	1	6	0	5	4
1910	3	3	4	3	0	4	11	2	3	1
1920	4	4	2	3	5	1	8	4	4	3
1930	2	2	6	10	6	5	7	6	3	3
1940	4	4	4	5	8	4	3	6	7	8
1950	11	8	6	6	8	9	4	3	7	8
1960	4	8	3	7	6	4	7	6	4	10
1970	3	5	3	4	–	–	–	–	–	–

Both the histogram and the bar chart below contain the same shape information about the data, but the bar chart (on the right) is preferred since it more clearly indicates the discrete nature of the data.

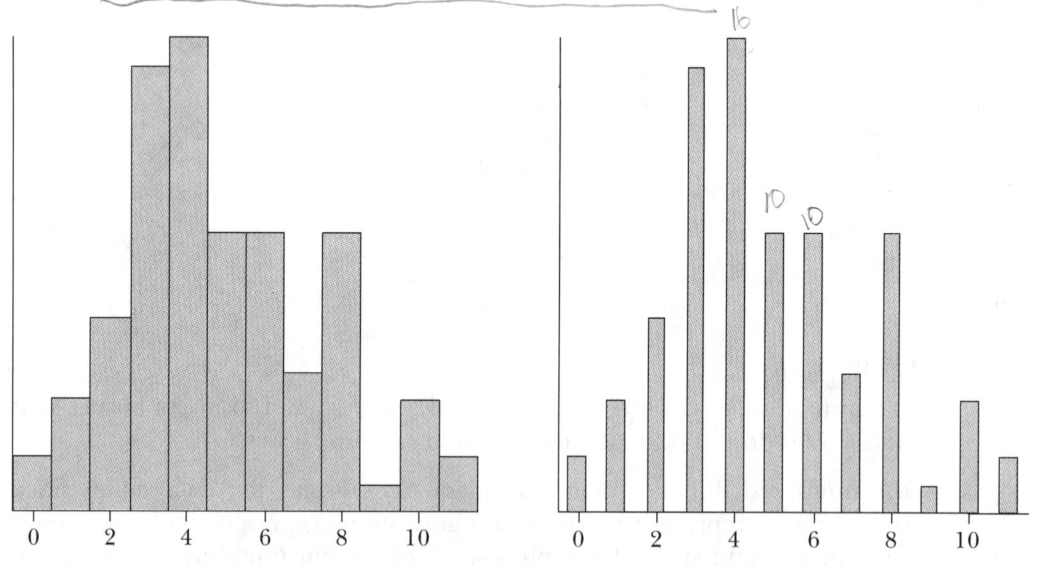

18. G. W. Cry, W. H. Haggard and H. S. White, *North Atlantic Tropical Cyclones*, Technical Paper no. 36, US Weather Bureau, Washington. In H. H. Lamb, *Climatic History and the Future*, Methuen, London, 1977, p. 640.

The height of each bar in a bar chart is proportional to the number of times that value occurs in the data set (the value's *frequency*). A bar chart is usually drawn from a *frequency table* in which the frequencies of all possible values are shown. (The frequency table also provides a convenient numerical data display.) The frequency table for the tropical hurricanes data is shown on the right.

Value	Frequency
0	2
1	4
2	7
3	16
4	17
5	10
6	10
7	5
8	10
9	1
10	4
11	2

EXERCISE 2.6

1. Explain why, for discrete data, a stacked dot plot and a bar chart will look essentially the same.
2. For the following data sets, would you prefer a histogram or a bar chart to display the distribution?
   (a) ages of patients in a hospital (with age rounded in the usual way)
   (b) number of brothers and sisters of children at a particular school
   (c) number of fire alarms in each month over the past five years, for a particular city

2.7 Comparing two or more batches

We have concentrated so far on examining the characteristics of a single batch of values. When we have several batches of related data, we might want to compare them. We want to know how the batches differ and, in particular, whether there are any differences in the location, spread and shape of the batches.

The methods of display we have looked at so far may all be used when comparing two or more batches of data, but certain modifications to the basic displays allow comparisons to be made more effectively. In particular, drawing the displays of the different batches against a common axis allows the eye to notice differences that might not be apparent otherwise.

One useful approach for two batches is to draw back-to-back stem-and-leaf plots (or histograms).

The following table describes leg measurements of common crossbills.[19] Approximately half of these birds have bills with the upper mandible crossing

19. A. J. B. Anderson, *Interpreting Data*, Chapman and Hall, New York, 1989, p. 45.

to the left (left-billed) and the remainder have bills with the upper mandible crossing to the right (right-billed). The data were recorded to determine whether the method of feeding places more strain on one leg than the other, depending on the crossing of the bill.

Left-billed				Right-billed			
Bird	Left leg (mm)	Right leg (mm)	Difference (left–right)	Bird	Left leg (mm)	Right leg (mm)	Difference (left–right)
1	19.0	17.4	1.6	13	17.8	18.0	–0.2
2	17.1	17.3	–0.2	14	15.6	17.1	–1.5
3	16.6	16.6	0.0	15	16.9	17.2	–0.3
4	17.0	16.7	0.3	16	16.9	17.1	–0.2
5	18.3	18.2	0.1	17	17.0	17.3	–0.3
6	17.5	17.3	0.2	18	16.9	17.1	–0.2
7	16.4	16.3	0.1	19	17.4	17.6	–0.2
8	17.1	16.9	0.2	20	17.4	17.7	–0.3
9	17.5	16.7	0.8	21	17.7	17.7	0.0
10	17.0	16.8	0.2	22	16.5	17.5	–1.0
11	18.6	17.5	1.1	23	17.1	17.3	–0.2
12	16.3	16.2	0.1	24	16.7	16.9	–0.2

A back-to-back stem-and-leaf plot of the differences in leg lengths for the two groups of birds is shown on the right. It is formed by drawing a standard stem-and-leaf plot for one group (the right-billed birds), then adding the leaves for the other group (the left-billed birds) to the left of the column of stems.

The stem-and-leaf plot clearly shows that left-billed birds usually have longer left legs than right legs (with positive differences) whereas right-billed birds usually have longer right legs (negative differences). Both distributions are skewed, but towards opposite ends of the axis.

This back-to-back display provides a quick and useful overall comparison of shape, spread and location, but a detailed comparison of the heights of leaf stacks for individual stems is not easy because the plots are drawn back-to-back. An overlaid display makes this aspect of the comparison easier.

When there are different numbers of individuals in the batches which are compared, comparisons of the two stem-and-leaf plots are particularly difficult. This problem is avoided if histograms are used, provided the two histograms are scaled to make their areas the same.

6	1	
	1	
	1	
1	1	
8	0	
	0	
	0	
3 2 2 2	0	
1 1 1 0	0	0
	–0	
2	–0	2 2 2 2 2 2 3 3 3
	–0	
	–0	
	–0	
	–1	0
	–1	
	–1	5

Stem unit: 1
Leaf unit: 0.1

In demography, back-to-back histograms are often used to display the age distributions of males and females in a population or to compare these age distributions at different periods or in different countries. They are then called *population pyramids*.

The population pyramids below show the age distributions of males and females from the major ethnic groups in New Zealand in 1989.[20]

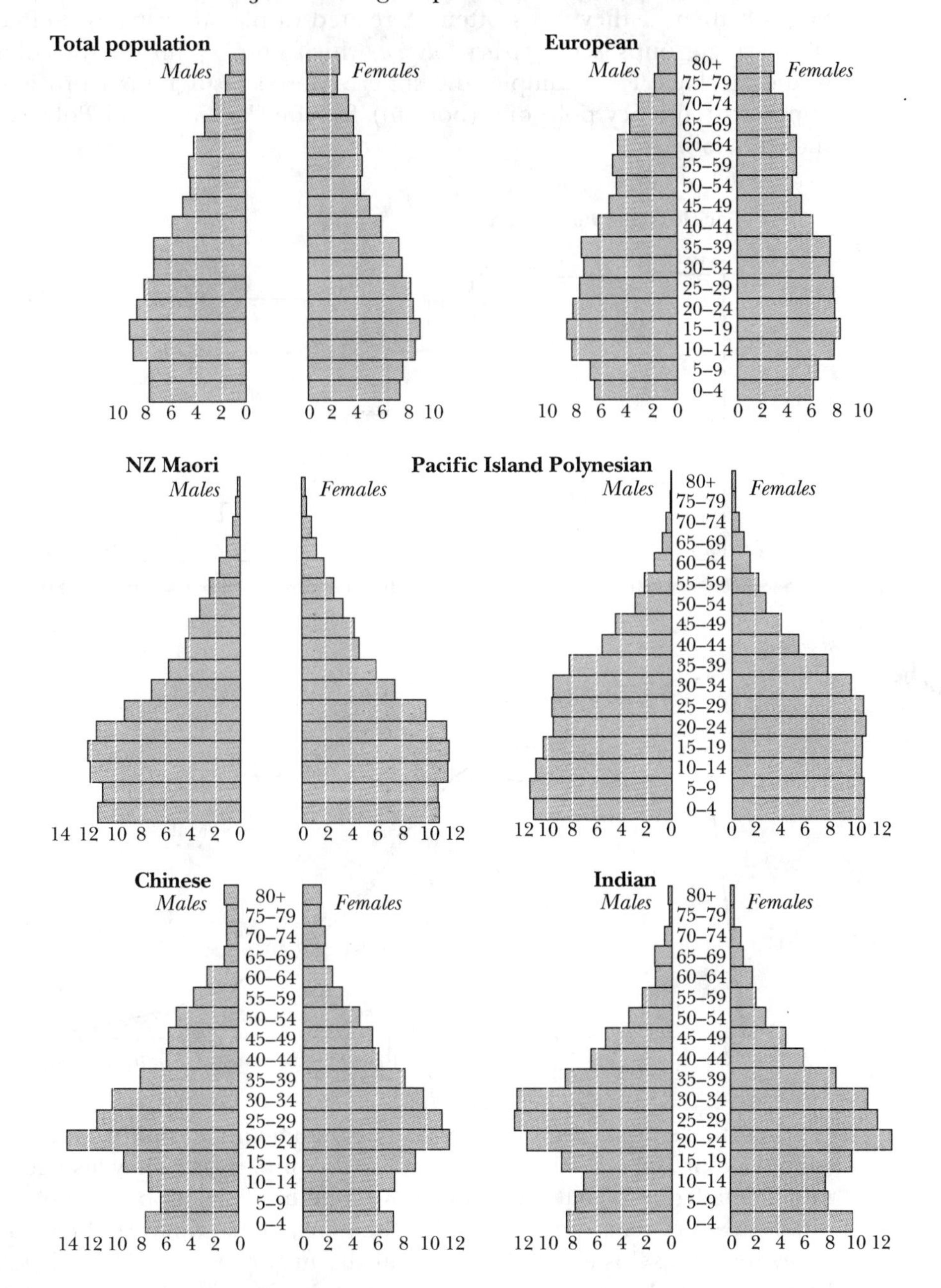

20. *New Zealand Official 1990 Yearbook*, Department of Statistics, Wellington, 1990, p. 155.

The Maori and Pacific Island population pyramids have wider bases than the others, indicating high birth rates and a relatively youthful population. Also note the asymmetry of the Chinese and Indian population pyramids. In particular, the distribution of Indian males peaks at a higher age than that of Indian females.

Two or more histograms may be superimposed, but it is often difficult to distinguish them if they cross often. A related display that is more effective for comparing groups is a *frequency polygon,* which simply joins the tops of adjacent histogram bars. For example, the superimposed histograms (top) and superimposed frequency polygons (bottom) for the Pacific Island Polynesians are shown below.

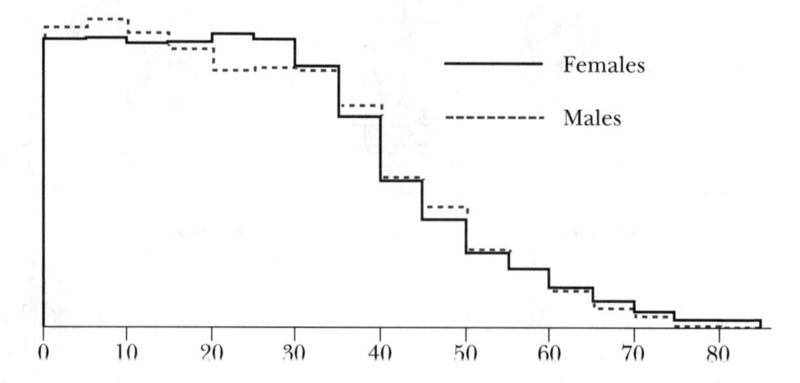

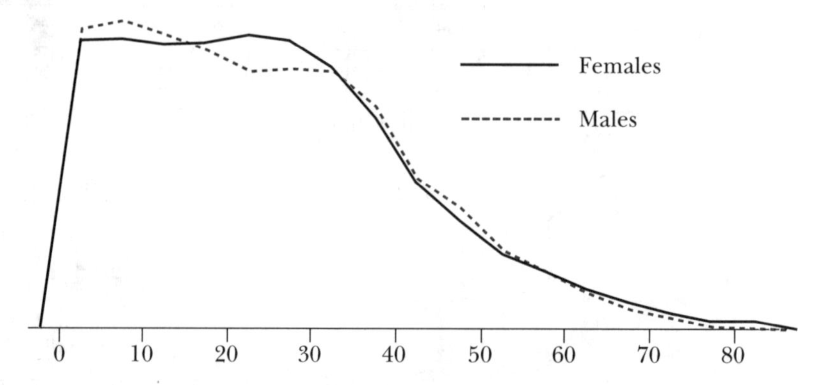

When there are three or more batches, stem-and-leaf plots and histograms cannot be linked effectively for easy visual comparison. They also contain too much detailed information that can overwhelm more important aspects of the differences between the batches. It is beneficial to summarise the data in such a way that 'noise' is eliminated, so that the major features of the batches may be compared. A more highly summarised display, called the box plot, allows better comparisons between several batches and will be described in section 2.8.

EXERCISE 2.7

1. The following table describes the age distribution of 123 males and 210 females in a nursing home.

Sex/Age	60–69	70–79	80–89	90–99	All
Males	23	46	32	12	123
Females	19	55	97	39	210
All	42	101	129	51	333

   (a) How might these numbers be adjusted to make it easier to compare the male and female age distributions?
   (b) How would you adjust the frequencies to enable the sex distribution over the various age classes to be compared?

2. Count the number of 'squares' (by adding portions of squares where necessary) in one of the population pyramids shown, for each sex separately. Do it again for another pyramid. Can you give a reason for the totals you get?

2.8 QUARTILES AND BOX PLOTS

When comparing several batches of values, the main problem with the graphical displays described in earlier sections is that they retain too much detailed information about the individual values. This detailed distributional information obscures the main differences between the batches.

Many batches can be effectively summarised with a few numerical values and such numerical summaries provide a concise description of the distribution of values within the batch. Numerical summaries of data are often useful adjuncts to graphical summaries and this is especially so when we are comparing batches of data; the more batches, the greater the need for numerical summaries.

In this section we will describe one particular set of numerical summaries called the *5-number summary*. This consists of the minimum and maximum values in the batch and three further summaries called the *median* and *quartiles*. The 5-number summary also forms the basis of an effective graphical display.

Median

The *median* is the middle value when the data are arranged in order of magnitude. The *rank* of a value indicates its position in the ordered list of values in the batch so, for example, the minimum value has rank 1 and the value with rank 10 is the 10th smallest value in the batch. The rank of the median is midway between the ranks of the minimum and maximum values. Thus, the rank of the median in a batch of n values is $\frac{n+1}{2}$.

If n is *odd*, this calculation gives an integer for the rank of the median and the median is *the* middle value. For example, if the batch is:

$$68, 27 \text{ and } 39 \quad (n = 3)$$

then the rank of the median is $\frac{3+1}{2} = 2$, so the median is the 2nd ordered value, 39. However if n is *odd*, the rank of the median is not an integer; we therefore define the median to be the *mean* (or *average*) of the two middle values. For example, if the batch is:

$$68, 27, 39 \text{ and } 42 \quad (n = 4)$$

then the rank of the median is $\frac{4+1}{2} = 2\frac{1}{2}$, so the median is halfway between the 2nd and 3rd ordered values, $\frac{39+42}{2} = 40.5$.

For the contamination data, $n = 20$, so the median has rank $10\frac{1}{2}$ and is the average of the 10th and 11th values in order of magnitude. Since a stem-and-leaf plot effectively sorts the data into order, the median can easily be determined from it by counting up leaves in columns until the 10th and 11th leaves are found.

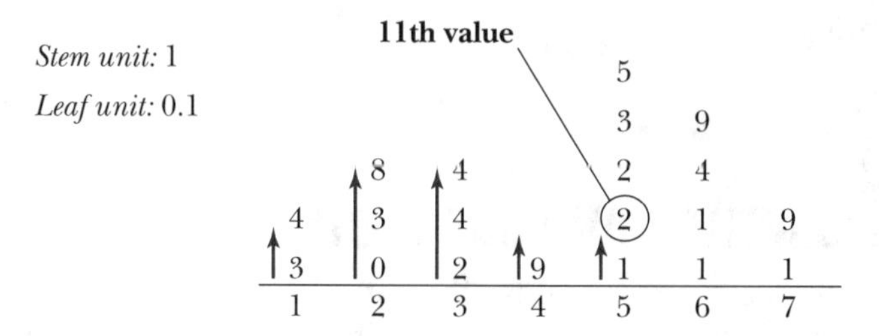

Thus: Median $= \frac{5.1 + 5.2}{2} = 5.15\%$.

Quartiles

The median divides the whole data set into two halves, so half the data lie above the median and half below it. The *quartiles* divide the data into quarters. The middle quartile is therefore the median.

The rank of the lower quartile is midway between the ranks of the median and the minimum values. It therefore has rank:

$$\frac{\left(1 + \frac{n+1}{2}\right)}{2} = \frac{n+3}{4}.$$

Similarly, the rank of the upper quartile is midway between the ranks of the median and maximum values in the batch:

$$\frac{\left(n + \frac{n+1}{2}\right)}{2} = \frac{3n+1}{4}.$$

As with the median, the ranks of the quartiles may be fractional and a complete definition should specify what is meant by fractional ranks. For example, if $n = 8$, the lower quartile has rank $\frac{11}{4} = 2\frac{3}{4}$. The precise definition of where the '$2\frac{3}{4}$th' ordered value lies between the 2nd and 3rd values is not, however, of practical importance and the details have therefore been relegated to a footnote.[21]

Box plots

The 5-number summary is defined in the following box.

> Smallest value, lower quartile, median, upper quartile, largest value.

Together with the sample size, these constitute a set of useful landmarks which are also the basis for an extremely useful graphical display — the *box plot*, which displays the five values graphically.

A box plot is drawn to the right of a jittered dot plot of the contamination data in the diagram below. Note that the five horizontal lines in the box plot split the data set into four groups of five data values. Depending on the number of values in a batch, the median and/or quartiles will coincide with one or more of the data values, but the values will still approximately be split into quarters.

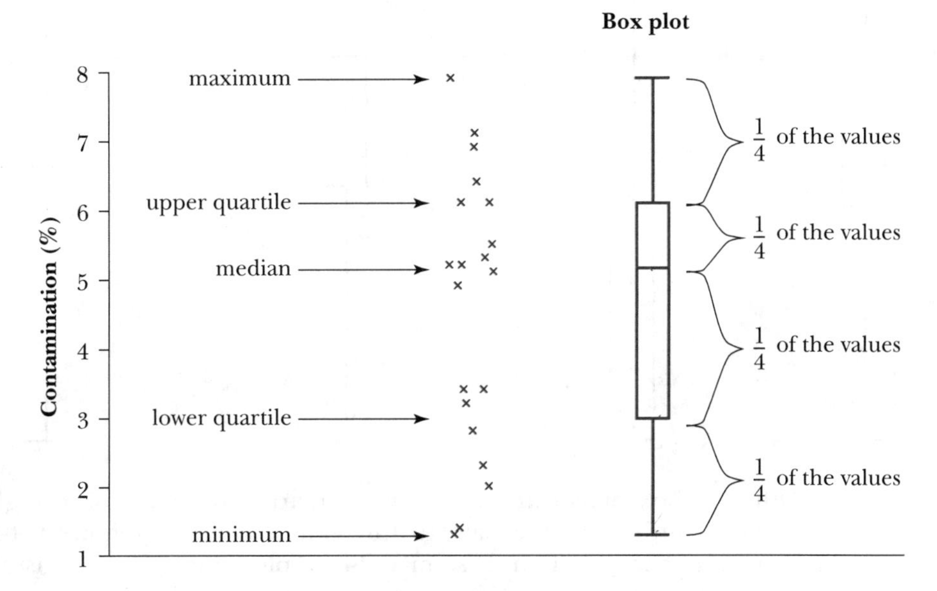

The shape of a box plot gives a lot of information about the distribution of values in a batch. It not only indicates the middle of the batch but also shows

21. To compute a value having a fractional rank, just use the two closest integral rank values in the natural way. For example, to compute the value with rank 2.75, start with the rank 2 value and add to it 0.75 of the difference between the rank 3 value and the rank 2 value. If the rank 2 value is 5 and the rank 3 value is 7, the rank 2.75 value would be $5 + 0.75 \times (7 - 5) = 6.5$.

which of the four 'quarters' of the data set have values close to each other and which 'quarters' have widely differing values. With a little experience, a box plot can therefore give a quick indication of the shape of a distribution of values.

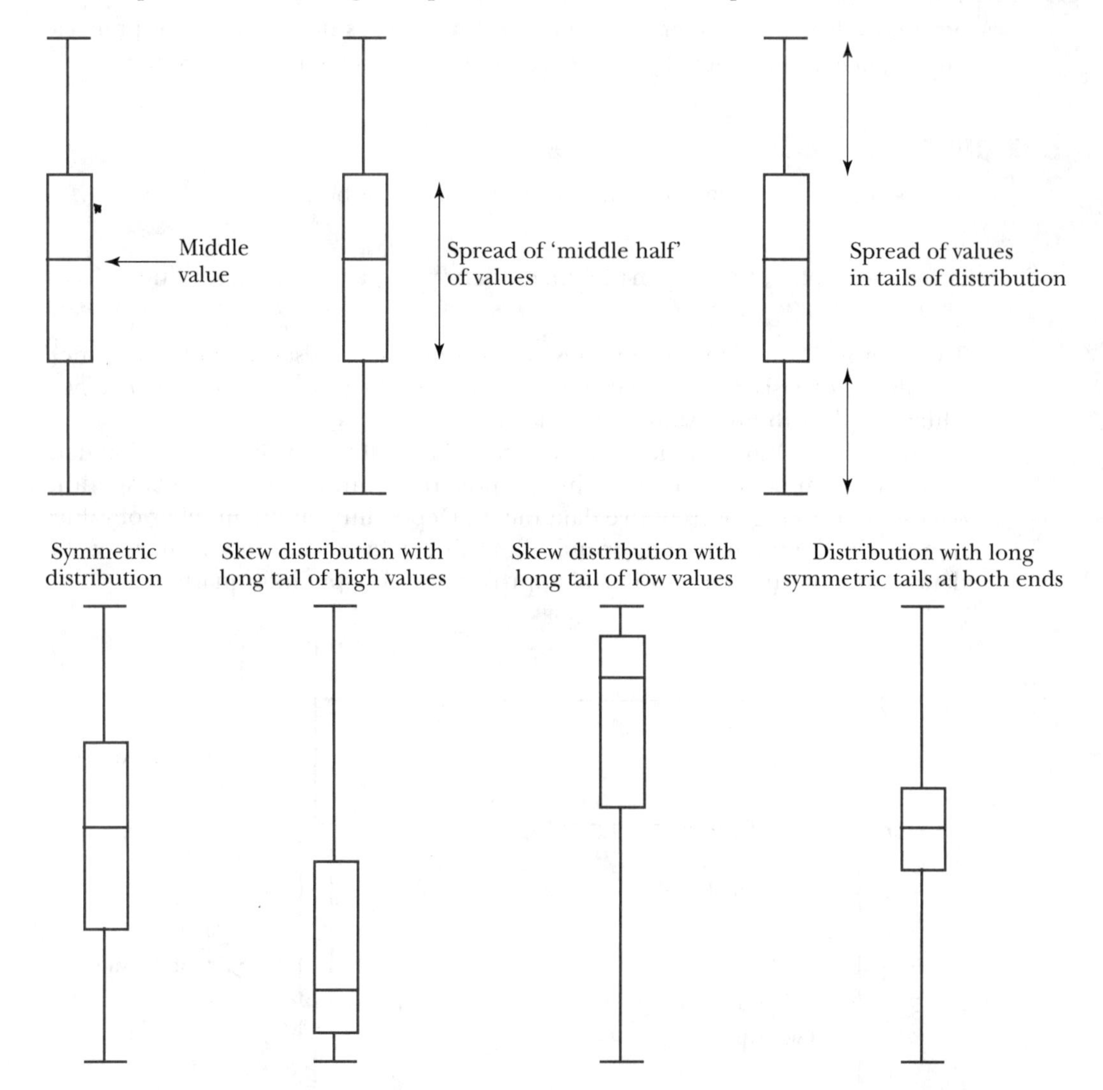

Although box plots can effectively summarise the shape of a single batch of values, condensing the data to this extent is rarely required; the extra information contained in a stem-and-leaf plot, dot plot or histogram is rarely distracting and can show features that cannot be seen in a box plot. However, box plots provide an extremely effective way to compare two or more batches of data. For example, the following table describes the voting turnout in each of the 65 electorates at the October 1996 general election in New Zealand.[22]

22. *The Dominion* (Wellington, New Zealand), 26 October, 1996, p. 2. Reproduced by permission.

Electorate	Party	Turnout %	Electorate	Party	Turnout %
Albany	National	87.42	Ohariu-Belmont	United NZ	87.67
Aoraki	Labour	87.95	Otago	National	87.58
Auckland Central	Labour	88.08	Otaki	Labour	88.22
Banks Peninsula	National	89.71	Owairaka	Labour	86.74
Bay of Plenty	National	85.23	Pakuranga	National	87.51
Christchurch Cent.	Labour	83.68	Palmerston North	Labour	86.73
Christchurch East	Labour	86.10	Port Waikato	National	83.98
Clutha-Southland	National	84.75	Rakaia	National	86.75
Coromandel	National	86.47	Rangitikei	National	86.45
Dunedin North	Labour	88.48	Rimutaka	Labour	88.32
Dunedin South	Labour	88.87	Rodney	National	86.99
Epsom	National	89.15	Rongotai	Labour	87.38
Hamilton East	National	86.89	Rotorua	National	86.05
Hamilton West	National	85.09	Tamaki	National	88.05
Hunua	National	85.50	Taranaki-King Cntry	National	83.91
Hutt South	Labour	87.85	Taupo	Labour	82.81
Ilam	National	89.22	Tauranga	NZ First	87.19
Invercargill	Labour	86.51	Tukituki	Labour	86.46
Kaikoura	National	86.56	Waimakariri	Labour	87.62
Karapiro	National	84.23	Waipareira	National	85.15
Mahia	Labour	84.97	Wairarapa	National	87.02
Mana	Labour	87.67	Waitakere	National	85.78
Mangere	Labour	79.78	Wellington Central	ACT NZ	91.49
Manukau East	Labour	84.11	West Coast-Tasman	Labour	86.88
Manurewa	Labour	81.16	Wanganui	Labour	86.37
Maungakiekie	National	85.28	Whangarei	National	85.83
Napier	Labour	86.95	Wigram	Alliance	86.35
Nelson	National	86.88	Te Puke/Whenua	NZ First	71.36
New Lynn	Labour	85.28	Te Tai Hauauru	NZ First	69.91
New Plymouth	Labour	86.86	Te Tai Rawhiti	NZ First	74.09
North Shore	National	88.34	Te Tai Tokerau	NZ First	73.15
Northcote	National	86.76	Te Tai Tonga	NZ First	73.56
Northland	National	84.31			

The data can be displayed with jittered dot plots, as shown in the diagram below.

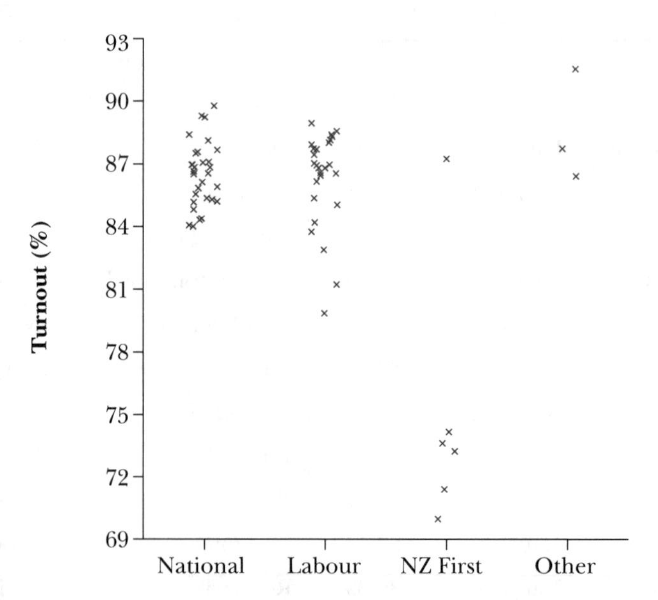

The dot plots clearly show the relatively narrow spread of voter turnouts in National seats and the group of NZ First seats with low turnouts (all Maori electorates). However, the dot plots retain a lot of detailed information about the individual National and Labour seats, which distracts from the comparisons between the parties. The following box plots show only the 'major' features of the batches.

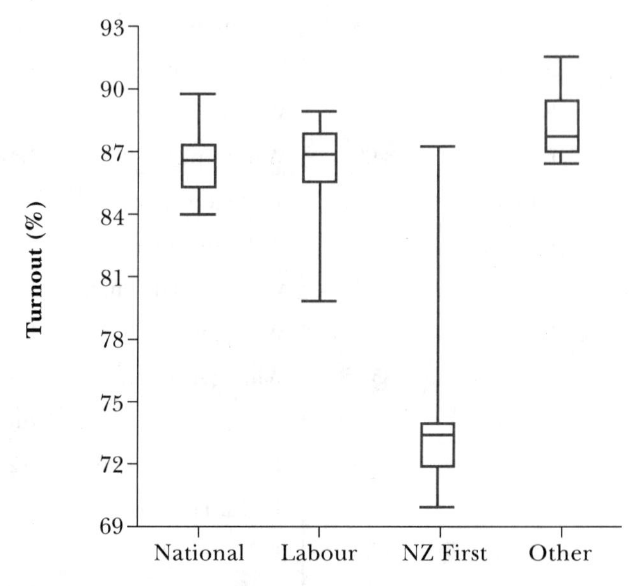

Although box plots are effective summaries of the National and Labour electorates, dot plots describe the small batches of NZ First and Other turnouts better.

The basic box plot can be beneficially modified or annotated with other relevant information, as discussed on page 57.

Outliers

It is useful to identify outliers on a box plot. A somewhat arbitrary but frequently used definition of an outlier is as follows. The difference between the two quartiles is called the *interquartile range* (IQR) and is a measure of the spread of the data (see section 2.9). Any observation more than 1.5 IQRs beyond the upper or lower quartiles may be regarded as an outlier. Any such outlier is worthy of more detailed examination.

Outliers
1.5 × IQR
IQR
1.5 × IQR
Outliers

The above rule defining outliers is most appropriately applied to data which are not skew. For symmetric bell-shaped distributions, such outliers are rare.[23] However, the outlier rule should be used with caution. If the data are markedly skew, it may be desirable to transform them to a shape more nearly symmetric and then to apply the rule. It does not make statistical sense to call a data value an outlier on one scale, but not on another transformed scale, whatever the visual impression may be.

If some values have been identified as outliers, the lines from the central box are drawn only as far as the most extreme (low and high) data values which are not outliers. The outliers themselves are shown as crosses on the plot.

Mean

Sometimes the mean (or average) of the batch is marked on a box plot with an asterisk (*), since two measures of centre may be better than one. (The mean of a batch of numbers will be formally defined in section 2.9.)

The box plot provides a very effective overall summary but, through summarising, it loses what may be important information. The most important example of this is with bimodal or multimodal data (where the shape of the histogram or stem-and-leaf plot has two or more distinct peaks). A box plot would hide the clustering in such data and would therefore be misleading. A stem-and-leaf plot or histogram would give a truer representation of the data.

EXERCISE 2.8

1. Since a box plot loses information about clusters, why is it sometimes preferred to a stem-and-leaf plot or a histogram?
2. Why is a box plot usually preferred to a histogram or a dot plot when batches from several groups are to be compared?
3. An index used by some researchers to describe the different body shapes of men and women is the ratio of the shoulder diameter to the hip diameter. If you had a measurement of this ratio for a group of adults, but did not have the information relating the sex of the individual to the measured ratio, which of the following methods would most likely reveal the expected presence of two clusters in the distribution: a box plot, a histogram or a dot plot?

23. In particular, for the normal distribution which will be described in chapter 8, approximately 1% of values would be classified as outliers by this rule.

2.9 NUMERICAL SUMMARIES

In the previous section, we defined five numerical summaries of a batch of numbers (the minimum, maximum, median and two quartiles) that together provide a great deal of information about the distribution of values in the batch. For many batches of values, two numerical summaries are sufficient to describe the main features of the data.

Mathematical notation

In defining some numerical summaries, it is useful to introduce some mathematical notation. For mathematical convenience, variables are commonly given one-letter names:

$$x, y, \ldots$$

Because there is only a finite alphabet, it is often useful to use subscripts to identify the individual values in a set of data:

$$x_1, x_2, \ldots, x_n$$

where n represents the size of the data set.

Some numerical summaries involve summation, which is represented by the symbol Σ (Greek, upper case 'sigma'). For example, the total of all values in a data set is denoted by:

$$\sum_{i=1}^{n} x_i \quad \text{or simply} \quad \sum x.$$

Measures of centrality or location

A *measure of centrality* describes, in some sense, the centre of a data set or the location of a 'typical value'. There are various such measures, including the median (which was defined in section 2.8).

The units for all measures of location are the same as for the 'raw' data (for example, if the data are measured in miles, the median is also measured in miles).

The most commonly used measure of centrality is the *mean*. For a data set $x_1, x_2, \ldots, x_n$ the mean is:

$$\bar{x} = \frac{x_1 + x_2 + \ldots + x_n}{n} = \frac{\sum_{i=1}^{n} x_i}{n}$$

or, in further abbreviation:

$$\bar{x} = \frac{\sum x}{n}.$$

For example, the mean of 68, 27, 39 and 42 is:

$$\frac{68 + 27 + 39 + 42}{4} = 44.$$

For the contamination data:

$$\sum x = 91.5$$

$$n = 20$$

$$\bar{x} = \frac{91.5}{20} = 4.575\%.$$

A physical interpretation can be given to the mean which helps to explain its properties. Imagine a horizontal dot plot where all dots are equally sized and each dot is a miniature sphere of lead that is attached to a rigid axis, and that the axis itself has negligible mass. The mean is the position on the axis where the beam will balance, as illustrated in the diagram below.

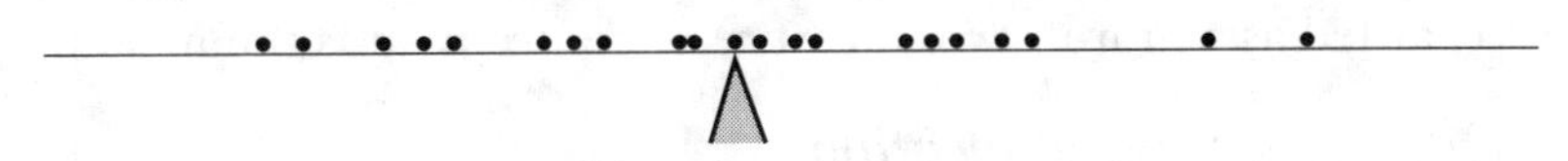

Mean = Point of balance

Comparison of mean and median

Do the mean and median measure the same thing? Typically, they do not.

For symmetric data, the mean and median both measure the same centre, but you will almost never come across a data set which is exactly symmetric. The mean and median normally do not have the same numerical value (although, by chance, this might happen).

When the data units are dollars, the mean is often thought of as an economic indicator and the median as a social indicator. The mean income of a group, for example, being the sum of the incomes divided by n, reflects the total wealth of the individuals, whereas their median income is the income that exactly half the group exceeds, representing the income of a 'typical' person. Similar differing interpretations may be placed on the mean and median in other contexts.

It is important to recognise that, as summary statistics, each offers a numerical summary of only one aspect of a data set. The mean or median should not be looked at in isolation, but must be used in conjunction with other numerical and graphical summaries. When the mean and median differ substantially, these other summaries will point to why this is so.

For data which are skewed with a long upper tail, the mean is larger than the median since, unlike the median, it is influenced by the numerical values of *all* observations in the data set, as illustrated below. Likewise, the mean is less than the median for data sets which are skewed with a long lower tail.

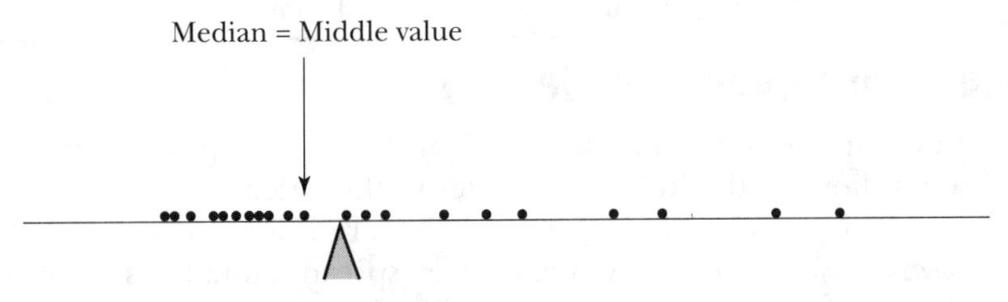

Mean = Point of balance

Similarly, a single 'outlier' will exert high leverage on the axis and therefore the mean will be considerably increased. For this reason, the mean is more sensitive than the median to outliers. To illustrate this, consider the following data set of size 4:

6.8, 2.7, 3.9, 4.2 $\bar{x} = 4.4$ median = 4.05.

If the first value was erroneously recorded as 68 instead of 6.8, the mean and median would be as follows:

68, 2.7, 3.9, 4.2 $\bar{x} = 19.7$ median = 4.05.

As illustrated by this example, a single outlier can have a large effect on the mean, but may not change the median at all. The median is therefore regarded as a more *robust* measure of centre than is the mean.

Other measures of location

Various other numerical summaries have been proposed to measure the centre of a batch of values. These are far less commonly used than the median and mean, but some will be described below since the terms are used elsewhere.

The *mode* is the most common value in a data set. It is sometimes suggested as a reasonable measure of location; however, it is so inferior to the mean and median for this purpose that we do not recommend it. The concept of mode is useful in other contexts in statistics, but these applications are beyond the scope of this book.

The median 'trims' off 50% of the values from each tail of the data set (leaving the middle value). *Trimmed means* are based on deletion of a smaller proportion of each tail (for example, 5%) and the value of the trimmed mean is the mean of the remaining data values.

Sometimes it is useful to use a numerical summary to describe the location of some part of a batch of values other than the centre. Quartiles are one example. More generally, a *percentile* indicates a point below which a chosen percentage of data values fall.

The main location measures used in statistics are summarised in the table below. Each measure of location has the property that adding the same constant to each data value shifts the location measure by that constant.

Measures of centre	Other measures
Mean	Quartiles
Median	Percentiles
Mode	Minimum
Trimmed mean	Maximum

Measures of spread or variability

Knowing the centre of a batch of values does not provide us with information about how far the individual values in the batch are from this centre. Thus, a second type of useful numerical summary is a *measure of spread* (also called a *measure of variability*). A measure of spread should distinguish between the three data sets illustrated on page 61, all of which have the same centre.

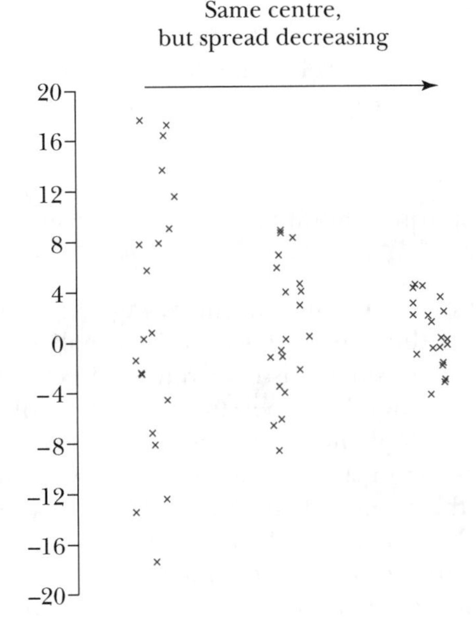

Whatever measure of spread we choose, it should not be affected by adding a constant to each value in a data set to change its location.

Consider the following two data sets of size 4:

Data set *A*	68, 27, 39, 42
Data set *B*	168, 127, 139, 142

Each observation in B is 100 more than the corresponding observation in A, and all measures of location will therefore be 100 higher for B than for A. However, the 'spread of values' is the same for both and, hence, so should be the value of any measure of spread. We will consider several definitions of spread which meet this requirement.

The *range, R*, of a batch of values is the difference between the largest and smallest values in the sample; that is:

$$R = x_{\text{max}} - x_{\text{min}}$$

The range is easy to compute, but it has some undesirable features:

- It is very sensitive to outliers (whether they are errors or just atypical values).
- It is insensitive to variation in the bulk of the data (since it is based on only the most extreme upper and lower values).
- Large data sets tend to have a bigger range than small data sets of the same type (since adding values to a data set can only increase the range).

These undesirable features suggest that the range should not be used as a measure of spread. The suggestion is valid, but in one important area of statistical practice — control charts — the norm in industry is to use the range as *the* measure of spread.[24] The practice is expected to diminish as the use of Statistical Process Control in industry increases.

A second measure of spread that was defined in section 2.8 is the *interquartile range* (IQR).

Interquartile range = Upper quartile – Lower quartile.

Like the range, the interquartile range is easy to compute and easy to interpret; it is the range of the middle half of the data. It also is a *robust* measure of spread since it is relatively unaffected by outliers.

Spread must always be measured in terms of differences between values. The range and IQR measure differences between selected values. The most popular measures of spread are defined in terms of differences (deviations) of values from the centre of the batch. The deviation from the mean, $\bar{x}$, of an individual value x is $x - \bar{x}$. We now want to find an overall measure of spread which aggregates these individual deviations.

Suppose we define, as a measure of spread, the average deviation from the mean:

$$\frac{\sum(x - \bar{x})}{n}.$$

Is this useful? Consider the data set of four values (–4, 3, –2, 5). The mean of these values is $\bar{x} = \frac{2}{4} = 0.5$ and the deviations are shown in the table below.

	x	$x - \bar{x}$
	–4	–4.5
	3	2.5
	–2	–2.5
	5	4.5
Total	2	0.0

Some of the deviations are positive and some are negative, and they always balance out to give a total of *exactly zero*. That is, the average deviation from the mean of any batch of values is always zero:[25]

There are two ways to overcome this problem, both of which make all deviations positive before averaging them.

24. Section 3.4 will briefly describe this use of the range in quality control.

25. $\sum_{i=1}^{n} (x_i - \bar{x}) = \sum_{i=1}^{n} x_i - \sum_{i=1}^{n} \bar{x} = \sum_{i=1}^{n} x_i - n\bar{x} = \sum x - n\frac{\sum x}{n} = \sum x - \sum x = 0.$

We can ignore the signs of the individual deviations and define *mean absolute deviation* to be:

$$\frac{1}{n}\sum|x-\bar{x}|.$$ [26]

	x	$x-\bar{x}$	$\lvert x-\bar{x}\rvert$	$(x-\bar{x})^2$
	–4	–4.5	4.5	20.25
	3	2.5	2.5	6.25
	–2	–2.5	2.5	6.25
	5	4.5	4.5	20.25
Total	2	0.0	14.0	53.00

The mean absolute deviation for our data set is therefore $\frac{14.0}{4} = 3.5$.

Although the mean absolute deviation is a sensible measure of spread, it is rarely used (for a variety of reasons which relate to computational, mathematical and theoretical statistical issues). Another measure of the spread of values in a batch is based on their squared deviations rather than their absolute values. The *variance* is an average of these squared deviations.

$$\text{Variance} = \frac{1}{n-1}\sum(x-\bar{x})^2.$$

For technical reasons, which we will not go into here, the divisor $(n - 1)$ is used rather than n. When the data are concentrated near their mean, the individual deviations will be small and therefore the variance will also be small; when the data are more variable, the deviations, and hence the variance, will be larger. In the artificial example above, the variance is $\frac{53.00}{3} = 17.67$.

Although the variance does reflect the spread of values in a batch, it is not an easily interpreted quantity, and should not be used as a descriptive numerical summary of the batch. The problem is that the variance is defined in terms of squared deviations and therefore its units are the square of the units of the sample values. For example, the variance of a batch of values which are lengths in metres is measured in square metres.

The units of variance are often unpleasant. (Square metres are fine, but how do we interpret a squared year?) For this reason, the square root of variance is used as a more easily interpreted measure of spread. This measure, called the *standard deviation* and denoted s, is the most commonly used measure of spread. The variance is therefore denoted by s^2. The formula for the standard deviation is therefore:

$$s = \sqrt{\frac{\sum(x-\bar{x})^2}{n-1}}.$$

26. The symbols | | mean 'take the absolute value of'. In other words, 'turn negative values into positive and leave positive alone'.

The units of measurement of a standard deviation are the same as those of the sample values and of the mean. The standard deviation of a batch of values should always be used as a numerical summary of spread in preference to its variance.

An alternative formula for the standard deviation is:

$$s = \sqrt{\frac{\sum x^2 - n\bar{x}^2}{n-1}}.$$

This formula is easier to evalute by hand than the earlier formula and is used by hand-held calculators to evaluate standard deviations. However, it obscures what the standard deviation is actually measuring! In this computer age, it is more important to understand concepts than the mechanics of evaluation, so do not waste time memorising this formula. The equivalence of the two formulae can be demonstrated through simple mathematical maniplations.[27]

Interpreting the standard deviation

Unfortunately, although the standard deviation is the most commonly used measure of spread, its value is also the hardest to interpret. A useful guide to interpreting the standard deviation is given by the following rule of thumb called the *70-95-100 rule*.

For many data sets:

- about 70% of observations lie within $1s$ of the mean
- about 95% of observations lie within $2s$ of the mean
- almost all observations lie within $3s$ of the mean.

This rule of thumb is a most useful guide for data sets which are reasonably symmetric and bell-shaped. For asymmetric distributions, and also for unusual distributional shapes such as bimodal or multimodal distributions, the approximation is less accurate; there are often different proportions of observations in the central region and tails of the distribution. However, even for these distributions, we can guarantee a lower bound for the fraction of the data that will be within a fixed number of standard deviations from the mean.[28]

---

27. $\sum(x-\bar{x})^2 = \sum(x^2 - 2\bar{x}x + \bar{x}^2) = \sum x^2 - 2\bar{x}\sum x + n\bar{x}^2 = \sum x^2 - 2\bar{x}(n\bar{x}) + n\bar{x}^2 = \sum x^2 - n\bar{x}^2$.

28. The guaranteed rule is known as *Chebychev's inequality*. It guarantees that for *all* data sets:
    - at least 0% of observations lie within $1s$ of the mean (not of profound value!)
    - at least 75% of observations lie within $2s$ of the mean
    - at least 88.9% of observations lie within $3s$ of the mean.

    This guarantee will usually substantially understate the actual percentages, which are typically much closer to the rule of thumb percentages.

Mean and standard deviation for discrete data arranged in a frequency table

The calculations for evaluating the mean and variance of a batch of discrete values can be greatly simplified if the data are presented in a frequency table. Consider the following frequency table, the first two columns of which show the household sizes for a sample of 60 households.

Household size	**Number of families**		
x	f_x	xf_x	x^2f_x
1	13	13	13
2	18	36	72
3	6	18	54
4	10	40	160
5	7	35	175
6	4	24	144
7	2	14	98
Total	60	180	716

The mean household size is:

$$\bar{x} = \frac{\overbrace{1+1+\ldots+1}^{13} + \overbrace{2+2+\ldots+2}^{18} + \overbrace{3+3+\ldots+3}^{6} + \ldots}{60}$$

$$= \frac{13\times 1 + 18\times 2 + 6\times 3 + \ldots + 2\times 7}{60}$$

$$= \frac{180}{60} = 3.0.$$

These calculations can be generalised to give the formula:

$$\bar{x} = \frac{1}{n}\sum xf_x$$

where f_x denotes the frequency of occurrence of the value x in the data set.

In a similar way, the formula for the variance can be expressed in the following way which, again, is easier for hand calculations:

$$s^2 = \frac{1}{n-1}\left[\sum(x^2 f_x) - n\bar{x}^2\right].$$

The raw data in the first two columns of the frequency table are augmented by two further columns to aid these calculations:

$$s^2 = \frac{716 - 60\times 9.0}{59}$$

$$= \frac{176}{59} \approx 2.98.$$

So, the standard deviation is:

$$s \approx 1.73.$$

EXERCISE 2.9

1. Refer to the 'point of balance' graphic illustrating that the mean is this point. Now imagine what would happen if the right-most point were moved further to the right. Is it possible that all the other points would eventually be to the left of the point of balance?
2. Under what circumstances will the median be less than the mean?
3. It is sometimes said that the median does not depend on the extreme values in a distribution. Is this correct? (*Hint*: Consider what would happen to the median if the largest value in a data set were removed or, alternatively, if the largest value were made even larger.)
4. In discrete data sets (as defined in section 2.6), the most commonly occurring value is called the mode. Is this a good measure of location in comparison with the mean or median? Give a reason for your opinion.
5. Consider the hurricane data. As you proceed from 1886 through to 1973, what happens to the range of the data as you successively add the number of hurricanes in each year to the batch? Does this suggest a problem with the range as a measure of spread of a distribution?
6. Describe the advantages and disadvantages of the interquartile range and the standard deviation as descriptive measures of the variability of a batch of values.
7. Is the variance a useful descriptive measure of the variability in a single batch of data?
8. Consider the following distribution of the number of children per household in a small community.

Number of children	Number of households
0	7
1	4
2	8
3	4
4	2
5 or more	0

Compute the mean number of children per household using the frequency formula. Verify that the ordinary average of the 25 values displayed as follows gives the same answer:

0, 0, 0, 0, 0, 0, 0, 1, 1, 1, 1, 2, 2, 2, 2, 2, 2, 2, 2, 3, 3, 3, 3, 4, 4.

9. Compute the standard deviation of the data in question 8. Suppose each household has two parents, or two adults in the case of the 0 children households, and no others in the household. Compute the standard deviation of the household size.

1. The body lengths of 1116 plaice (a kind of flatfish) are recorded in the following frequency table.

Lower end of interval	Frequency
17.5	4
18.5	21
19.5	80
20.5	100
21.5	110
22.5	40
23.5	12
24.5	24
25.5	45
26.5	91
27.5	152
28.5	170
29.5	140
30.5	63
31.5	25
32.5	24
33.5	9
34.5	2

Source: D. W. Thompson, *On Growth and Form,* Vol. 1, Cambridge University Press, London, 1959, p. 213.

(a) Why is a histogram better than a bar chart for these data?
(b) Construct a histogram for the data using 1 cm class intervals. (*Hint*: Lowest class should be 17.5 to 18.5.) Construct another histogram using 4 cm intervals.
(c) Comment on the possible interpretation of the histogram based on 1 cm intervals.
(d) Comment on the relative merits of the two histograms of part (b).

2. Compare dot plots and histograms under the headings below. You may assume that both kinds of display are available from computer programs.
(a) usefulness for descriptive displays
(b) usefulness for detection of anomalies in the data
(c) simplicity of understanding
(d) ease of construction
(e) ease of detection of bimodality (two humps)

3. As mentioned in section 2.9, the mean can be interpreted as a balance point; if the 'dots' of a dot plot have equal weights, and the axis is a bar of negligible weight, then the balance point of this bar is at the mean. There is a sense in which the median has an interpretation in terms of these dot-weights. Apply the definition of the median to discover this interpretation.

4. A biologist measures a sample of maple trees on the local university campus. The trees have trunk circumference (cm) as shown in the following data set:

   92 55 83 49 51 79 75 93 72 67 60 78 57
   65 66 91 33 77 94 97 73 79 68 74 92.

   Provide a statistical summary of these data that is likely to be useful to the biologist. (*Hint*: Investigate this data set using graphical methods. Do you see any features that might interest a biologist?)

5. In a large statistics class there are 324 students: 182 men and 142 women. The average height of the men in this class is 170 cm and the standard deviation of their heights is 8 cm, while for the women heights average 160 cm and the standard deviation is also 8 cm.
   (a) What is the average height of all 324 people?
   (b) Is the standard deviation of all 324 heights larger than 8 cm, or smaller than 8 cm, or just about 8 cm?
   (c) Use the 70-95-100 rule to determine very approximately what proportion of the women are over 170 cm tall in the statistics class (that is, the proportion who are taller than the average male in the class).

6. Draw a freehand sketch of the histogram of the age distribution of students currently enrolled in your course. (This should be based on your general knowledge, not on actual data.) Based on your sketch, estimate the average age, the median age and the standard deviation of the age. Include a brief explanation of your sketch. (A correct answer would be one that is reasonable based on general knowledge and with approximate consistency between the sketch and the summary numbers.)

7. Construct a stem-and-leaf plot of the compressive strength data in section 2.1. What does this tell you about the precision of the measurement process?

8. Consider the aircraft emissions data in section 2.1. We noted earlier that the raw CO_2 emissions are larger for heavier aircraft, suggesting that we instead examine CO_2 emissions per unit weight. However, these values, which are also shown in the table, tend to be larger for lighter aircraft, suggesting that this standardisation might also be improved. A fairer adjustment of the emissions would show no systematic trend as the weight of the aircraft decreases. Explore a few other adjusted pollution measures, such as:

$$\text{Pollution index} = \frac{CO_2 \text{ emissions}}{\sqrt{\text{Weight}}}.$$

   Report your best standardisation scheme, and describe whether your conclusions from analysing this index differ from the earlier analysis of CO_2 emissions per unit weight.

9. Summarise the maximum isometric strength data using both graphical and tabular techniques. You should select one table and one graph that portray the result of your analysis as clearly as possible.

10. From the data on mammal brain weights, select any 10 species with which you are familiar. List them in order of their brain weight.
    (a) What transformation of brain weight might be related to the diameter of the brain (the length across the brain that is as large as possible)? Justify your choice.
    (b) Is the distribution of the transformed data easier to interpret than that of the raw brain weights? (Use any graphical summaries to make the comparison.)

11. Summarise the major differences in the age structure of the six populations portrayed by population pyramids in the text. In your comparison, include a consideration of what the passage of time will do to each pyramid.

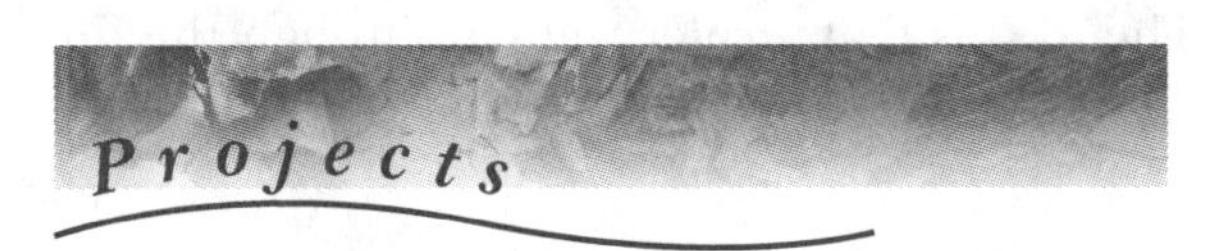

1. **Olympic medals**
   There are many published sources of data on performance at Olympic Games — check with your library. Media coverage usually includes the number of medals won by each country's athletes and these tables are published in most sources. However, large countries invariably receive more medals than smaller countries and it is hard to judge how a country has done after considering its population. The aim of this project is to find a way to compare countries' medal counts in a way that is fair to the smaller countries. To do this, we must examine the relationship of the number of medals a country receives to its population. It may be possible to devise a way to assess a country's medal performance that takes account of its population. If you find a good way, send it to the Olympic Committee — the small countries will thank you!
   (a) Choose at least 20 countries to include in your project. (They should have a wide range of populations.)
   (b) For the countries you choose, compile a table of the medals' count from the most recent Olympics for which published data are available and include the populations of these countries.
   (c) Re-express the number of medals as medals per million population and examine the distribution of these values.
   (d) Report your conclusion about whether this standardisation is effective.
   Note: This project may be extended by considering the 'wealth' of the countries as an alternative index of their 'size' to adjust the medal counts. A measure of wealth that is widely available in published sources is the Gross Domestic Product, or GDP.

2. **Paper planes, distance**

The purpose of this project is to determine the effect size has on the performance of a certain model of paper aeroplane.

(a) Construct five paper aeroplanes according to your own design. Each aeroplane should be made with the same design and type of paper, but the sizes of the sheets from which they are folded should vary — the biggest being at least twice as big as the smallest.

(b) Measure the aeroplanes' performances by the straight-line distance, in any direction, from the launching point to the point where the aeroplane touches down. You should launch all the aeroplanes using the same launching technique to eliminate this factor as a source of variation.

(c) Repeat the experiment with the five aeroplanes enough times that the average distance flown for each is well established. Report the results of your experiment in both tabular and graphical form. Include a written conclusion as well.

3. **Reaction times**

An ordinary ruler approximately 30 cm long (or 1 foot long) can be used to measure reaction time. The idea is that a ruler will fall through the fingers and thumb a distance that depends on the time you take to clasp your finger and thumb together.

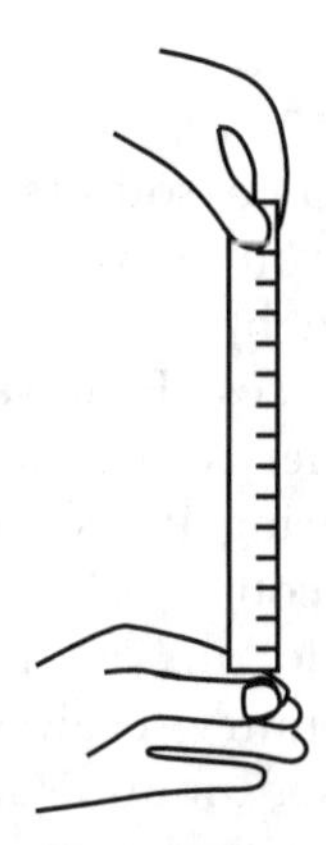

The measurement requires a helper and cannot be self-administered. The helper holds the ruler so that it is suspended vertically, with the free end being the 0 cm end and at the bottom. The 0 cm end is placed just above your thumb and fingers, and the helper drops the ruler at an unannounced moment. You catch the ruler as quickly as you can and read the distance the ruler travelled before you caught it. This is a measure of your reaction time. Actually, it is a measure whose square root is proportional to your reaction time. A typical reaction time is about $\frac{1}{5}$ of a second, which corresponds to just about 19 cm or 7.5 inches drop. It is advisable to do a few practice measurements before you start to record data.

The goal of the project is to determine your own reaction time and that of your helper, and to measure the variability of each. Your report should comment on any anomalous data. It should also report the data and provide appropriate summaries of them (numerical, graphical and textual!)

4. **The letter 'e'**

The most commonly used letter in the English language is the letter 'e'. For example, this paragraph contains more e's than any other letter. Of course, the frequency of e in any particular selection will vary from one line to the next but, in a large selection, the proportion of e's is fairly predictable. The object of this project is to study the variation in the number of e's in a particular selection of English prose.

Choose a book from your bookshelf, or this textbook if you wish, and select some prose containing about 25 lines. The number of e's per line will vary and will have a certain frequency distribution.

(a) Count the number of e's in each complete line in your selection. Ignore the incomplete lines.

(b) From your counts in (a), summarise the distribution of the number of e's per line.

The number of e's counted in a given line of prose can also vary in the sense that we can count them imperfectly — we call this 'measurement error'. Some variation is due to measurement error, while some variation is not.

(c) Repeat your counts from part (a) carefully and note any differences. Summarise the distribution of your counting errors.

(d) Again summarise the distribution of the number of e's per line, using the corrected counts. Compare the summary from this part with your summary from part (b).

(e) Comment on your results — in particular, compare the size of the two kinds of variation mentioned above.

CHAPTER 3

Exploring change — time series

Time reveals all things.

Erasmus (1465–1536)

3.1 TIME SERIES DATA

In chapter 2, we investigated the distribution of a single batch of univariate data and saw how to extend the exploratory analysis to compare several batches. In some such batches there is a further structure which we ignored in chapter 2 — the values in the raw data were collected in order. For example, the hurricane data in section 2.6 were ordered in time, as were the ages at death of the US presidents in section 2.1. In this chapter, we will use the time ordering of values in a data set to assess change.

A *time series* is a collection of univariate data in which the values are recorded at successive (and usually equally spaced) time intervals. We are interested in detecting patterns of change in the measurements over time.[1]

The following table gives information about the Academy Awards for Best Actress between 1928 and 1991.[2] Is there any pattern to the ages of the actresses? Were there any periods where older or younger actresses were chosen?

Year	Actress	Movie	Age
1928	Janet Gaynor	*Seventh Heaven*	22
1929	Mary Pickford	*Coquette*	36
1930	Norma Shearer	*The Divorcee*	26
1931	Marie Dressler	*Min and Bill*	62
1932	Helen Hayes	*The Sin of Madelon Claudet*	32
1933	Katharine Hepburn	*Morning Glory*	26
1934	Claudette Colbert	*It Happened One Night*	29
1935	Bette Davis	*Dangerous*	27
1936	Luise Rainer	*The Great Ziegfield*	24
1937	Luise Rainer	*The Good Earth*	25
1938	Bette Davis	*Jezebel*	30
1939	Vivien Leigh	*Gone with the Wind*	26
1940	Ginger Rogers	*Kitty Foyle*	29
1941	Joan Fontaine	*Suspicion*	24
1942	Greer Garson	*Mrs Minever*	34
1943	Jennifer Jones	*The Song of Bernadette*	24
1944	Ingrid Bergman	*Gaslight*	29
1945	Joan Crawford	*Mildred Pierce*	41
1946	Olivia de Havilland	*To Each His Own*	30
1947	Loretta Young	*The Farmer's Daughter*	34
1948	Jane Wyman	*Johnny Belinda*	34
1949	Olivia de Havilland	*The Heiress*	33
1950	Judy Holliday	*Born Yesterday*	28
1951	Vivien Leigh	*A Streetcar Named Desire*	38

1. Such a pattern may be considered as a relationship between the variable of interest and time, a special type of *bivariate* relationship. More general bivariate relationships between variables will be explored in chapter 4.
2. S. P. Gordon and F. S. Gordon, *Contemporary Statistics: A Computer Approach*, McGraw-Hill, New York, 1994, p. 103.

Year	Actress	Movie	Age
1952	Shirley Booth	*Come Back, Little Sheba*	45
1953	Audrey Hepburn	*Roman Holiday*	24
1954	Grace Kelly	*The Country Girl*	26
1955	Anna Magnani	*The Rose Tattoo*	48
1956	Ingrid Bergman	*Anastasia*	41
1957	Joanne Woodward	*The Three Faces of Eve*	27
1958	Susan Hayward	*I Want to Live*	40
1959	Simone Signoret	*Room at the Top*	38
1960	Elizabeth Taylor	*Butterfield 8*	28
1961	Sophia Loren	*Two Women*	27
1962	Anne Bancroft	*The Miracle Worker*	31
1963	Patricia Neal	*Hud*	37
1964	Julie Andrews	*Mary Poppins*	30
1965	Julie Christie	*Darling*	24
1966	Elizabeth Taylor	*Who's Afraid of Virginia Woolf?*	34
1967	Katharine Hepburn	*Guess Who's Coming to Dinner*	60
1968*	Katharine Hepburn	*The Lion in Winter*	61
1968*	Barbra Streisand	*Funny Girl*	26
1969	Maggie Smith	*The Prime of Miss Jean Brodie*	35
1970	Glenda Jackson	*Women in Love*	34
1971	Jane Fonda	*Klute*	34
1972	Liza Minnelli	*Cabaret*	26
1973	Glenda Jackson	*A Touch of Class*	37
1974	Ellen Burstyn	*Alice Doesn't Live Here Anymore*	42
1975	Louise Fletcher	*One Flew over the Cuckoo's Nest*	41
1976	Faye Dunaway	*Network*	35
1977	Diane Keaton	*Annie Hall*	31
1978	Jane Fonda	*Coming Home*	41
1979	Sally Field	*Norma Rae*	33
1980	Sissy Spacek	*Coal Miner's Daughter*	30
1981	Katharine Hepburn	*On Golden Pond*	74
1982	Meryl Streep	*Sophie's Choice*	33
1983	Shirley MacLaine	*Terms of Endearment*	49
1984	Sally Field	*Places in the Heart*	38
1985	Geraldine Page	*Trip to Bountiful*	61
1986	Marlee Matlin	*Children of a Lesser God*	21
1987	Cher	*Moonstruck*	41
1988	Jodie Foster	*The Accused*	26
1989	Jessica Tandy	*Driving Miss Daisy*	80
1990	Kathy Bates	*Misery*	43
1991	Jodie Foster	*The Silence of the Lambs*	29
1992	Emma Thompson	*Howards End*	33
1993	Holly Hunter	*The Piano*	35
1994	Jessica Lange	*Blue Sky*	45

*There was a joint award of Best Actress in 1968.

A time series may consist of measurements of an underlying process that is either continuous or discrete. Some time series describe discrete processes where measurements are undefined at times between those when the data were recorded. The time series of ages of Academy Award winning actresses is an example of such a process. A single data value, the age of the actress who wins the Oscar, is added to the series at the time of the annual Academy Awards.

A continuous process is one for which the value of the variable being studied (for example, temperature or exchange rates) could be measured at any moment in time. Most commonly, a continuous time series is sampled at predetermined and, at least approximately, regularly spaced times (for example, noon temperatures, stock market prices at the close of a day's trading).

The table below describes changes in the Consumer Price Index (CPI) in New Zealand between 1950 and 1989.[3] Note that the CPI is based on the prices of a basket of goods and services. Continual changes in these prices underlie the process which is measured, even though data acquisition is only carried out at regular intervals, perhaps monthly. So, the process which generates the CPI can be regarded as continuous.

Year	CPI	Year	CPI	Year	CPI
1950	55.1	1964	94.0	1978	300.9
1951	61.1	1965	97.2	1979	342.2
1952	65.9	1966	99.9	1980	400.8
1953	68.9	1967	106.0	1981	462.4
1954	72.0	1968	110.5	1982	537.2
1955	73.9	1969	116.0	1983	576.9
1956	76.4	1970	123.5	1984	611.9
1957	78.1	1971	136.4	1985	706.6
1958	81.6	1972	145.8	1986	800.1
1959	87.7	1973	157.7	1987	925.8
1960	85.3	1974	175.3	1988	984.8
1961	86.8	1975	201.0	1989	1041.0
1962	89.1	1976	235.0		
1963	90.9	1977	268.8		

We have seen examples of discrete and continuous processes. Time series data may vary structurally in other ways. Some data sets are aggregate measurements of a quantity over some time period, such as total rainfall in a month or total annual petroleum imports in a country. Although aggregate measurements could be broken down into finer time periods (daily petroleum imports, for example), the corresponding measurements from the finer time period would be of greatly different magnitudes.

Just as the process which generates a time series may be discrete or continuous, the measurements may also be either discrete (counts) or continuous.

3. *New Zealand Official 1990 Yearbook*, Department of Statistics, Wellington, p. 614.

The two time series examples we have seen so far represent continuous measurements, although the ages of the actresses are truncated to whole years.

An example of a time series of discrete counts is given in the table below, showing annual road traffic deaths in New South Wales from 1950 to 1993.[4] Here, the process itself is continuous in that it is updated whenever a road traffic death occurs; this may happen at any time.

Year	Deaths	Year	Deaths	Year	Deaths	Year	Deaths
1950	634	1961	918	1972	1092	1983	966
1951	728	1962	876	1973	1230	1984	1037
1952	700	1963	900	1974	1275	1985	1067
1953	704	1964	1010	1975	1288	1986	1029
1954	754	1965	1151	1976	1264	1987	959
1955	820	1966	1143	1977	1268	1988	1037
1956	801	1967	1117	1978	1384	1989	960
1957	765	1968	1211	1979	1290	1990	797
1958	824	1969	1188	1980	1303	1991	663
1959	859	1970	1309	1981	1291	1992	649
1960	978	1971	1249	1982	1253	1993	581

We will not further distinguish between discrete and continuous processes, or between discrete and continuous measurements in our analysis of time series in this chapter. Our basic approach to exploring time series is the same as in chapter 2. We use graphical and numerical summaries to find patterns and study deviations from those patterns.

EXERCISE 3.1

1. Which data sets from chapter 2 are time series?
2. Refer to the Best Actress data set. Use a plotting technique of your choice (dot plot, stem-and-leaf plot, histogram, frequency polygon or box plot) to compare actresses' ages in the following periods: 1928–1950, 1951–1973, 1974–1994. (Before you begin, think about how you can make the three plotted distributions easily comparable.) Do you see any trend over time from your three plots?
3. (a) For each of the following time series, state whether the time variable is discrete or continuous.
   - (i) number of loaves sold by the supermarket daily
   - (ii) number of customers in the supermarket throughout a day
   - (iii) total purchases ($) of successive customers
   - (iv) inventory of two-litre containers of milk during a day

   (b) For each of the series just described, state whether the quantity measured at each point in time is discrete or continuous.

4. *Road Traffic Accidents in New South Wales*, Roads and Traffic Authority, Sydney, 1994.

3.2 DISPLAYING TIME SERIES

In this section, we will introduce some simple tools for monitoring time series. Often, it is useful to ignore the sequential nature of the data at first and use tools from chapter 2 (especially stem-and-leaf plots or histograms of the relevant variable) to display the data.

The data in the example which follows are 70 consecutive yields from a batch chemical process (ordered by column).[5]

Batch yield of a chemical production process

47	44	50	62	68
64	80	71	44	38
23	55	56	64	50
71	37	74	43	60
38	74	50	52	39
64	51	58	38	59
55	57	45	59	40
41	50	54	55	57
59	60	36	41	54
48	45	54	53	23
71	57	48	49	
35	50	55	34	
57	45	45	35	
40	25	57	54	
58	59	50	45	

Arranging the data in a stem-and-leaf plot (on the right) shows that the yield data are roughly symmetrically distributed around a median of 51.5. However, the true nature of the variability in a time series can only be observed when the time ordering of the data is used explicitly. Even the suggestion of symmetry from the stem-and-leaf plot must be confirmed from the time series itself.

To discover whether some of the variability in yields is caused by systematic changes in time, we plot the observations against time. A time series plot displays time along a horizontal axis and the relevant variable on a vertical axis, as illustrated on page 79.

```
8 | 0
7 |
7 | 1 1 1 4 4
6 | 8
6 | 0 0 2 4 4 4
5 | 5 5 5 5 6 7 7 7 7 7 8 8 9 9 9 9
5 | 0 0 0 0 0 0 1 2 3 4 4 4 4
4 | 5 5 5 5 5 7 8 8 9
4 | 0 0 1 1 3 4 4
3 | 5 5 6 7 8 8 8 9
3 | 4
2 | 5
2 | 3 3
```

Stem unit: 10
Leaf unit: 1

5. G. E. P. Box and G. M. Jenkins, *Time Series Analysis: Forecasting and Control*, Holden-Day, San Francisco, 1970, p. 530.

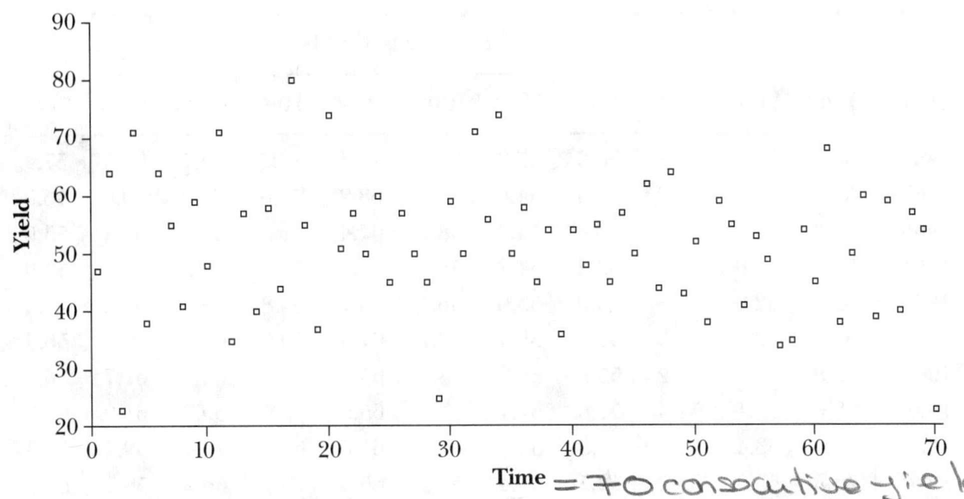

Observations are usually made at known, equally spaced times. Sometimes, however, data are collected in chronological order, but the exact time of collection may vary and is not recorded. The commonly used methods of plotting and analysing time series, described in this chapter, assume equispaced observations, but can also be used for a rough analysis of other ordered data sets, such as the chemical yield data. A time series plot is usually improved if successive points are joined by lines (called a *line graph*). Generally, this makes it easier to detect patterns.

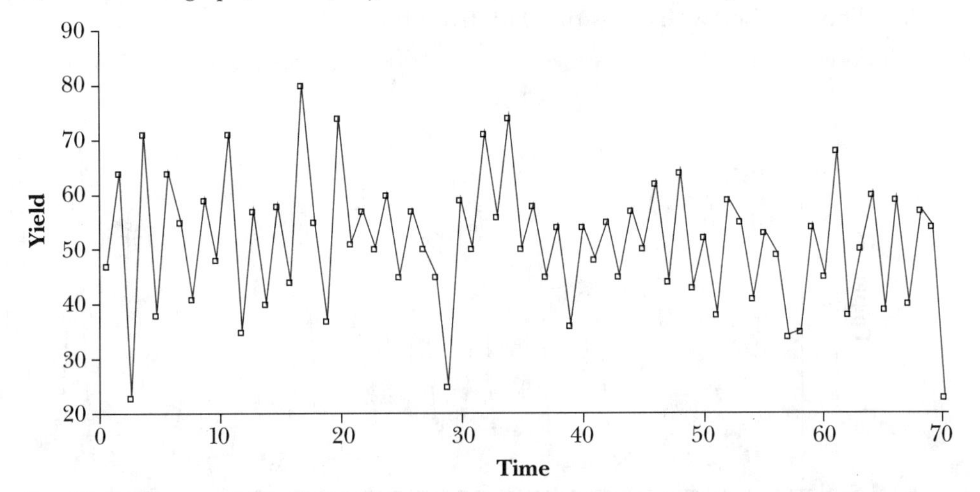

The plot above shows considerable variability in the yields of the batches, but no clear *trend* towards either higher or lower yields over time. However, note that there is a tendency for high yields to follow low yields and vice versa.

The next data set gives monthly numbers of employees ('000s) involved in the production of food and kindred products in the State of Wisconsin over 178 months from January 1961 through to October 1975.[6]

6. R. B. Miller and D. W. Wichern, *Intermediate and Business Statistics: Analysis of Variance, Regression and Time Series*, p. 440, Copyright © 1977, Holt, Rinehart and Winston, New York. By permission of Brooks/Cole Publishing Company, a division of International Thomson Publishing Inc.

Year	Employees ('000s)											
	Jan.	Feb.	Mar.	Apr.	May	June	July	Aug.	Sept.	Oct.	Nov.	Dec.
1961	56.3	55.7	55.8	56.3	57.2	59.1	71.5	72.2	72.7	61.5	57.4	56.9
1962	55.3	54.9	54.9	54.9	54.6	57.7	68.2	70.6	71.0	60.0	56.0	54.4
1963	53.3	52.8	53.0	53.4	54.3	58.2	67.4	71.0	69.8	59.4	55.6	54.6
1964	53.4	53.0	53.0	53.2	54.2	58.0	67.5	70.1	68.2	56.6	54.9	54.0
1965	52.9	52.6	52.8	53.0	53.6	56.1	66.1	69.8	69.3	61.2	57.5	54.9
1966	53.4	52.7	53.0	52.9	55.4	58.7	67.9	70.0	68.7	59.3	56.4	54.5
1967	52.8	52.8	53.2	55.3	55.8	58.2	65.3	67.9	68.3	61.7	56.4	53.9
1968	52.6	52.1	52.4	51.6	52.7	57.3	65.1	71.5	69.9	61.9	57.3	55.1
1969	53.6	53.4	53.5	53.3	53.9	52.7	61.0	69.9	70.4	59.4	56.3	54.3
1970	53.5	53.0	53.2	52.5	53.4	56.5	65.3	70.7	66.9	58.2	55.3	53.4
1971	52.1	51.5	51.5	52.4	53.3	55.5	64.2	69.6	69.3	58.5	55.3	53.6
1972	52.3	51.5	51.7	51.5	52.2	57.1	63.6	68.8	68.9	60.1	55.6	53.9
1973	53.3	53.1	53.5	53.5	53.9	57.1	64.7	69.4	70.3	62.6	57.9	55.8
1974	54.8	54.2	54.6	54.3	54.8	58.1	68.1	73.3	75.5	66.4	60.5	57.7
1975	55.8	54.7	55.0	55.6	56.4	60.6	70.8	76.4	74.8	62.2		

The plot below displays the data from the table.

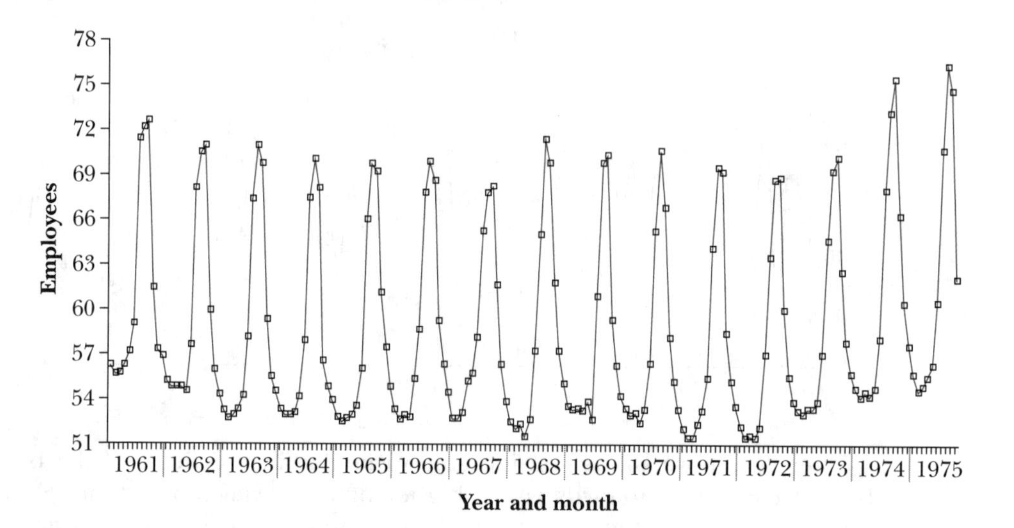

In this time series, there is a pattern that repeats in a similar form each year. Numbers of employees in food production are relatively high during the northern hemisphere summer months and relatively low in the northern hemisphere winter. Superimposed on this pattern there also seems to be a general drop in the numbers employed in the initial years of the series, a fairly constant pattern during the middle period and an increase from late 1972 onwards.

Our final example in this section deals with monthly numbers (in thousands) of international airline passengers, from January 1949 through to December 1960.[7]

	Passengers ('000s)											
Year	**Jan.**	**Feb.**	**Mar.**	**Apr.**	**May**	**June**	**July**	**Aug.**	**Sept.**	**Oct.**	**Nov.**	**Dec.**
1949	112	118	132	129	121	135	148	148	136	119	104	118
1950	115	126	141	135	125	149	170	170	158	133	114	140
1951	145	150	178	163	172	178	199	199	184	162	146	166
1952	171	180	193	181	183	218	230	242	209	191	172	194
1953	196	196	236	235	229	243	264	272	237	211	180	201
1954	204	188	235	227	234	264	302	293	259	229	203	229
1955	242	233	267	269	270	315	364	347	312	274	237	278
1956	284	277	317	313	318	374	413	405	355	306	271	306
1957	315	301	356	348	355	422	465	467	404	347	305	336
1958	340	318	362	348	363	435	491	505	404	359	310	337
1959	360	342	406	396	420	472	548	559	463	407	362	405
1960	417	391	419	461	472	535	622	606	508	461	390	432

The time series is displayed below.

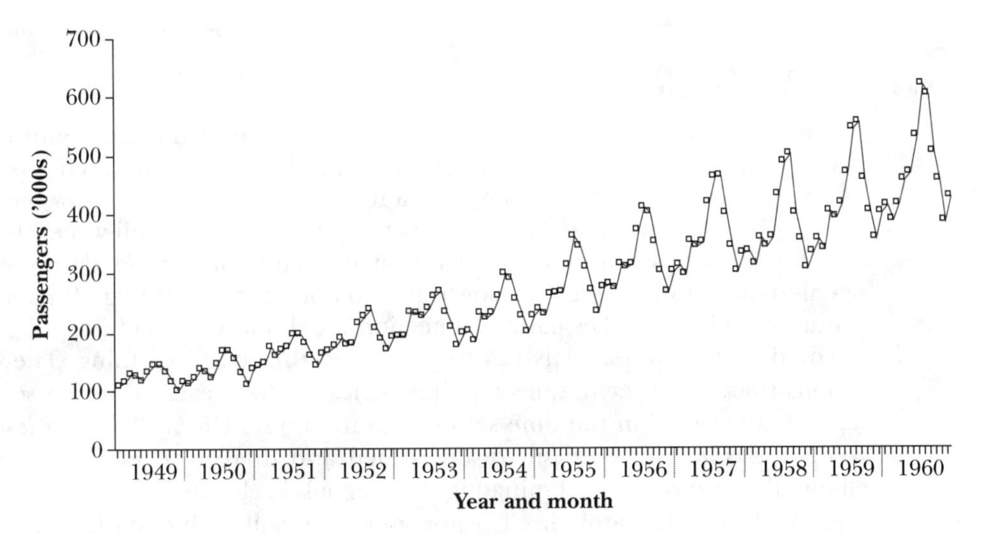

This time series also has a repeating annual pattern (a cycle), with higher numbers of airline passengers in the northern hemisphere summer months and lower numbers in the northern hemisphere winter. The series is, however, much less stable than the previous one. There is a general increase in passenger numbers between 1949 and 1960 and the month-to-month fluctuations become larger.

7. G. E. P. Box and G. M. Jenkins, op. cit., p. 531.

Provided the time series is sufficiently long, a plot of the data against time, drawn to an appropriate scale, will often reveal any strong features such as long-term trends, *seasonal* patterns (such as annual, weekly or daily cycles), outlying observations and any breaks or discontinuities in the series.

If the scale of a time series plot is badly chosen, it is sometimes difficult to see patterns, and the plot should be redrawn. It has been recommended that the dimensions of the two axes should be chosen, if possible, to make trends or other patterns have an angle of approximately 45° to the axes on the plot.[8] It will often be impossible to achieve this, due to different parts of the trend (for example, seasonal and long-term) having different angles, but the guideline is well worth keeping in mind.

EXERCISE 3.2

1. Examine the time series for Wisconsin food workers and for airline passengers. Both series have an annual pattern and a long-term trend, but there is one feature of the airline passenger series that is not present in the food workers series. What is it?
2. How much of the pattern you described in question 1 would be visible using the techniques discussed in chapter 2?

3.3 SMOOTHING

While some patterns in time series are very clear from the basic plot of the observations against time, this is not always so. The extent to which a pattern can be readily detected is substantially a function of the *signal-to-noise ratio.* If a data set has a clear trend or cyclical pattern that can be readily seen by eye, and the deviations about the trend are small, then signal dominates noise (the signal-to-noise ratio is high). A low signal-to-noise ratio with large fluctuations about a trend line makes pattern detection much more difficult.

The detection of patterns can be enhanced through smoothing. There are various smoothing techniques for time series analysis, each of which seeks to separate the signal in the time series from the noise. Unfortunately, these two components cannot be easily distinguished and there is a trade-off between eliminating noise and eliminating the signal itself when a time series is smoothed — if the signal itself is not smooth, it will be hidden by smoothing because part of the signal will be indistinguishable from noise. The smoothing techniques that we will describe are based on an assumption that the signal is smooth and that the noise is not smooth, an assumption that is often justified.

Smoothing can display trends in time series very effectively. All smoothing techniques involve some sort of centring or averaging, using medians and/or

8. W. S. Cleveland, *Visualising Data*, Hobart Press, New Jersey, 1993, p. 89.

means.[9] All such methods are applied locally so that each point in the original time series is replaced in the smoothed series by an 'average' of adjacent points.

In order to describe the methods that are commonly used for smoothing data, we will denote the n ordered observations in the time series by $x_1, x_2, \ldots, x_n$.

End effects

With all smoothing techniques, end points in the time series (at one end or both) have to be treated differently to other values, since adjacent values are not available on one side.

Running means

The simplest form of smoothing is to take running means. A 3-point running mean replaces each observation in the series by the mean of that observation and the two adjacent ones. A k-point running mean replaces x_i by the average of k observations centred on x_i. Since this can only be done for k odd, running means are usually not used for k even.[10]

For a $(2k + 1)$-point running mean, the smoothed value $\tilde{x}_i$ replacing x_i is:

$$\tilde{x}_i = Mean\,\{x_{i-k}, x_{i-k+1}, \ldots, x_{i+k}\}.$$

Calculation of 5-point running means for the chemical yield data is illustrated in the table below.

Raw data	Running mean of 5
47	47*
64	$(47 + 64 + 23)/3 \approx 44.7$ *
23	$(47 + 64 + 23 + 71 + 38)/5 = 48.6$
71	$(64 + 23 + 71 + 38 + 64)/5 = 52.0$
38	$(23 + 71 + 38 + 64 + 55)/5 = 50.2$
64	.
55	.
.	.
.	.
.	.

* Note that end points are treated differently.

9. Sometimes, weighted means are used, as in exponential smoothing.
10. A closely related smoothing method is called 'hanning'. This smoothes the value x_t with the weighted mean:

$$\frac{\frac{1}{2}x_{i-1} + x_i + \frac{1}{2}x_{i+1}}{2}.$$

This is intermediate between the 1-point running mean, x_i and the 3-point running mean and, so, can be thought of as a '2-point' running mean.

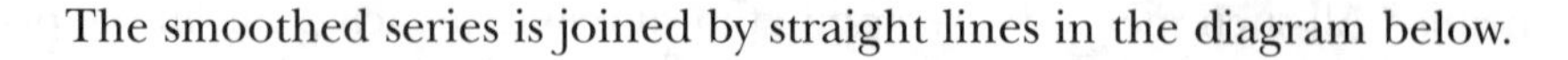

The smoothed series is joined by straight lines in the diagram below.

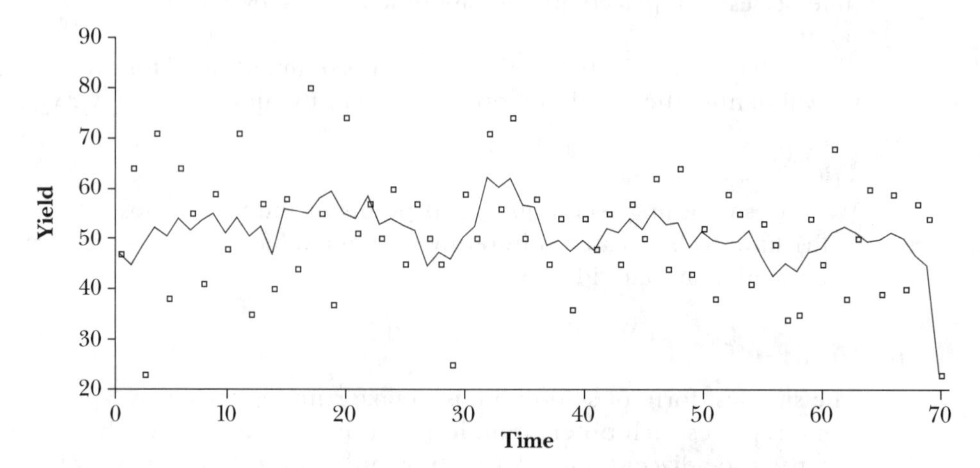

Note that the first (and last) two points are smoothed by lower-order running means than the interior points in the series. As a result, the end points are less highly smoothed, which should be kept in mind when interpreting fluctuations at the ends of a smoothed time series. For example, the third observation of the chemical yield data is smoothed by the adjacent values but the last observation is not, even though both have the same value (23).

Most commonly, a value of k between 3 and 13 is used for running means.

A disadvantage of using running means as a smoothing method is the sensitivity of means to outliers. As we discovered in chapter 2, medians are more robust than means; they are less influenced by outliers. This suggests that running medians may provide a useful alternative smoothing method which is more robust than taking running means.

Running medians

Running medians are defined in a similar way to running means. A $(2k+1)$-point running median replaces x_i by the median of x_i and the k observations on either side:

$$\tilde{x}_i = Median\,\{x_{i-k}, x_{i-k+1}, \ldots, x_{i+k}\}.$$

As with running means, the running median for points near either end of the time series must be modified by taking a running median of fewer points to avoid 'going beyond the data'.

Except at either end of the series, running medians are not affected by outliers. However, a time series smoothed by running medians tends to look rather jagged, with some values repeated because successive running medians are often the same; a sequence of repeated values is often followed by a relatively large jump.

Calculation of 5-point running medians for the chemical yield data is illustrated in the following table.

Raw data	Running median of 5
47	47*
64	47*
23	47
71	64
38	55
64	.
55	.
.	.
.	.
.	.

* As with running means, the end points are treated differently. For the 1st and 2nd points, medians of one data value and three data values respectively are used.

The smoothed series resulting from this process is shown below. As the end values in the series are again less highly smoothed than the central values, care should be taken not to over-interpret fluctuations at the ends of the smoothed series.

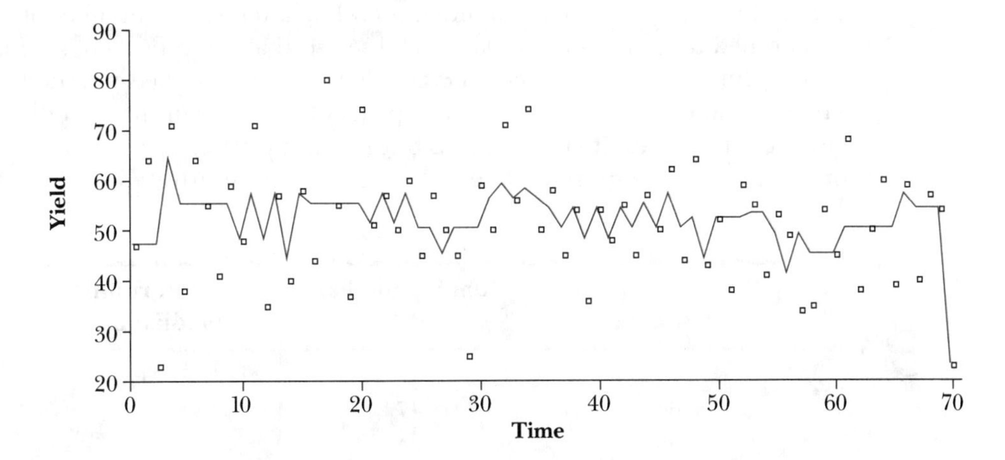

Running medians, then means

A smoothing procedure, which has the advantages of both running means and running medians but avoids many of their disadvantages, is to use running means and running medians in combination. This is done in two stages: firstly smoothing the data by taking *k*-point running medians to remove outlier effects and then further smoothing by taking *p*-point running means. As a general rule, small values of *k* and *p* are used for this two-stage process. Each stage of this process must be modified at either end of the time series, as described previously.

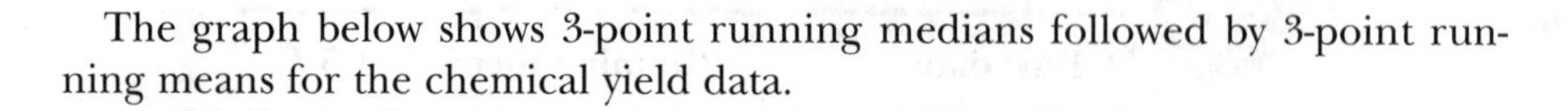
The graph below shows 3-point running medians followed by 3-point running means for the chemical yield data.

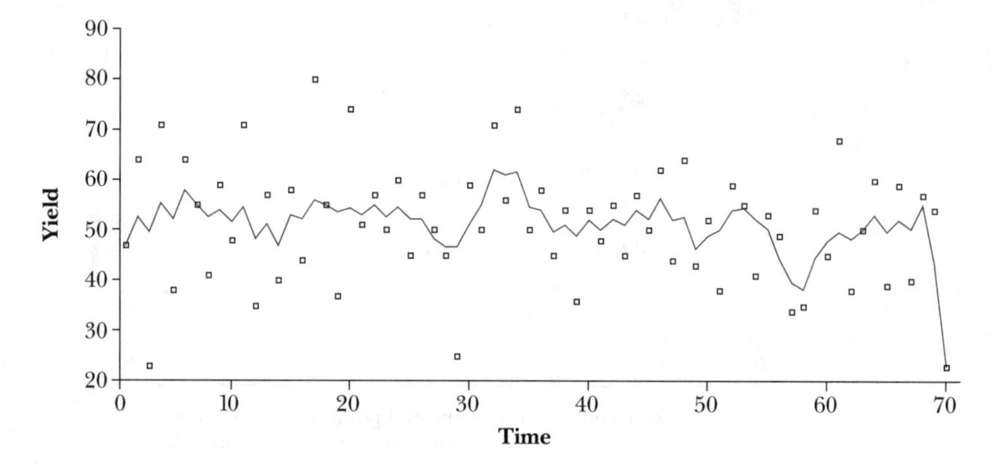

Repeated running medians

Repeated k-point running medians with small k are sometimes used in preference to running medians with a larger value of k. The series is first smoothed with k-point running medians. Then, k-point running medians are applied to this smoothed series to smooth it further, and the process is repeated several times. The repetition can be carried out indefinitely but, after a small number of cycles, the smoothed time series has usually stabilised so that the process of smoothing through a further 'running median cycle' does not change the smoothed series.

For the chemical yield data, nine repeats of 3-point running medians are required before stability is achieved, but for many other series only three or four repeats are required. The calculations are illustrated in the table below.

Raw data	Running median of 3	Second running median of 3
47	47	47
64	47	47
23	64	47
71	38	64
38	64	55
64	55	55
55	55	55
41	55	55
59	48	55
48	59	48
71	48	.
35	.	.

Even the second iteration of the procedure alters very few of the smoothed values and fewer still are modified in subsequent iterations. The graph below shows the final smoothed series.

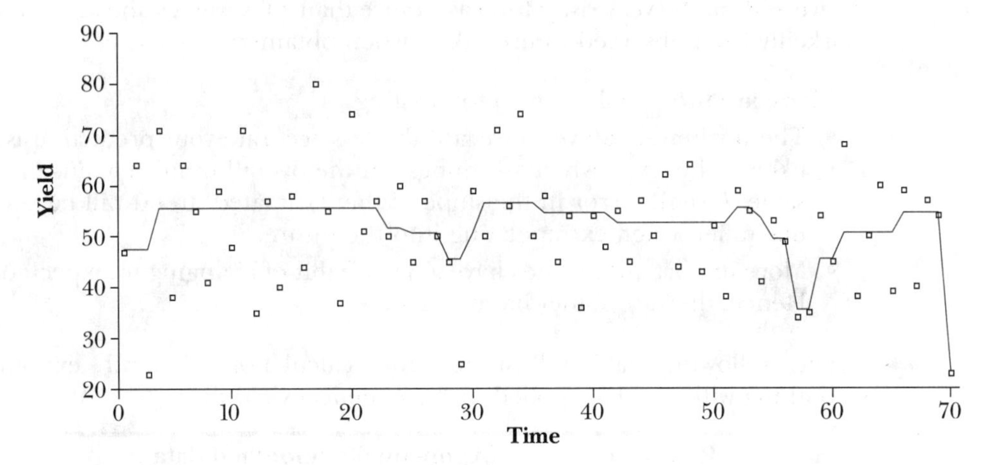

Note that the smoothed series has several flat regions where many successive smoothed values are the same. The use of repeated running medians can, as here, oversmooth the series.

None of the running smoothing methods described above has any effect on the final value of the series since each operates symmetrically on the values on both sides of each value being smoothed. If an aim of the smoothing is to help forecast future values, a different smoothing method must be used to smooth the end of the series.

Exponential smoothing

The methods discussed so far are mainly used for smoothing time series in an exploratory way to detect trends and cycles. Smoothing can also be used to forecast future values. Because it is used for forecasting, *exponential smoothing* only uses previous observations in calculating a smoothed value rather than taking an average centred on the value being smoothed. Because previous observations far back in time are less relevant and less reliable for forecasting than the value being smoothed and those in the immediate past, exponential smoothing uses 'weighted' averages. It is best to choose the weights so that they are a smooth and decreasing function of the time lag. Exponential smoothing uses a particular form of 'weight function', according to the following computational formula:

$$\tilde{x}_i = ax_i + (1-a)\tilde{x}_{i-1}.$$

This is essentially equivalent (except for end effects) to giving weight a to x_i, $a(1-a)$ to x_{i-1}, $a(1-a)^2$ to x_{i-2}, and so on. Therefore:

$$\tilde{x}_i = ax_i + a(1-a)x_{i-1} + a(1-a)^2 x_{i-2} + \dots.$$

Exponential smoothing is a crude forecasting method which (unless modified) only works well in the absence of trend and seasonality. In the presence of trend or seasonality, the exponentially smoothed series tends to lag behind the actual series and forecasts likewise lag behind the (eventually observed) series of future values.[11]

Forecasting involves extrapolating the pattern of the observed data into the future and can be very risky; forecasts more than a few values ahead often differ markedly from observed future values when obtained.

> There are two problems with forecasting:
> - The further ahead you forecast, the less accurate your prediction is likely to be, even when assuming that the overall trend remains the same. A slight error in the slope of the 'estimated' trend will be magnified when extrapolating into the future.
> - More importantly, trends have a nasty habit of changing unexpectedly. Hence, they are somewhat unreliable.

The following table illustrates the calculations behind exponential smoothing with $a = 0.3$, applied to the chemical yield data.

Raw data	Exponentially smoothed data
47	47*
64	$0.3 \times 64 + 0.7 \times 47 = 52.1$
23	$0.3 \times 23 + 0.7 \times 52.1 = 43.37$
71	$0.3 \times 71 + 0.7 \times 43.37 = 51.66$
38	$0.3 \times 38 + 0.7 \times 51.66 = 47.56$
64	$0.3 \times 64 + 0.7 \times 47.56 = 52.49$
55	.
41	.
.	.
.	.

* As there are no earlier values, exponential smoothing has no effect on the first data value.

The smoothed series is shown on the graph below.

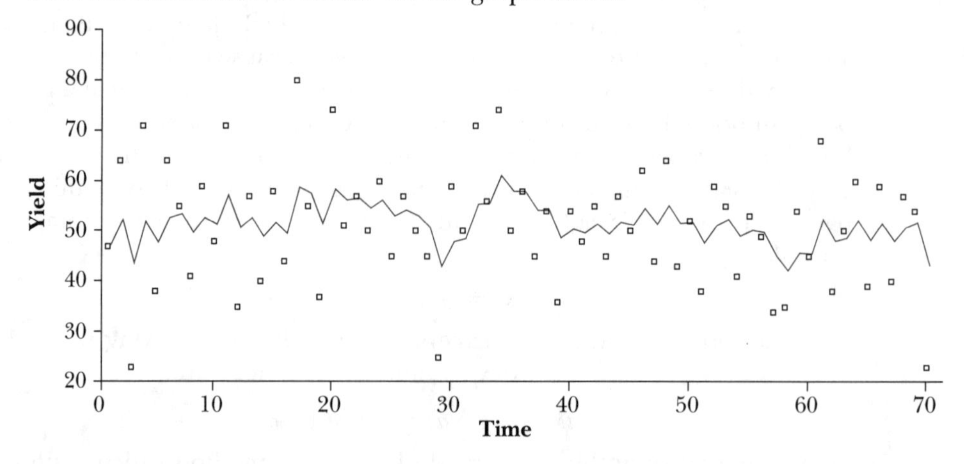

11. There are, however, enhancements to the basic method that avoid these problems. Exponential smoothing is the basis of many popular forecasting methods.

For these data, exponential smoothing does effectively smooth the series, including the final observation. A value of a closer to zero would smooth the series even further.

For the chemical yield data which we have used to illustrate the various methods of smoothing, all of the techniques suggested have led to broadly similar smoothed trends. This is commonly the case, provided the smoothing methods are used sensibly (neither too much nor too little smoothing). Better techniques for smoothing time series have been developed but are beyond the scope of this book.[12]

Transformation

Quite commonly, time series exhibit substantial skewness in the 'marginal' distribution of the variable of interest; that is, the distribution of the measurements ignoring the time ordering of the observations. For example, a stem-and-leaf plot of the airline passenger data is shown below — the distribution is skewed with a long upper tail.

Stem unit: 100
Leaf unit: 10

Stem	Leaves
6	0 2
5	5
5	0 0 3 4
4	6 6 6 6 6 7 7 9
4	0 0 0 0 0 0 1 1 1 2 2 3 3
3	5 5 5 5 6 6 6 6 6 7 9 9 9
3	0 0 0 0 0 1 1 1 1 1 1 1 1 3 3 4 4 4 4 4 4
2	5 6 6 6 6 7 7 7 7 7 7 8 9
2	0 0 0 0 1 1 2 2 2 2 3 3 3 3 3 3 3 3 4 4 4
1	5 5 6 6 6 7 7 7 7 7 7 7 8 8 8 8 8 8 9 9 9 9 9 9 9 9 9
1	0 1 1 1 1 1 1 2 2 2 2 3 3 3 3 3 4 4 4 4 4 4 4

In such a case, transformation of the values in the time series often makes interpretation easier and smoothing more effective. The marginal distribution of the measurements is a useful guide in the choice of a transformation. For example, a square root or logarithmic transformation often helps to spread out a cluster of low values and to compress a long tail of high values in the marginal distribution, as discussed in section 2.5. However, trends and seasonal patterns should also be taken into account in the choice of a transformation. A suitable transformation should also make the spread of the 'scatter' about the smoothed line more nearly equal at different times.

For the airline passenger data, a square root or logarithmic transformation might initially be tried to remove the skewness of the distribution. The effect of a logarithmic transformation is shown on the top of page 90 by box plots of the untransformed data (left) and logarithmically transformed data (right).

12. For example, a computer-intensive technique called 'loess' is very popular and effective. It is described in W. S. Cleveland, op. cit., pp. 93–101.

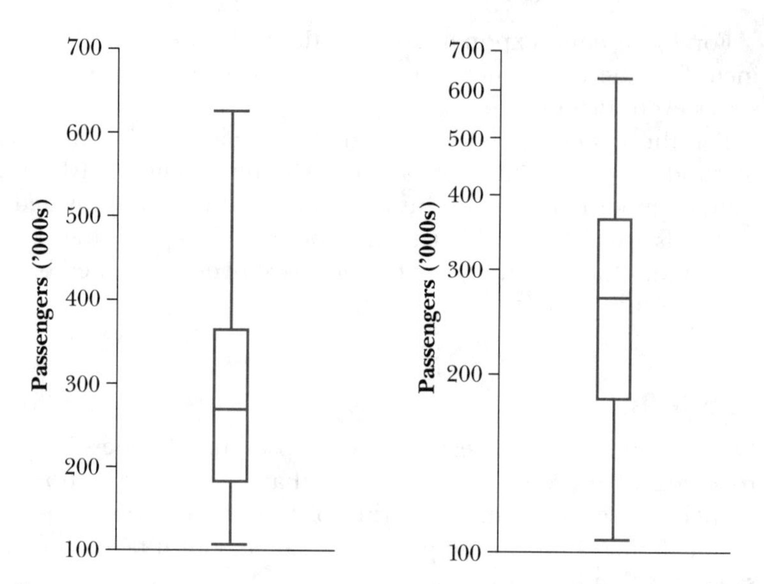

The diagram below shows the time series on a logarithmic scale.

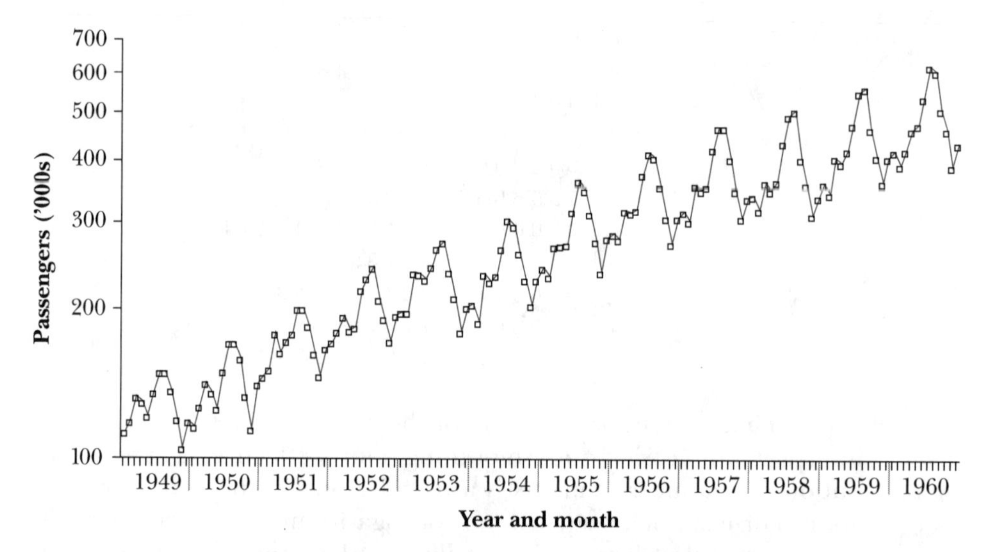

This time series plot of the logarithmically transformed data demonstrates an additional benefit. On a logarithmic scale, the seasonal variability is fairly consistent from year to year. In contrast, the amplitude of the seasonal fluctuations of the original time series increased markedly during the 12 years.[13]

13. Note that constant percentage changes are transformed by a logarithmic transformation into constant absolute changes. For example, the change in airline passengers from 136 000 in September 1949 to 119 000 in October 1949 — a decrease of 12.5% — is similar on a logarithmic scale to the change from 404 000 to 359 000 between September and October 1958 — a decrease of 11.1%.

EXERCISE 3.3

1. Explain why exponential smoothing has a different end-effects problem than do the techniques of running means or medians.
2. In exponential smoothing, what range of values of the smoothing constant, a, should be tried to obtain an appropriate amount of smoothing? What happens when a is close to 0 or 1?
3. Contrast running means and running medians with respect to the smoothness of the result they produce and also with respect to their sensitivity to extreme values.
4. What are the objectives of transforming data before they are plotted as a time series?
5. In this question, we will examine whether repetition of the 3-point running median procedure eventually leaves the smoothed series unchanged.
   (a) Find the 3-point running medians for the series {1, 2, 3, 4, 5}. Remember to leave the end values unchanged, so that the smoothed series still has five values in it.
   (b) Do the same thing for the series {1, 2, 3, 2, 1}. Does the smoothed series change if running medians are repeated?
   (c) Write down a series of eight values for which running medians must be repeated twice to obtain a series that does not change on further applications of running 3-point medians.
6. The CPI time series increases from 55.1 to 1041.0, with much of the increase occurring in the last ten years of the 40-year series. Graph the series after transforming the CPI to log(CPI). Does the log series have the same feature?

3.4 SEASONAL TIME SERIES

Many time series, such as the Wisconsin food employees data from section 3.2, exhibit a regular pattern of seasonal variation which repeats itself regularly (for example, every year). The seasonal pattern in the series is constant from year to year but may occur concurrently with a long-term trend. The logarithmically transformed airline passenger data set (see sections 3.2 and 3.3) provides another example of such a time series. Regular cyclical variation with periods of one day or one week are also common in time series. In this section, we will examine data with cyclical patterns that repeat at regular *known* intervals.[14]

---

14. Other time series show cyclic variation where the cycle length is unknown and may vary from cycle to cycle. Various time series in the world of commerce are presumed to be driven by business cycles. Other phenomena that have been investigated for cyclic but non-seasonal behaviour include sunspot counts and counts from trappings of Canadian lynx. When the cycle is not of fixed length, it is potentially very misleading to analyse these data in the same way as seasonal time series. More appropriate analyses which go beyond simple smoothing are not within the scope of this book.

For annual seasonal data, observations are usually made quarterly, monthly, every four weeks or at weekly intervals. For many types of quarterly or monthly data (for example, economic and count data), an adjustment needs to be made for the different numbers of days in different quarters or months. For example, monthly sales can be converted from $ to $/day. Likewise, monthly totals of traffic fatalities can be converted to fatalities per day. A similar adjustment may be needed in other circumstances where, for example, a business may be closed for some part of a period. Such adjustments should be made before other analyses are carried out.

Decomposing seasonal time series

Because of the regularity of seasonal variations from year to year, a useful general approach to exploratory analysis of seasonal time series is to decompose them into several components.

These components are:

- an overall mean (a constant)
- a long-term trend which will include cyclical non-seasonal fluctuations. This component, added to the mean, produces a smooth de-seasonalised time series.
- a seasonal pattern, constant from year to year
- residuals.

Different ways to decompose a series have been proposed. The details of decomposition are beyond the scope of this book. In practice, a computer and specialist software would be used to perform the calculations. In this section, we will only demonstrate what can be achieved by such a decomposition. The analysis below shows the decomposition implemented in the statistical program *Models'n'Data.*[15]

Models'n'Data decomposes the Wisconsin food employees data into the four components shown in the following four graphs. Adding together the four components gives the original time series:

Data = Mean + Trend + Seasonal pattern + Residual.

The diagram on the top of page 93 overlays the mean of the process over all time periods, a constant, on the time series plot. This is the first of the four components of the time series.

15. W. D. Stirling, *Statistical Exercises Using Models'n'Data,* John Wiley and Sons, Brisbane, 1995.

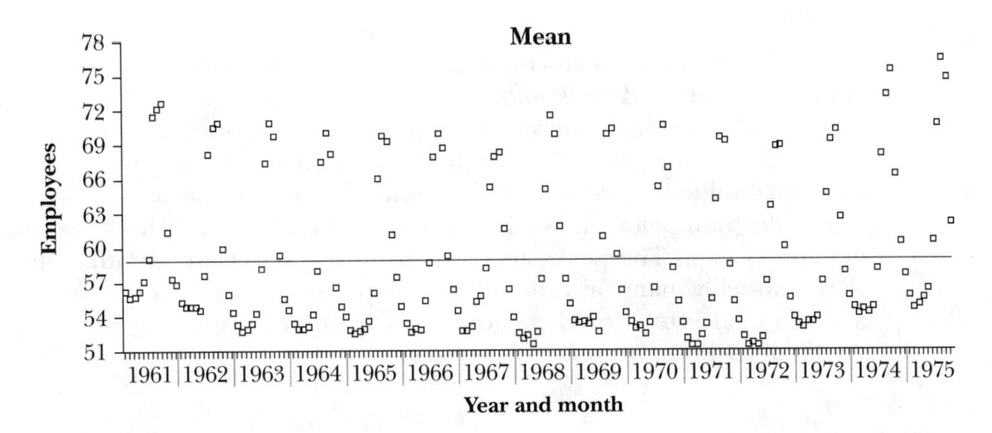

The 'trend' component (see diagram below), when added to the mean, is a smoothed version of the data (after eliminating the seasonal pattern). It provides a useful overall summary of movements in the underlying trend of the series.

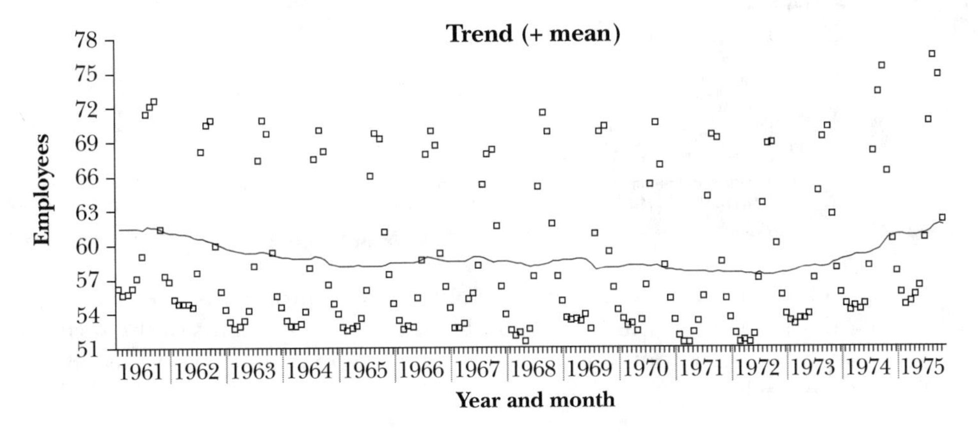

The seasonal component (again added to the mean in the diagram below) shows the repeating annual cycle, but does not follow any overall trend in the series. It is an average over the whole time series of the annual pattern of seasonal variation (after eliminating trend).

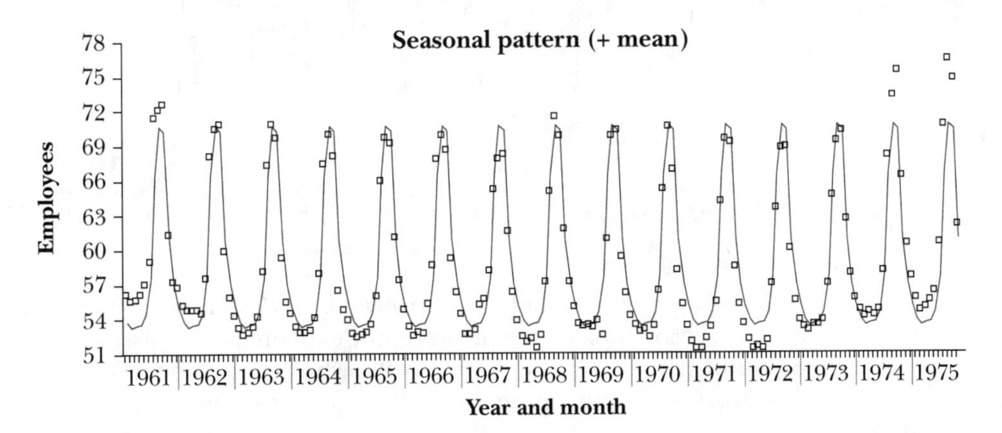

The residuals are what is left in the series after removing (subtracting) the mean, trend and seasonal components. They show variation in the series that is neither explained by a smooth long-term trend nor by a regular seasonal pattern. The residuals therefore represent other influences on the measurements. These are often, but not necessarily, 'special causes' which affect only individual values or a small number of neighbouring values.

The diagram below shows the residuals (plus mean) for the Wisconsin food employees data. The peaks and troughs in this series are months when there were unusually many or unusually few employees (after adjusting for the trend and seasonal effect for that month). It is often informative to look for the causes of such peaks and troughs.

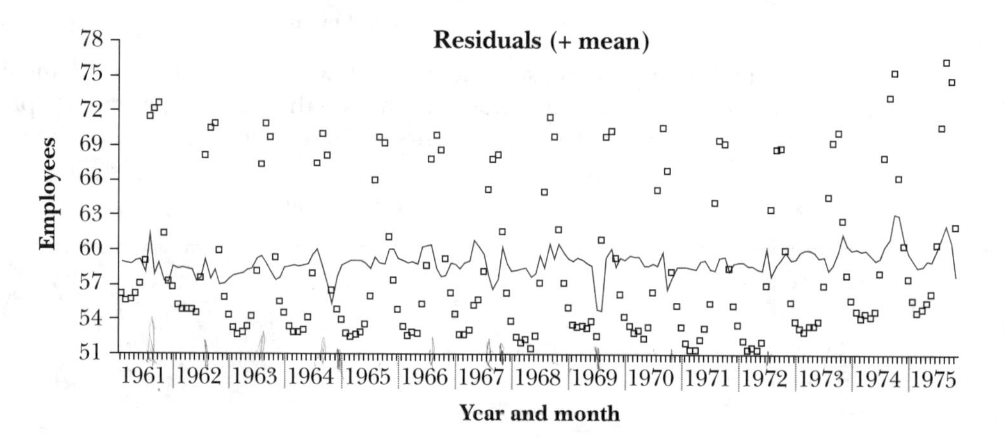

Other combinations of the four basic components are also meaningful. Adding together the first, second and third components of the series produces an effective smoothed version of the original time series, as shown in the graph below.

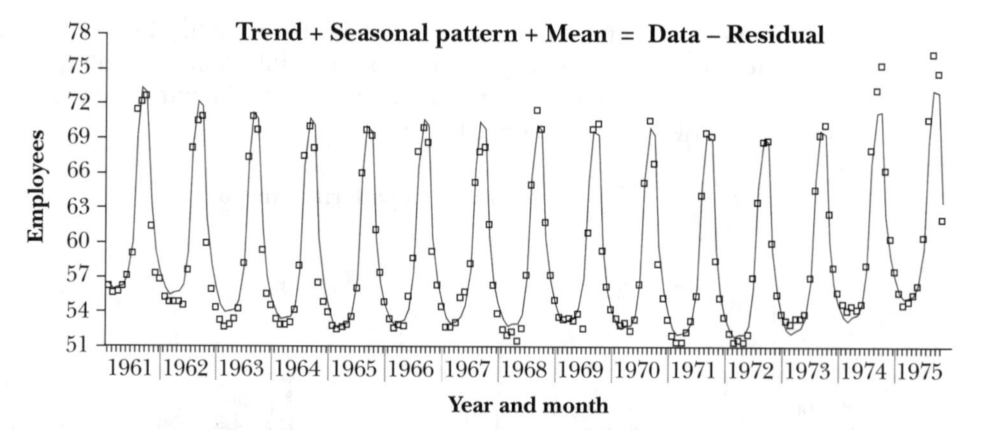

Looking at this graph, we might tentatively conclude that there are subtle variations in the seasonal pattern which are not explained if we assume a seasonal component which remains the same for the whole time series. For example, in several years the trough appears to show a consistent variation from the smoothed series; likewise the summer peaks during the last three years are higher than indicated by the smoothed series.

Adding the fourth component to the first and second, as illustrated below, *deseasonalises* the original series, producing a 'seasonally adjusted' series. Many economic (and some other) time series are published in this deseasonalised form.

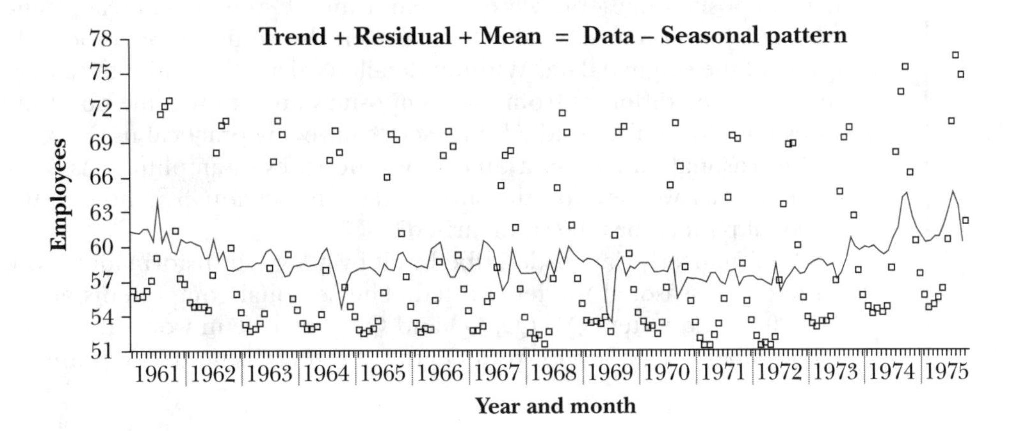

Sharp peaks and troughs in this series can be interpreted as months where the number of employees was higher or lower than would be expected from the long-term trend and seasonal effect. A search for the cause of such unusual months often gives the data analyst considerable insight into the data.

Transformation of seasonal time series

As discussed in section 3.3, time series often exhibit substantial skewness in the 'marginal' distribution of the relevant variable that can be removed by transforming the measurements. In particular, a logarithmic transformation is often effective in giving a more symmetric distribution of values. As was illustrated for the airline passenger data, such transformations often also make the seasonal pattern in the data more regular. After transformation, the fluctuations are of a similar magnitude in each cycle.

> When a logarithmic transformation has been used, the standard decomposition:
>
> $$\text{Ln(Data)} = \text{Mean} + \text{Trend} + \text{Seasonal pattern} + \text{Residual}$$
>
> can be written in the form:
>
> $$\text{Data} = \text{Exp(Mean)} \times \text{Exp(Trend)} \times \text{Exp(Seasonal pattern)} \times \text{Exp(Residual)}.$$

In other words, the raw data are *multiplicatively* decomposed into components rather than *additively* decomposed.

EXERCISE 3.4

Questions 1–5 below relate to the Wisconsin food employees data.

1. Does the seasonal component remain the same in each year?
2. Does the trend component include any regular seasonal changes?
3. If a composite time series were formed from the sum of the mean, trend and seasonal components, would the graph of the resulting data look like the graph of the original data? Without detailed calculations, identify a raw value that would be different from the composite value at the same point in time.
4. Does the plot of the residual time series have any practical use?
5. If the seasonal pattern in a time series increases in amplitude as the trend increases, as we saw for the airline data in section 3.3, how would the seasonal pattern best be summarised?
6. A set of quarterly data is described well by a $\log_{10}$ transformation, to which an additive seasonal model is fitted. The seasonal components are {−1, 0, 1.5, −0.5} in quarters Q1, Q2, Q3 and Q4. Describe in words the change in the number of employees from Q3 to Q4 if there is no trend in this period.

3.5 CONTROL CHARTS

Although many time series have an obvious trend (such as steady upward or downward movement), a fundamental question for many other time series is whether there is a trend or whether the series is stable over time. Commonly, time series data are collected for the purpose of monitoring stability.

This is the case in clinical medicine where physiological measurements are taken on a patient, perhaps after surgery or prescription of a drug (used to control blood pressure, for example). It is desirable that the physiological measurement remains within predetermined limits and steady over time, apart from relatively small fluctuations about a constant level.

Variation is inevitably present in time series, but the resulting fluctuations may mask an important shift or systematic trend. It is a key role of time series analysis to 'unmask' the trend and other information in the data. In fact, it is often the timing of observed trends or shifts that provides a clue to the causes of our process veering out of control.

Another important example of the search for instabilities in time series is in statistical process control, which began in manufacturing industry where production processes are monitored. Measurements of the process are made regularly and analysed so that if quality changes, corrective action can be taken quickly. Statistical process control is now applied widely to many processes outside of manufacturing, for example in service industries as a key element in the modern 'total quality management' revolution.

The purpose of control charts is to continuously monitor a process and assess whether that process appears to be operating normally. The regular measurements form a time series containing both signal (reflecting changes in performance of the underlying process) and noise (which is 'random' variation in the measurements). As each measurement (or group of measurements) is recorded, a decision is made about process stability; if it is decided that the process has changed, action would be taken to seek the cause for that change (and to correct it). However, the noise in the measurements means that the decision will occasionally be wrong. Statistical procedures try to minimise decision errors, which may occur either through not detecting changes in the process or through taking action on the basis of false signals.

Before control charts were first used in industry, it was (and, in some industries, still is) common practice to inspect most or all of the product to see that measurements on that product conformed with requirements. Control charting focuses on frequently monitoring the process for stability but avoids inspection of the majority of product.

Run charts

The following table concerns an industrial process manufacturing insecticide dispensers.[16] It shows successive measurements of sampled dispensers over a period of two days, in the order 476, 478, 473, 459, 485, ...

Date	Charge weight of insecticide dispenser (g)			
Dec. 13	476	478	473	459
	485	454	456	454
	451	452	458	473
	465	492	482	467
	469	461	452	465
	459	485	447	460
	450	463	488	455
	Lost	478	464	441
	456	458	439	448
	459	462	495	500
Dec. 14	443	453	457	458
	470	450	478	471
	457	456	460	457
	434	424	428	438
	460	444	450	463
	467	476	485	474
	471	469	487	476
	473	452	449	449
	477	511	495	508
	458	437	452	447
	427	443	457	485
	491	463	466	459
	471	472	472	481
	443	469	462	479
	461	476	478	454

16. I. W. Burr, *Elementary Statistical Quality Control*, Marcel Dekker, New York, 1979, p. 130.

It was specified that the dispensers should be of average weight 454 g with a minimum of 427 g and a maximum of 481 g.

In monitoring a continuous process, a first step may be to record individual values which may either be all the output of a process or, more usually, values sampled at regular intervals. In process control, the plot of such a time series is called a *run chart.* A run chart should be produced 'dynamically' so that, as data are accumulated, the values can be assessed to see if they are consistent with the previously established level and variability of the process.

The simplest type of control chart is a control chart for individual measurements.[17] A control chart for individual measurements adds control limits to a run chart. The control limits represent values beyond which it would be unusual to obtain data values, given the established level and spread of the process.

The diagram below shows a run chart of the charge weight data with the specification limits (the minimum and maximum charge weights from the process specifications) superimposed, plus a further line at the specification average weight.

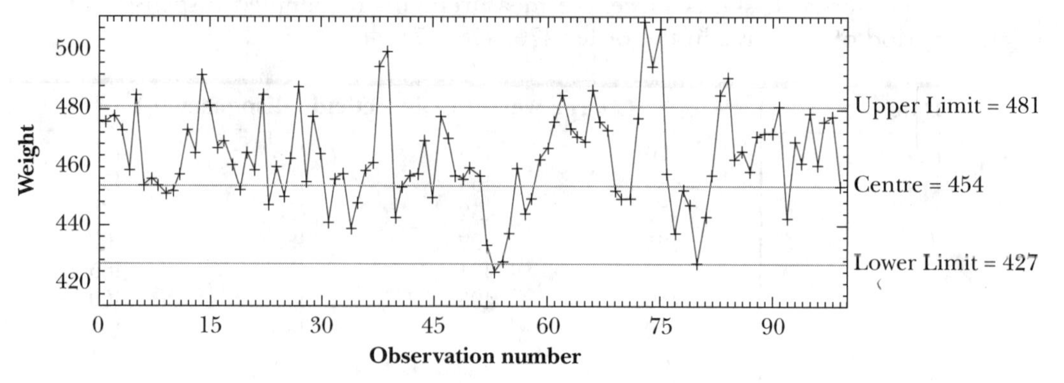

The chart shows that there is greater variation in the dispenser weights than the specifications require. The process is not satisfactory, because too much of the output *does not conform with specification limits* (it is 'out of spec').

The variability in dispenser weights may simply result from natural variability in the filling process. Process capability measures the extent to which the natural variation of a stable (*in control*) process is sufficiently small that the output is satisfactory. The process may be stable over time but, if it is not capable of producing output of sufficient quality (that is, most output within specification limits), the process must be changed in some fundamental way (for example, new equipment, retraining personnel).

Fluctuations in the output of a process can, however, often be attributed to specific causes (sometimes called *special, or assignable, causes*); such causes need to be sought while the process is being monitored. Deciding when a process is going 'out of control' as a result of assignable causes is one of the main uses of control charts. For this purpose, control charts do not use specification limits; instead, they use limits defined by the performance of the system when it is in control.

17. We will describe other control charts, such as control charts for sample means and for measures of sample spread later in the section.

In chapter 2, we noted that, irrespective of the shape of a batch of numbers, it was commonly the case that:

- about 70% of values lie within one standard deviation of the mean
- about 95% of values lie within two standard deviations of the mean
- almost all lie within three standard deviations of the mean.

This was called the *70-95-100 rule.*

For this reason, it is common to determine the standard deviation of the process when it is in control, then define the control limits to be three standard deviations on either side of a centre line (the mean of the process). These control limits are used to monitor the future performance of the process. Values outside these limits are very unlikely if the process is in control. Any value outside these limits indicates a possible change in the process which may be due to a sudden shift, a trend in the mean, an increase in variability, or some other 'special cause' (or simply an unusual value with no attributable cause).[18]

In addition to plotting upper and lower control limits, it is common also to plot a centre line. Although neither common nor recommended for the routine production of control charts, it is instructive to also draw lines representing distances of one and two standard deviations either side of the centre, to establish six zones within the control limits, as illustrated below. Individual values outside the control limits are indicative of the process being out of control, but various patterns of points within the six zones also indicate problems with process stability.

Upper control limit	————————————	$\bar{x} + 3s$
	Zone A	
	- - - - - - - - - - - - - -	$\bar{x} + 2s$
	Zone B	
	- - - - - - - - - - - - - -	$\bar{x} + s$
	Zone C	
Centre line	————————————	$\bar{x}$
	Zone C	
	- - - - - - - - - - - - - -	$\bar{x} - s$
	Zone B	
	- - - - - - - - - - - - - -	$\bar{x} - 2s$
	Zone A	
Lower control limit	————————————	$\bar{x} - 3s$

18. The ideas behind control limits and the decision that a process is out of control are essentially the same as those of testing a hypothesis about a process parameter (see section 9.3), except that in statistical process control, the process is assessed for change as each of the data values is recorded and included in the updated analysis.

The simplest statistical indication that the process is out of control is provided by:

- a single point beyond the upper or lower control limits (that is, outside all six zones), as shown in the diagram below. This suggests that the process mean has changed or the process spread has increased.

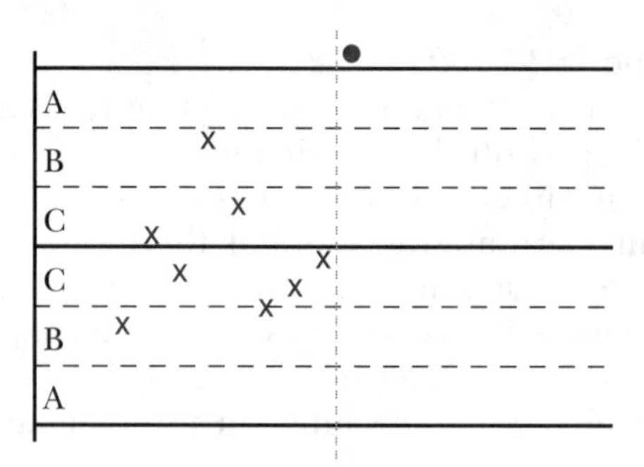

However, there are other statistical indicators that the process is out of control. In all of these, there are too many neighbouring points which exhibit a particular pattern. There is no fixed practice determining the number of points in such patterns which constitute a signal that the process is out of control. The practice varies, but the principle remains the same; the pattern is unlikely if the process is in control. The patterns include:[19]

- too many successive points all on one side of the centre line (often nine points). This suggests that the process mean has changed.

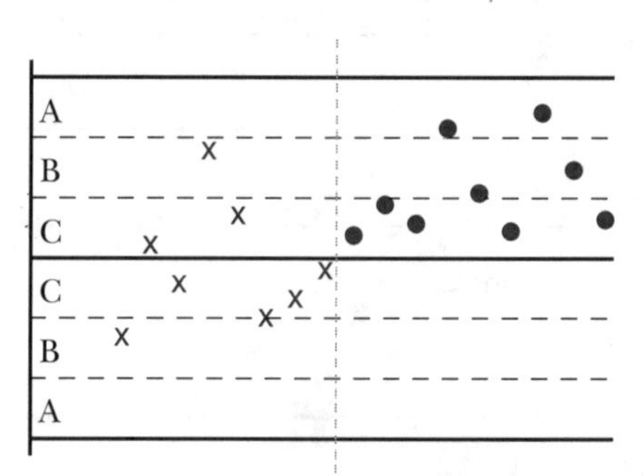

19. L. S. Nelson, 'The Shewhart Control Chart — tests for special causes', *Journal of Quality Technology*, 16, pp. 237–39.

- too many neighbouring points all in Zone A, the zone nearest the control limits (often two out of three successive points). This suggests that the process mean has changed or the process variability has increased.

- too many neighbouring points all in zones A and B (often four out of five successive points). This suggests that the process variability has increased.

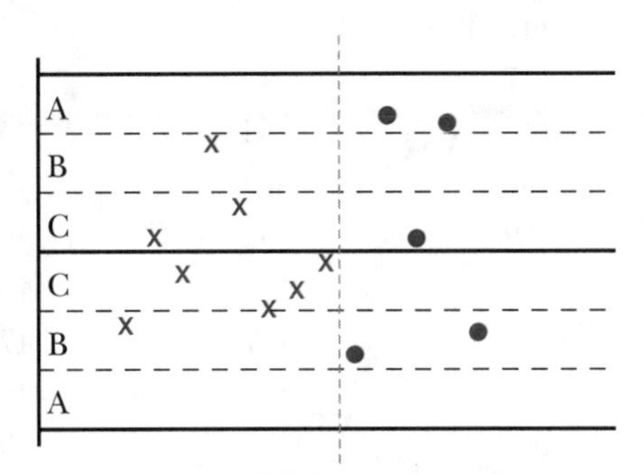

- too many points in a row steadily increasing or decreasing (often six points). This suggests that the process mean has changed.

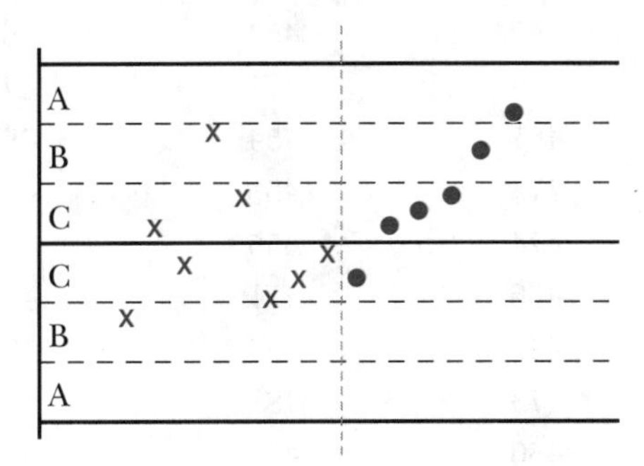

Any such patterns should be regarded as evidence that the process may be out of control and should be followed by an examination of the process to try to find (and correct) an assignable cause.

The process should be monitored carefully (to avoid special causes) for a period to obtain good estimates of the mean and standard deviation of the output when the process is under control. We will use the mean (463.66) and standard deviation (16.94) of the 100 observations of insecticide dispenser charge weight as our estimates. (Despite an indication that the process was not conforming to specifications, we will, for the moment, assume that the process was in control.) Control limits are therefore placed at:

$$463.66 - 3 \times 16.94 = 412.84 \text{ and } 463.66 + 3 \times 16.94 = 514.48.$$

We will use these control limits to assess the subsequent 100 measurements that were made of insecticide dispenser weights following the initial study. These are shown in the table below.

Date	Charge weight of insecticide dispenser (g)			
Dec. 17	450	441	444	443
	454	451	455	460
	456	463	Lost	445
	447	446	431	433
	447	443	438	453
	440	454	459	470
	480	472	475	472
	449	451	463	453
	454	455	452	447
	474	467	477	451
	459	457	465	444
	465	475	456	468
	458	450	451	451
	447	417	449	445
	453	442	456	453
Dec. 18	471	467	461	455
	462	454	462	468
	474	471	471	463
	461	454	468	452
	473	453	465	475
	474	455	486	490
	466	471	482	474
	447	454	476	486
	473	488	482	475
	460	450	461	445

The control chart for these data is shown below. Although nonc of thc observations fall outside the upper and lower control limits, several of them fail the test for nine consecutive points on the same side of the mean; these are marked with solid circles in the chart.

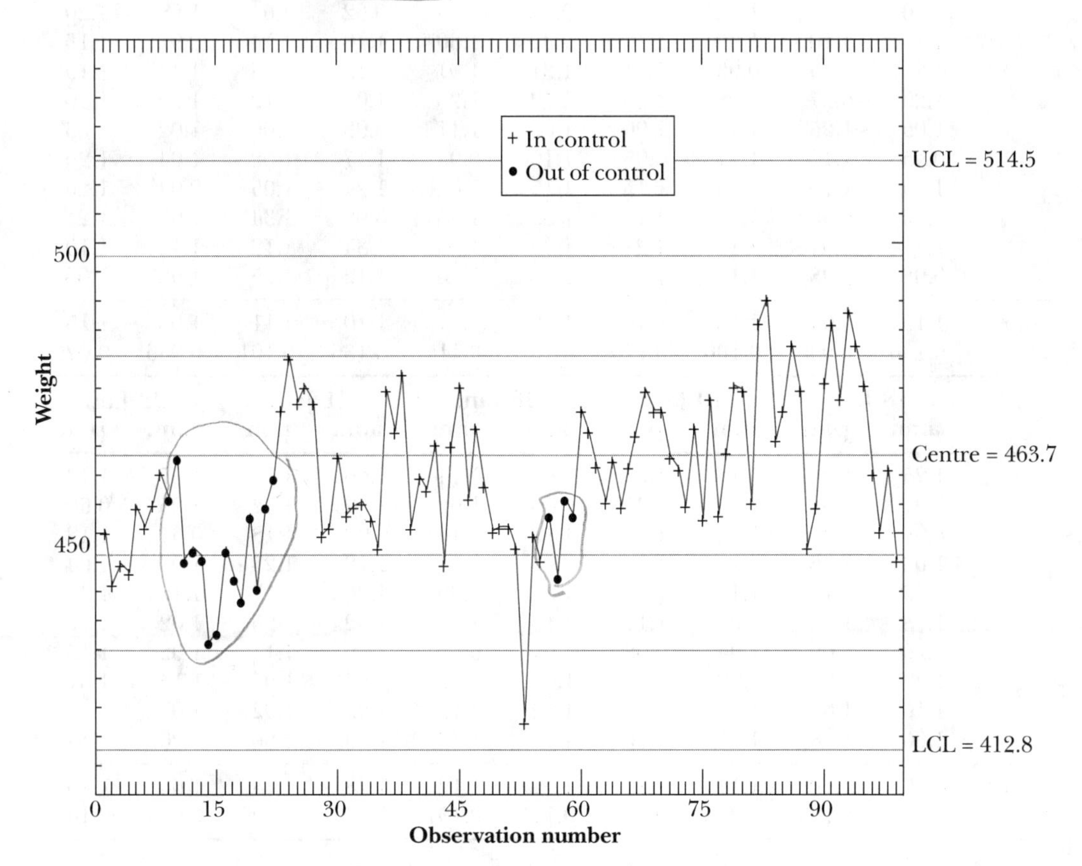

As the control chart would be constructed observation by observation while the process is being monitored, the process would be carefully scrutinised when the two runs of 'low' weights were recorded to try to find an attributable cause. At other times, there is no evidence of the process being out of control.

Control charts for means

Although control charts for individual measurements are the simplest type of control chart, in practice measurements are usually recorded in groups, not individually. For example, the following table shows paint primer thickness, which was recorded twice a day from 10 consecutive samples on each of 10 working days.[20]

20. *Continuing Process Control and Process Capability Improvement,* Statistical Methods Office, Ford Motor Company, Dearborn, Michigan, 1984.

Paint primer thickness (mils*)

	11 Jan.		12 Jan.		13 Jan.		14 Jan.		15 Jan.	
	a.m.	p.m.	a.m.	p.m.	a.m.	p.m.	a.m.	p.m.	a.m.	p.m.
	1.30	1.01	1.22	1.08	0.98	1.12	0.92	1.04	1.08	1.20
	1.10	1.10	1.05	1.12	1.30	1.30	1.10	1.14	0.92	1.13
	1.20	1.15	0.93	1.11	1.31	1.01	1.13	1.18	1.14	1.19
	1.25	0.97	1.08	1.28	1.12	1.20	1.02	1.12	1.20	1.16
	1.05	1.25	1.15	1.00	1.08	1.11	0.93	1.00	1.02	1.03
	0.95	1.12	1.27	0.95	1.10	0.93	1.17	1.02	1.04	1.25
	1.10	1.10	0.95	1.15	1.15	1.02	1.24	1.05	0.94	1.20
	1.16	0.90	1.11	1.14	1.35	1.25	0.98	1.34	1.05	1.24
	1.37	1.04	1.12	1.28	1.12	1.05	1.34	1.12	1.12	1.10
	0.98	1.08	1.10	1.31	1.26	1.10	1.12	1.05	1.06	1.03
$\bar{x}$	1.15	1.07	1.10	1.14	1.18	1.11	1.10	1.11	1.06	1.15
s	0.136	0.098	0.106	0.120	0.121	0.115	0.136	0.101	0.086	0.079
	18 Jan.		**19 Jan.**		**20 Jan.**		**21 Jan.**		**22 Jan.**	
	a.m.	**p.m.**	**a.m.**	**p.m.**	**a.m.**	**p.m.**	**a.m.**	**p.m.**	**a.m.**	**p.m.**
	1.25	1.24	1.13	1.08	1.08	1.14	1.06	1.14	1.07	1.13
	0.91	1.34	1.16	1.31	1.26	1.02	1.12	1.22	1.05	0.90
	0.96	1.40	1.12	1.12	1.13	1.14	0.98	1.18	0.97	1.12
	1.04	1.26	1.22	1.18	0.94	0.94	1.12	1.27	1.05	1.04
	0.93	1.13	1.12	1.15	1.30	1.30	1.20	1.17	1.16	1.40
	1.08	1.15	1.07	1.17	1.15	1.08	1.02	1.26	1.02	1.12
	1.29	1.08	1.04	0.98	1.07	0.94	1.19	1.15	1.02	1.15
	1.42	1.02	1.28	1.05	1.02	1.12	1.03	1.07	1.14	1.01
	1.10	1.05	1.12	1.00	1.22	1.15	1.02	1.02	1.07	1.30
	1.00	1.18	1.10	1.26	1.18	1.36	1.09	1.36	1.00	1.14
$\bar{x}$	1.10	1.19	1.14	1.13	1.14	1.12	1.08	1.18	1.06	1.13
s	0.170	0.125	0.070	0.107	0.111	0.137	0.074	0.099	0.059	0.141

* A mil is an imperial measurement equivalent to one-thousandth of an inch.

For such data, the most commonly used control chart is the $\bar{x}$ chart, in which the means of groups of values are plotted against time. For reasons which will be elaborated on in chapter 8, the *70-95-100* rule is especially useful when applied to means.

The procedures for using $\bar{x}$ charts are essentially the same as those for charts for individual values, but the control limits should now be based on the standard deviation of the $\bar{x}$ values, not on the standard deviation of the individual values. There are two ways to estimate the standard deviation of the $\bar{x}$ values.

1. Directly evaluate the standard deviation of the observed means from the first few samples.

From the first 10 samples above, the standard deviation would be estimated by the standard deviation of the 10 values 1.15, 1.07, . . ., 1.15, $s_{\bar{x}} = 0.0371$.

2. Estimate the standard deviation of $\bar{x}$ from the first k samples of size n with the following formula:[21]

$$s_{\bar{x}} = \sqrt{\frac{\sum_{i=1}^{k} s_i^2}{k \times n}}$$

Using the first $k = 5$ samples, we get the estimate $s_{\bar{x}} = 0.0361$.

Both formulae should give similar estimates of $s_{\bar{x}}$ provided the process is in control. The second method gives a more accurate estimate if the estimate must be made from a small number of preliminary samples.

We will use the estimate $s_{\bar{x}} = 0.0361$ from the second method and the average primer thickness from the first 5 samples, 1.1182, to obtain control limits at $1.1182 + 3 \times 0.0361 = 1.226$ and $1.1182 - 3 \times 0.0361 = 1.010$. These are shown in the control chart below.

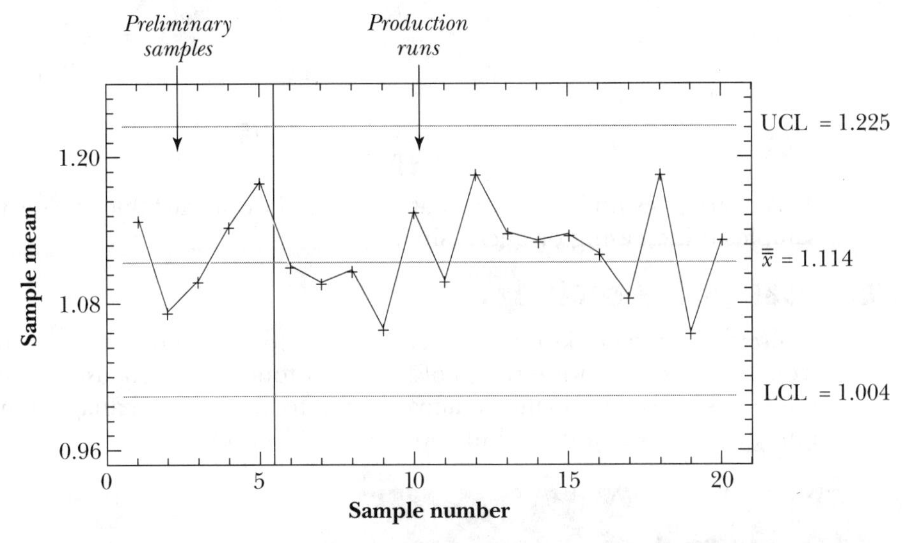

This $\bar{x}$ chart is interpreted in the same way as control charts for individual values. It does not give any indication that the process is out of control since it exhibits none of the patterns of points that we described earlier.

Control charts for process spread

While an $\bar{x}$ chart assesses stability of the mean, or level, of a process, it is also common practice to simultaneously use an accompanying chart to assess stability of the spread as measured by the range (an R chart) or standard deviation (an s chart) of successive groups of values. R charts are the most commonly used control charts for process spread, both in specialist statistical process control monographs and in industry. The R chart (or s chart) should be used in conjunction with the $\bar{x}$ chart.

21. This formula will be explained in chapter 8.

As with other control charts, the *R* chart and *s* chart are time series charts of either *R* or *s* with upper and lower control limits added. Determining the position of these control limits is harder than with individual value or $\bar{x}$ charts, but many statistical computer programs will perform the calculations, so the underlying mathematical detail does not need to be understood. The important characteristic of *R* and *s* charts is that values of *R* or *s* rarely cross the control limits when the process is under control. A range chart for the paint primer data is given below.

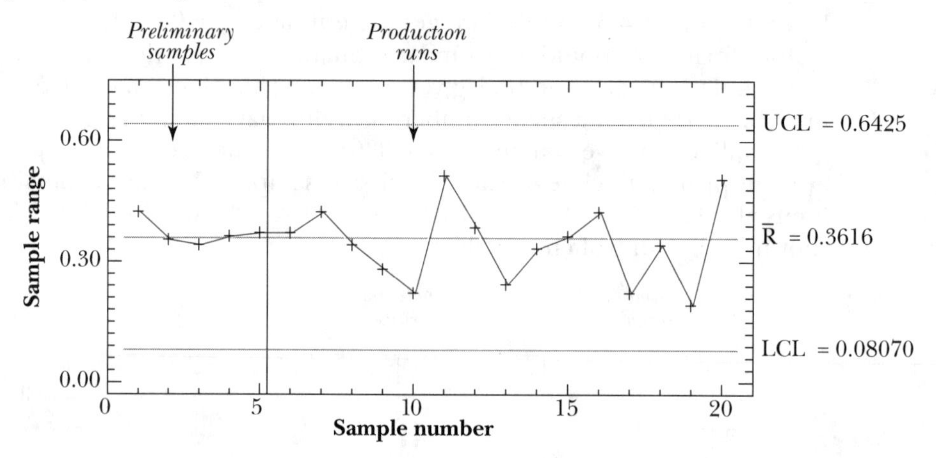

This chart gives no indication that the spread of primer thicknesses within the samples is increasing or decreasing.

Statistical quality control

There are many books on statistical process control and statistical quality control. These cover a wide range of statistical tools and their use in what is variously described as quality management, total quality management, quality improvement, continuous improvement, and so on.[22]

EXERCISE 3.5

1. Are all of the control charts in this section time series?
2. A common management strategy is 'management by exception'. Executives examine data on operations and further examine only those data which seem out of line with the rest — data indicating that either something good or something bad may have occurred. Then, some management action would follow this examination if indicated. Does this strategy have any relationship to control charts?
3. Why is the range used as a measure of variation for quality control applications when it has so many shortcomings as a descriptive measure of variation?

22. A good overview is provided in D. C. Montgomery, *Introduction to Statistical Quality Control* (2nd ed.), John Wiley and Sons, New York, 1991 and in K. Ishikawa, *Guide to Quality Control* (2nd ed.), Asian Productivity Organisation, Tokyo, 1986.

Problems

1. Winning times for the men's 100 metre race at the Olympic Games are noted below.

Year	Name	Country	Time (seconds)
1896	Thomas Burke	United States	12
1900	Francis Jarvis	United States	10.8
1904	Archie Hahn	United States	11
1908	Reginald Walker	South Africa	10.8
1912	Ralph Craig	United States	10.8
1916	(Games cancelled)		
1920	Charles Paddock	United States	10.8
1924	Harold Abrahams	Great Britain	10.6
1928	Percy Williams	Canada	10.8
1932	Eddie Tolan	United States	10.3
1936	Jesse Owens	United States	10.3
1940	(Games cancelled)		
1944	(Games cancelled)		
1948	Harrison Dillard	United States	10.3
1952	Lindi Remigino	United States	10.4
1956	Bobby Morrow	United States	10.5
1960	Armin Hary	Germany	10.2
1964	Bob Hayes	United States	10.0
1968	Jim Hines	United States	9.95
1972	Valeri Borzov	USSR	10.14
1976	Hasely Crawford	Trinidad	10.06
1980	Allan Wells	Great Britain	10.23
1984	Carl Lewis	United States	9.99
1988	Carl Lewis	United States	9.92
1992	Lynford Christie	Great Britain	9.96
1996	Donovan Bailie	Canada	9.84

Source: R. Famighetti (ed.), *The World Almanac and Book of Facts 1995*, Funk and Wagnalls, Mahwah, New Jersey, 1995, p. 858. Reprinted with permission.

(a) Plot the data from the table. Does a logarithmic transformation of the winning times make the trend more linear?

(b) Apply exponential smoothing to the transformed data (ignoring the gaps when there were no Olympic Games). Use the smoothed value in 1996 as a prediction of the time of the winner in the year 2000. Is it

likely to overestimate or underestimate the winning time? (*Note:* If the winning time in 2000 is already known, supplement the series and predict the winning time at the next Olympics.)
(c) Fit a straight line by eye to the transformed data, and use it to predict the winning speed in the year 2000.
(d) How would you attempt to predict the winning time in the year 2048?

2. In process control charts, why are control limits positioned three standard deviations on either side of the mean most commonly used as a trigger for a single value?

3. (a) For the Wisconsin food employees data, is there any pattern to the residual series (graphed as residuals + mean in section 3.4, page 94)? What does it tell you about the seasonal component?
(b) Why is question (a) an important question to ask while considering the forecast of the next year's monthly data?

4. Can you think of a situation where the stem-and-leaf plot (or the dot plot) of the time series values would suggest a symmetrising transformation that would make the residuals from the time series trend less symmetric? (*Hint*: Consider a time series with a nonlinear trend, but equal variability over the time interval of the data.)

5. For the chemical yields data graphed in section 3.2, the 70th value is a yield of 23 and is quite a bit lower than typical values. This suggests that the 71st value will be greater than 23, and a look at the typical yield values suggests that the 71st value may be about 50. Is there a reason why the 71st value would be expected to be *more than* 50, based on the graph? What about the 72nd value?

6. A manager wants to produce tyres whose diameter is within certain specification limits (for example, between 94.0 cm and 95.0 cm).
(a) Why would the manager want to monitor the mean of the process?
(b) Why would the manager want to monitor the standard deviation of the process?

7. Forecasts of time series rely upon the unchanging nature of the process generating the time series — although random variation is expected to make forecasts imperfect, the size of errors should have a distribution that is predictable. For example, consider a government budget announcement that specifies revenue and expenditure plans for the next three years. It would be based on an assumption about the rate of growth of the economy and the resulting tax revenues. However, these assumptions are likely to be wrong — in fact, given the interdependence of national economies, and the likelihood of unpredictable events like wars and earthquakes, the assumptions are likely to turn out to be very wrong. Does this mean that the forecast is of no use? Why is an unreliable forecast better than no forecast?

1. **Consumer Price Index**
   Use the CPI data from section 3.1 (page 76) to forecast the CPI for 1990–1994. (*Suggested steps*: First examine the raw trend; then consider whether a transformation would simplify this trend; do the transformation if appropriate, fit the trend, and then check the residuals; re-do the trend if necessary, and finally forecast the series to 1990–1994.) Then, find the actual data from your library and comment on the nature of your prediction errors over the five year forecast.

2. **Library circulation**
   Obtain data from your library on the circulation (number of books checked out) over a recent five week period, from the middle of a term. Put aside the most recent week of data and examine the earlier portion: plot the data and visually identify the nature of the trend and weekly periodicity. Note actual dates and events occurring during this period that might affect the series. Discuss the stability of the trend and periodicity, and the likely quality of a forecast of the time series for one week.

   If you have time series software available to do so, forecast the series for one week and compare with the actual data. If you do not have time series software, then simply plot the five weeks of data and note the degree to which the fifth week follows the patterns of the first four weeks.

3. **100 metres at the Olympics**
   Various sources compile the results of the Olympic Games which have occurred every four years since 1896. In particular, the winning times of the men's 100 metres event are recorded in problem 1 above, along with the reference showing where the data were published. Select another event and record the time series of winning times. Then answer the following questions:
   (a) Does the series have the behaviour shown for the 100 metres series, with improvement over time, but with a decreasing rate of improvement?
   (b) What standard in the event you chose will be necessary to win the event at the next Olympic games? (Although you cannot answer this precisely, the time series should allow you to make a reasonable estimate of this critical level.)

4. **Control chart for minimum and maximum**
   The control chart for means illustrated in section 3.5 for the paint primer data (page 105) plots one point for each batch of 10 measurements recorded at one time. However, when the maximum of the individual values is unusually high or the minimum unusually low, it would indicate that the process was veering out of control. The object of this project is to consider a new kind of control chart which uses individual measurements as well as the groups' means.

   If the individual data points were recorded graphically as they were measured, the control limits designed for the means, $\bar{x} \pm 3 \times \frac{s}{\sqrt{10}}$ would not be a useful warning level, since they are based on the distribution of the mean rather than the distribution of the individual values; individual values would often exceed these limits.

We could draw control limits based on individual values, $\bar{x} \pm 3s$, but there would be a reasonable chance of *at least one* value in a batch being outside these limits if the batch size is large. An alternative procedure that is less affected by the batch size is to add control limits for the maximum and minimum of each batch. It can be shown that appropriate control limits for the minimum and maximum are at $\bar{x} \pm 3.55 \times s$. (The value 3.55 is specific to the batch size being 10. It would be higher if the batch size was greater.) These limits can be overlaid on a plot of the individual values, and will be exceeded with the same frequency as the earlier limits for batch means.

Use estimates of the process mean and standard deviation from the first $k = 5$ samples of the paint primer data to obtain control limits, and apply them to the remaining samples. The chart below shows all data values; you should draw the control limits on this chart. Are the indications similar to the indications provided by the means control chart?

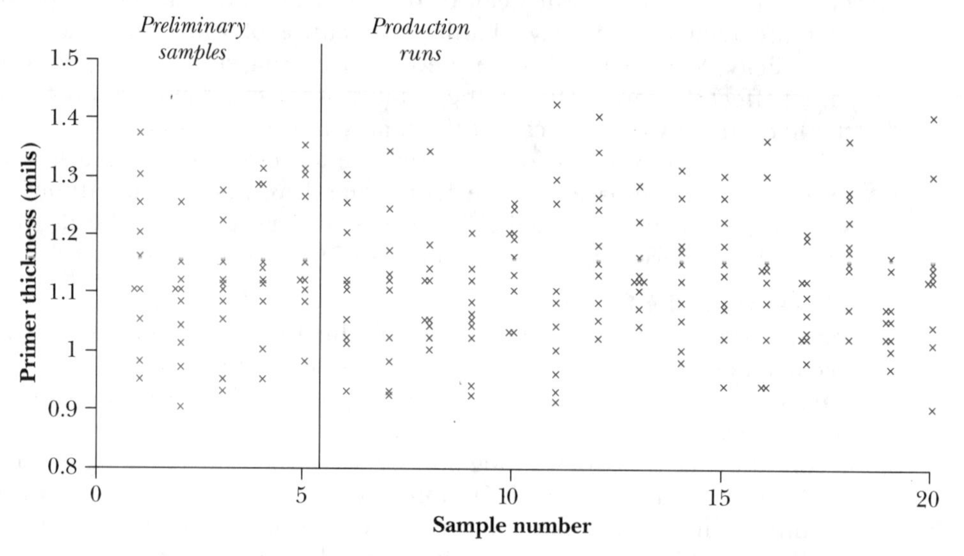

Does the use of both charts together give more useful information than either chart separately?

5. **Reaction times — diurnal trends**

Refer to the reaction time project of chapter 2 (page 70). This is an extension of it, although it is not necessary to have done the chapter 2 project to do this one.

Measure your reaction time each hour from when you first wake in the morning to when you retire in the evening. This will give you about 15 points. Is there any pattern to these reaction times? Use a smoothing method, and the residuals from it, to examine the trend. Any trend over a 24 hour day is called a diurnal trend. Can you predict what would happen if you continued the measurements later into the night? Do you wish to validate your forecast?

Compare your data with the data of another student doing the same project. Comment on the similarities and differences.

CHAPTER 4

Exploring relationships — bivariate data

Whatever moves is moved by another.

St Thomas Aquinas (AD 1227–1274)

4.1 INTRODUCTION

We discussed time series in chapter 3. In analysing time series, we are interested in investigating patterns of change of a variable over time. Since we are interested in the relationship between this variable and time, the pattern is a special type of *bivariate relationship*. In this chapter, we will study more general relationships between two variables.

For example, the table below describes the gross expenditures on Research and Development (R & D) by Business and, separately, by Government and Universities, in OECD countries in 1988.[1]

The whole range of graphical and numerical univariate summary statistical methods from chapter 2 can and should be applied to each variable separately. Such analyses would show the distribution of each type of research expenditure and would highlight any unusual features such as outliers, clusters or skewness.

Univariate methods do not, however, describe the *relationship* between the two variables. For example, do countries with high expenditures on Business R & D also have high expenditures on Government and University R & D? Which countries have unusually high Government and University R & D in relation to their Business R & D?

Country	Business (% GDP)	Government and Universities (% GDP)
Switzerland	2.24	0.55
Japan	1.89	0.85
Sweden	2.03	0.79
Germany	2.06	0.71
USA	1.87	0.72
Netherlands	1.37	0.87
UK	0.55	0.70
France	1.34	0.91
Norway	1.17	0.67
Finland	1.02	0.61
Belgium	1.14	0.38
Denmark	0.80	0.54
Canada	0.75	0.58
Austria	0.70	0.55
Australia	0.45	0.75
Italy	0.75	0.56
Ireland	0.63	0.35
New Zealand	0.21	0.75
Yugoslavia	0.42	0.47
Iceland	0.12	0.58
Spain	0.43	0.29
Turkey	0.42	0.13
Portugal	0.11	0.30
Greece	0.09	0.23

1. Australian Bureau of Statistics, *Australia — Working it Out*, AGPS, Canberra, 1990. Commonwealth of Australia copyright reproduced by permission.

Before we consider exploratory techniques to help answer such questions, we will briefly examine a few bivariate data sets with different characteristics.

In a stimulus–response relationship we are interested in the effect of one variable, the *stimulus,* on the value of another, the *response.* This is often a controlled experiment, where we are able to control the value of the stimulus and set it at different levels for different experimental units. Examples include the effects of fertiliser on crop yield, the dose of a drug (for example, an anti-hypertensive drug) on a physiological response (in this case, blood pressure) or hours of training (or study) on physical (or mental) performance. In the two examples which follow, we clearly distinguish between the stimulus, or *explanatory* variable, and the response variable.[2]

Amplitude of stress ($N\,m^{-2}$)	Cycles to failure (× 1000)
500	20
450	19
400	19
350	40
300	48
250	112
200	183
150	496
100	1 883
90	2 750
80	2 181
70	3 111
60	9 158
50	15 520
40	47 188

The data above were obtained from an experiment in which 15 steel objects with the same geometry were prepared and allocated to fluctuating loads of different amplitudes. The number of cycles to failure was recorded.[3]

The stress amplitude is under the experimenter's control and is the explanatory variable in this relationship. The number of cycles to failure is the response.

Sometimes the values of the explanatory variable cannot be set by the experimenter. When neither variable is controlled by the experimenter, it may still be possible to regard one variable (the explanatory variable) as affecting the other (the response), but not the reverse. Studies in which the values of the explanatory variable are not (and perhaps cannot be) controlled or designed are commonly referred to as *observational studies.*

2. It was historically common to use the terms 'independent' and 'dependent' to respectively describe explanatory and response variables, and this terminology is still seen in many recent books. These labels are unfortunate and should be avoided since the words 'dependent' and 'independent' have specific meanings in statistics which make their use inappropriate here.
3. A.V. Metcalfe, *Statistics in Engineering: A Practical Approach,* Chapman and Hall, New York, 1994, p. 225.

For example, we might be interested in the relationship between the final grade of graduating students at a university and their high-school grade. High-school grades should have an influence on final university grades, but the reverse is impossible. Often, as in this example, there is a temporal ordering of the explanatory and response variables, with the explanatory variable being measured 'prior' to the response.

The table below describes the lifespan and metabolic rate (measured by oxygen consumption) of various mammals.[4] Although we cannot be certain that either of the two variables will causally affect the other, metabolic rate is more likely to affect lifespan than the reverse. We may also be interested in *modelling* such a relationship and using it to *predict* the lifespan of other animals from their metabolic rate. Our analysis of these data treats lifespan as the response and metabolic rate as the explanatory variable.

Mammal	**Lifespan (year)**	**O_2 ($cm^3/g/h$)**
Echidna	50.0	0.22
N. American opossum	5.0	0.52
European hedgehog	6.0	0.75
E. American mole	3.5	1.90
Little brown bat	20.5	2.00
Owl monkey	12.0	0.51
Squirrel monkey	20.0	0.83
Rhesus monkey	29.0	0.43
Chimpanzee	50.0	0.25
Humans	100.0	0.22
Nine-banded armadillo	6.5	0.25
Rabbit	18.0	0.42
Ground squirrel	9.0	0.87
Golden hamster	3.9	0.93
Rat	4.7	0.86
Mouse	3.2	1.49
Guinea pig	7.6	0.74
Gray wolf	16.2	0.32
Red fox	9.8	0.52
Arctic fox	14.0	0.52
Raccoon	13.7	0.39
Cat	28.0	0.44
Asian elephant	69.0	0.07
Horse	46.0	0.15
Goat	20.0	0.19
Sheep	20.0	0.24
Cow	30.0	0.13

4. H. Zepelin and A. Rechtschaffen, 'Mammalian sleep, longevity and energy metabolism', *Brain Behaviour and Evolution*, 10, 1974, pp. 431–433. Reproduced with permission of S. Karger, AG, Basel, Switzerland.

It is not always necessary (or possible) to distinguish between response and explanatory variables. Some of the statistical methods described in this chapter require a distinction to be made; others do not.

Sometimes the two variables are of more-or-less equal status, and neither would automatically be considered to be the explanatory variable (or response). For example, the heights and weights of a group of individuals might be recorded, or the concentrations of two different chemicals in the soil adjacent to each plant in a tray of plant seedlings. In such situations, we are simply interested in exploring the relationship between the measurements of two different characteristics.

The following table describes emissions of gases by various jet aircraft in each takeoff/landing cycle.[5] Both types of emission have similar status, so neither should be regarded as a response. (Nor should either be regarded as a stimulus.)

	Emissions (kilograms per takeoff/landing cycle)	
Aircraft	**CO**	**NO_x**
B747-400	32.68	48.45
B747-200	126.90	56.89
MD-11	45.68	37.47
DC10-30	57.64	40.42
L1011-200	120.61	42.66
A300	38.43	26.95
DC8-63	128.15	19.15
A310	30.94	23.88
B707-320B	128.15	11.78
B767-300	32.18	27.23
B757-200	13.06	27.47
B727-200	26.18	13.08
A320	14.28	9.46
B737-300	13.50	8.10
B737-100	18.69	8.72
DC9-50	18.69	12.10
BAe-146	12.03	4.85
BAC 111-400	42.55	7.07
Fokker 28	10.01	5.85
Dassault Falcon 20	36.28	0.76
Gates Learjet 36	1.27	1.73
Gates Learjet 35	1.27	1.73
Gates Learjet 24D	42.02	0.71
Cessna Citation	11.37	0.53

5. B. G. Woodmansey and J. G. Patterson, 'New methodology for modeling annual aircraft emissions at airports, *Journal of Transportation Engineering*, 120, 1994, p. 345. Reproduced by permission of the American Society of Civil Engineers.

For all bivariate relationships, our aim is to use the data to graphically and numerically describe the relationship between the two variables. Various aspects of such relationships are of special interest. In particular, we will address the following questions.

- Are increases in one variable associated with increases or decreases in the other?
- Does change in one variable *cause* change to occur in the other?
- Can one variable (usually the response) be predicted from the other? This question is of particular interest in a stimulus/response relationship.
- How should the relationship be described quantitatively?
- Is there evidence of deviation from the relationship?
- Do all the data fit the one relationship, or are there distinct groups?
- Are there outliers?
- If there is no evident relationship, do the data show any other interesting patterns?

In particular, we will investigate graphical methods which are especially useful in describing the relationship and identifying outliers, and statistics which separately provide numerical summaries of the strength and structure of the relationship. In section 4.6, we will consider some extensions to these graphical methods which can be used to describe relationships between three or more variables.

EXERCISE 4.1

1. Consider the research expenditure data set introduced in section 4.1 (page 112). If we were to compare the dot plot of the expenditure data for Business with the dot plot of the expenditure data for Government and Universities, could the association between the sets of expenditures be gleaned?
2. Sort the research expenditure data with respect to the Business expenditure variable, but carry along the corresponding expenditure for Government and Universities. Does this make the association between the variables clear from the table itself?
3. Consider the stress loads data in section 4.1 (page 113). The natural analysis of this data set treats amplitude of stress as the explanatory variable and cycles to failure as the response. Are there circumstances where cycles to failure might be used as an explanatory variable to predict amplitude of stress?

4.2 SCATTERPLOTS

The bivariate analogue of the dot plot is the *scatterplot* (or scatter diagram) which represents the paired data values on a two-dimensional plot. In mathematics, this is commonly termed an *x–y* plot. Each point of the scatterplot reveals the values of one pair of numbers from the data set. The scatterplot is the basic tool of bivariate exploratory data analysis (EDA).

Scatterplots are among the most commonly used and most useful graphical techniques in statistics. Although there are numerical summaries of bivariate data that are useful descriptive summaries of some aspects of a relationship, such numerical summaries can sometimes be very misleading and should never be used on their own; a scatterplot should always be used in conjunction with them, at least for data analysis if not for data summary. A particularly useful feature of a scatterplot is that it presents an overview of the general pattern of association between two variables as well as highlighting unusual observations atypical of the relationship between the two variables.

The following scatterplot displays the research expenditure data from section 4.1.

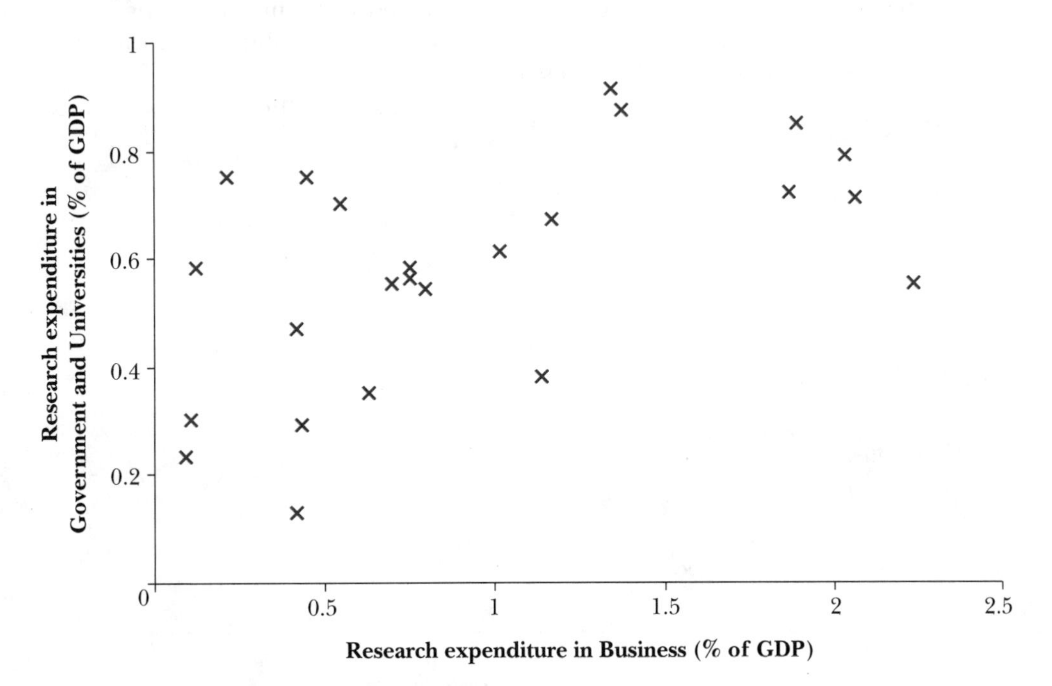

There is a very weak relationship between research expenditure in Government and Universities and that in Business. There is a slight tendency for countries with greater than average research expenditure in one category to also have greater than average research expenditure in the other category, but there are many pairs of countries where this does not hold.

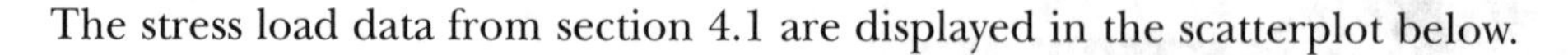
The stress load data from section 4.1 are displayed in the scatterplot below.

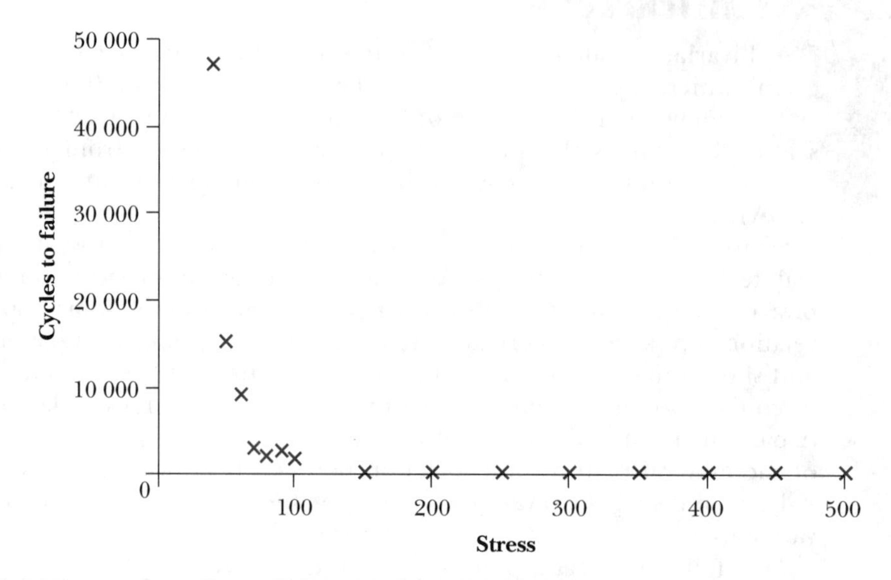

There is a very strong relationship between the cycles to failure of these steel objects and the amplitude of the stress applied to them. Increasing the stress seems to drastically reduce the number of cycles to failure, with the sharpest reductions occurring at low stress levels.

The relationship between animal lifespans and their metabolic rates is shown below.

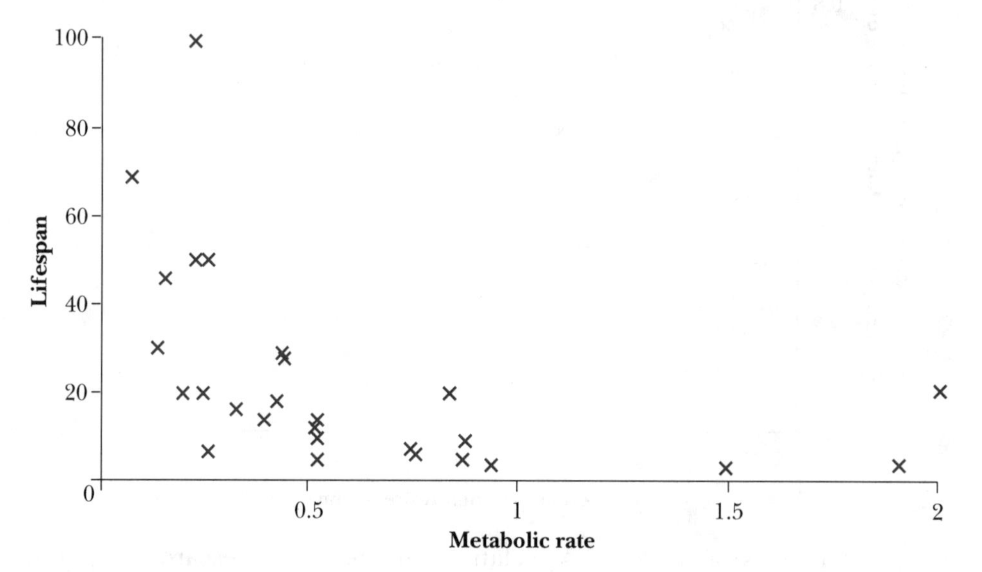

As in the previous scatterplot, high values of the explanatory variable (metabolic rate) correspond to low values of the response (lifespan), but the relationship is much weaker with a greater scatter of points around the relationship. In particular, there are two animals that have lifespans that seem unusually high in relation to their metabolic rates — Humans and the Little Brown Bat.

The aircraft emissions data are displayed in the following scatterplot.

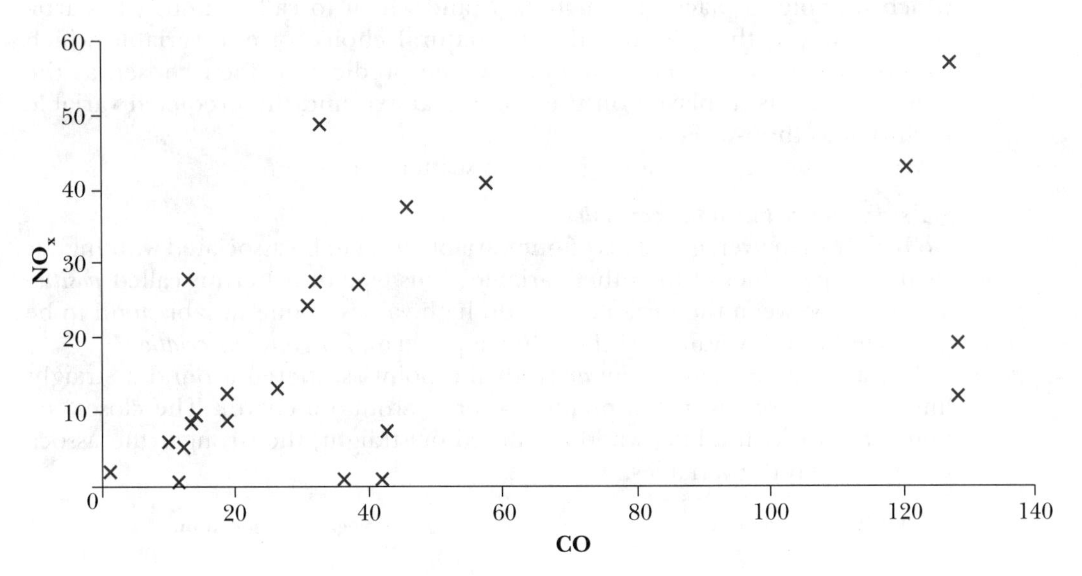

Like the research expenditure scatterplot, this scatterplot exhibits a weak relationship between the two measurements (NO_x and CO emissions from jet aircraft). There seems to be a separate cluster of four aeroplanes with high CO emissions (B747-200, L1011-200, DC8-63 and B707-320B), but they do not seem to exhibit a markedly different relationship between the variables than that suggested by data for the other jets.

As a matter of convention, when the two variables are identified as a response and an explanatory variable, we often denote the response by 'y' and the explanatory variable by 'x'. For time series, time is explanatory; the other variable of interest (for example, unemployment, the consumer price index) is the response. For stimulus–response relationships, we use x for the stimulus and y for the response. In both cases, the response is represented by the vertical axis and the explanatory variable by the horizontal axis, as shown in the diagram below.

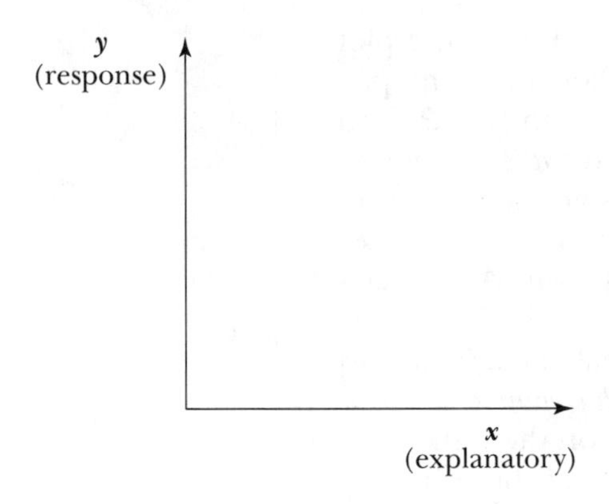

The main value in choice of these (or other) shorthand mathematical labels is convenience of mathematical and statistical representation (as was the case with the mathematical notation adopted in chapters 2 and 3).

Where there is no clear explanatory and response variable, the choice of which variable to place on which axis (and which to call 'x' and 'y') is arbitrary. However, there is sometimes a natural choice of one variable which we would like to predict from another: the predictor is then chosen as the x-variable and is displayed on the horizontal axis and the predicted variable is chosen as the y-variable.

What features are we looking for in a scatterplot?

- ***Association between the variables***

Do higher than average values of one variable tend to be associated with higher than average values of the other variable? This type of pattern is called *positive association* between the variables. Or, do high values of one variable tend to be associated with low values of the other, a pattern of *negative association*?

The association may be *linear* (with the points scattered around a straight line) or *nonlinear* (with the points scattered around a curve). The closer the points are to such a line, whether curved or straight, the stronger the association between the variables.

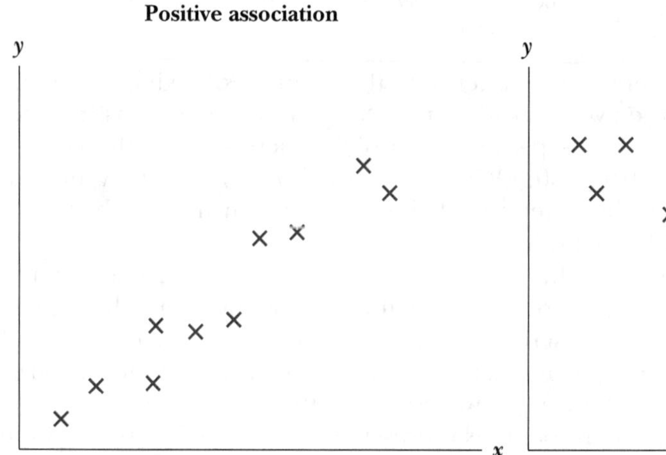

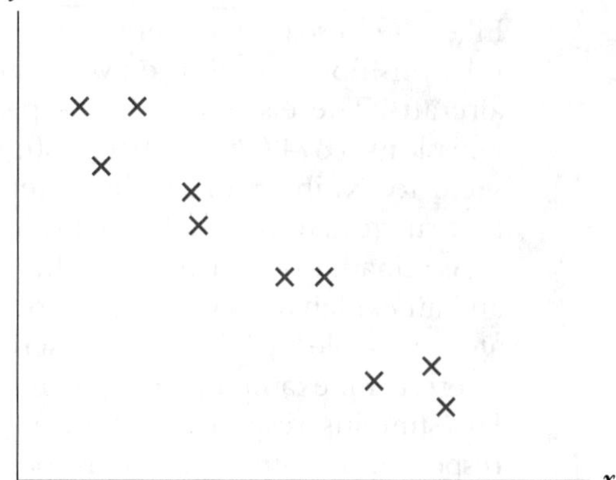

- ***Outliers***

Are there points in the scatterplot that do not follow the same pattern as the rest of the data? Such points are called *scatterplot outliers.* An outlier does not need to be an extreme measurement for either variable on its own. An observation is also an outlier if the y-value is unusual *in relation to other observations with similar x-values*, as in the diagram on the right.

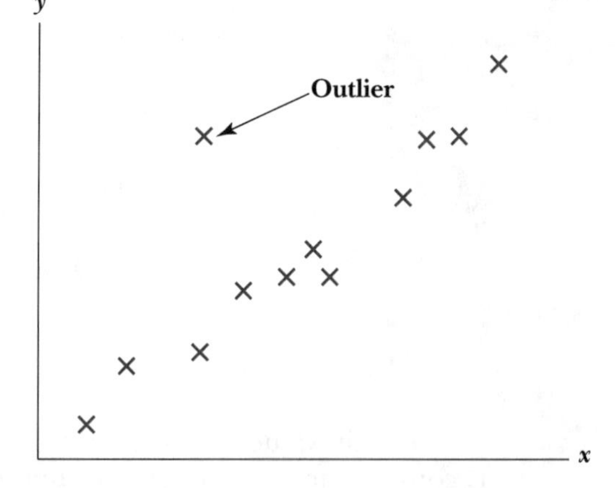

Outliers should be carefully examined. If possible, go back to the original data and check for recording or transcription errors.

If there were no such errors, look carefully for unusual features of the individual (or experimental) unit from which the measurements were made. Could they explain the observed values? Such detailed scrutiny of outliers can lead to new theory and discoveries and may well be the most important feature of the data.

- ***Clusters of points***

Do the data points separate into two or more *clusters* of points? Such clusters may indicate different groups that should be separately examined and which may indicate that another factor or variable is needed (and may be known) to explain the separate clusters.

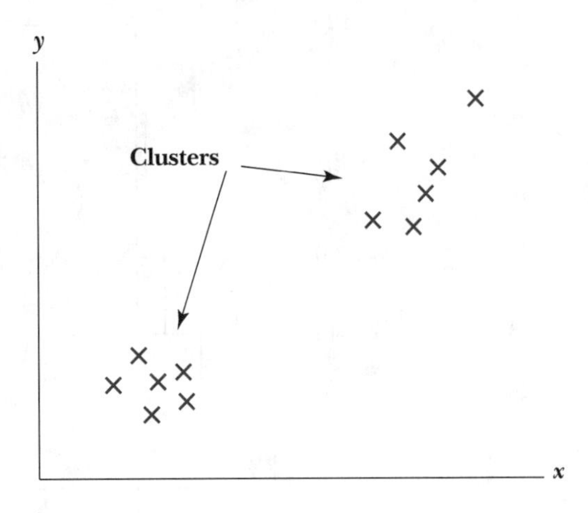

Scatterplots may be enhanced by adding univariate information about the distributions of each variable to the margins of the scatterplot. One economical way of doing this is to draw marginal box plots to the left and below the scatterplot.[6] The diagram below adds box plots to the margins of the mammalian lifespan scatterplot.

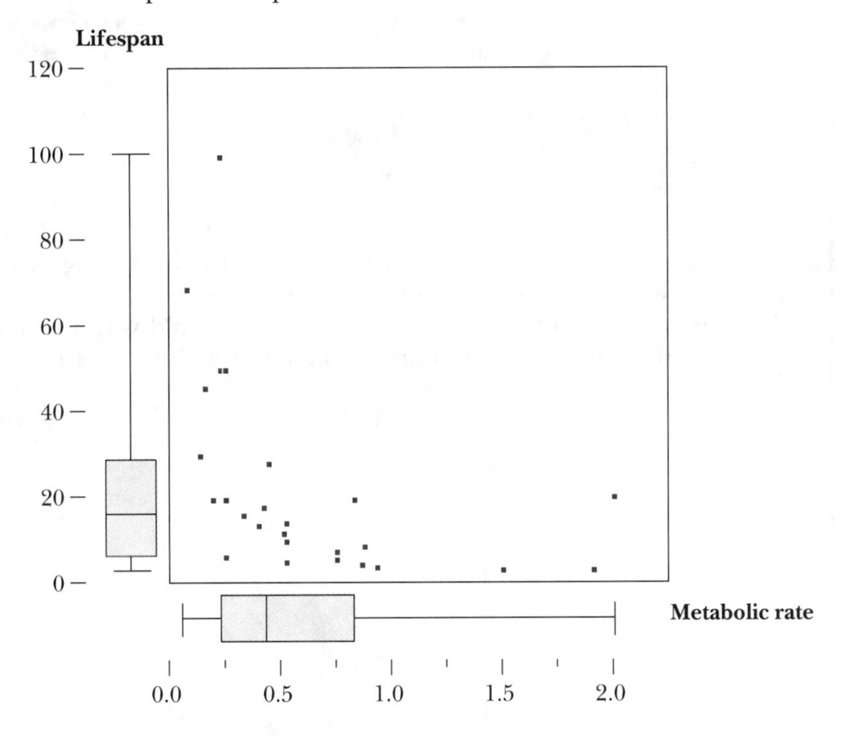

6. These are often called the *marginal distributions* of the two variables.

In this example, the location of 'zero' on each axis needs to be shown clearly, so the original horizontal and vertical lines at zero have been retained. For scatterplots of some data sets, the location of zero on each axis is unimportant, so the box plots may replace the horizontal and vertical axes completely.

The following diagram illustrates how the same scatterplot may be enhanced with histograms on its axes.

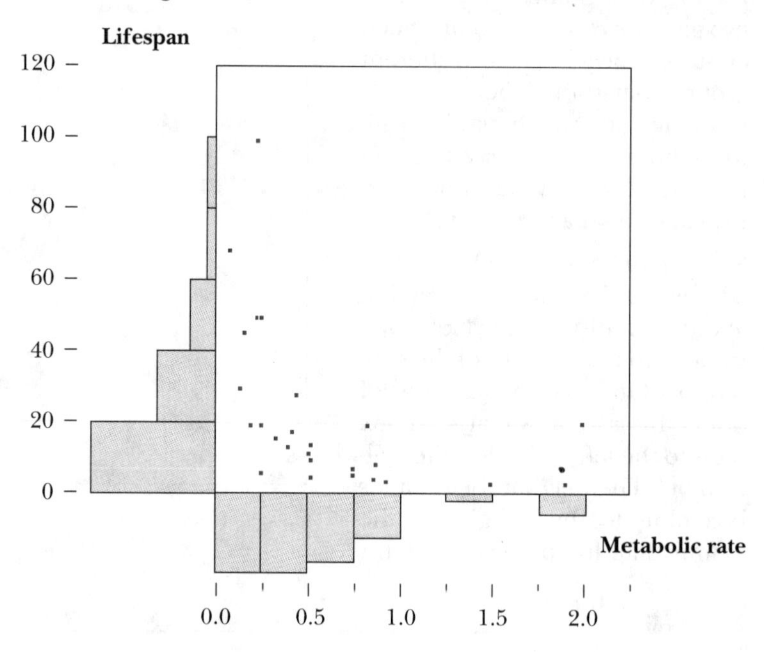

EXERCISE 4.2

1. Describe in words the nature of the relationship between the two variables displayed in each of the four scatterplots at the beginning of section 4.2.
2. Why are clusters useful to detect in a scatterplot?
3. Consider the scatter diagram shown below. Would you describe this as evidence of a positive association between X and Y, or a negative association?

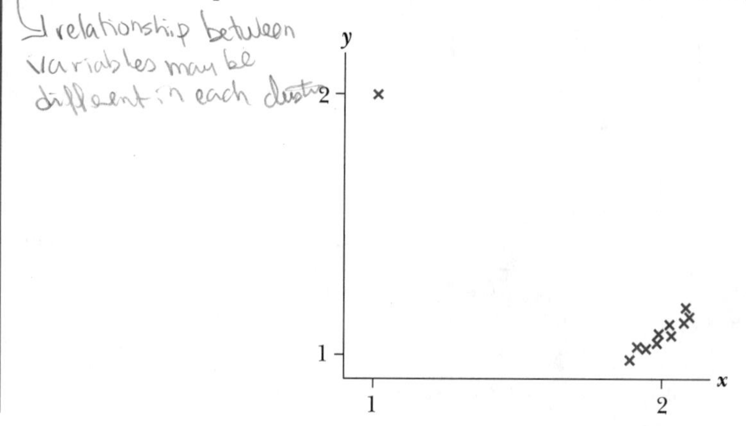

4.3 Strength of a relationship — correlation

Pearson's correlation coefficient

Correlation is a useful numerical summary of the strength of a relationship between two variables. There are three different correlation coefficients commonly used in statistics, each of which is named after a famous statistician (Kendall, Pearson and Spearman). We will only consider the most commonly used summary, Pearson's correlation coefficient.

The strength of a relationship should not depend on the scaling of the two variables. For example, the relationships exhibited in the two scatterplots below have the same strength, and their correlation coefficients should therefore be the same.

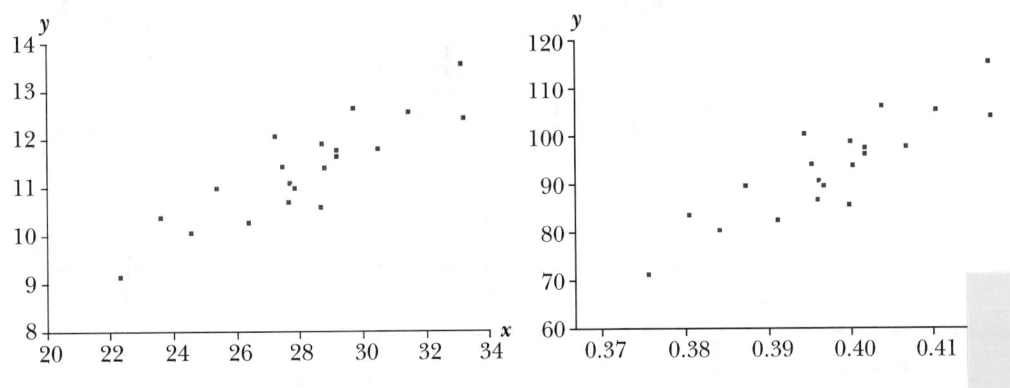

To remove the possibility of dependence on the scaling of the variable correlation coefficient is defined in terms of *standardised* versions of the variables, found by subtracting the mean and dividing by the standard deviation of each variable separately. For example, the table below illustrates standardisation of the data set on the left of the previous diagram.

	x	y	$z_x = \dfrac{x-\bar{x}}{s_x}$	$z_y = \dfrac{y-\bar{y}}{s_y}$
	31.462	12.517	1.172	1.159
	25.385	10.934	−0.975	−0.365
	27.264	11.993	−0.311	0.654
	26.381	10.204	−0.623	−1.068
	33.146	13.516	1.767	2.120
	24.590	10.000	−1.255	−1.264
	.	.	.	.
	.	.	.	.
	.	.	.	.
	30.486	11.748	0.827	0.418
	27.855	10.934	−0.102	−0.366
Mean	28.144	11.314	0.000	0.000
St. devn	2.831	1.039	1.000	1.000

A scatterplot of the standardised variables is shown below. Note that the standardised values are the same for both of the original scatterplots.

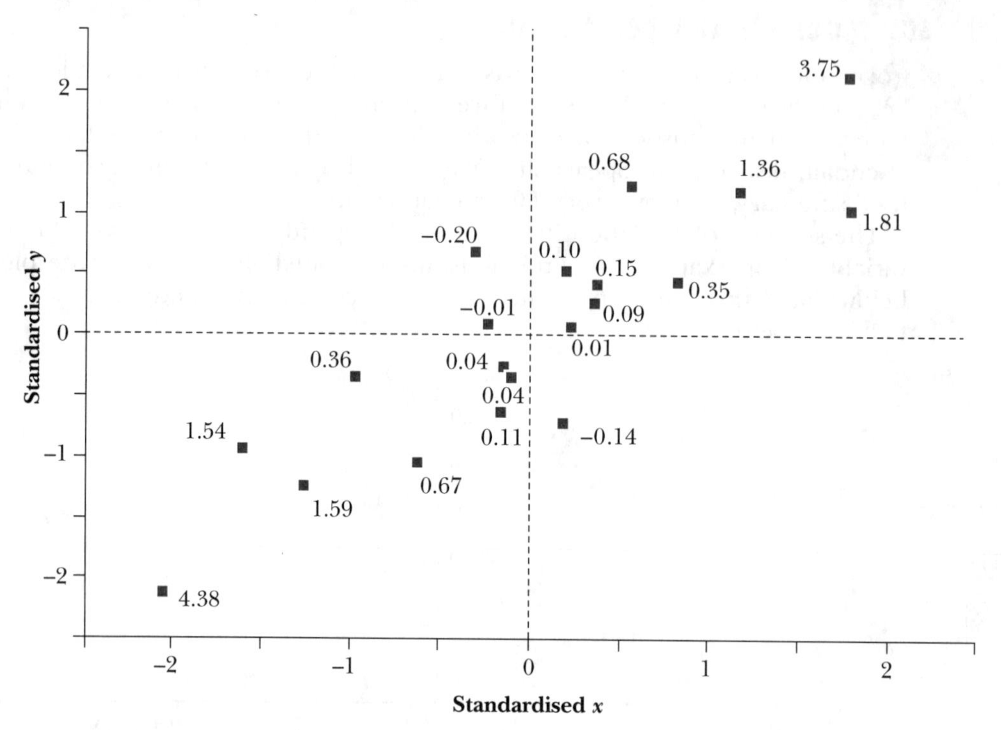

Pearson's correlation coefficient, *r*, can be defined in terms of the standardised variables, z_x and x_y with the formula:[7]

$$r = \frac{\sum z_x z_y}{n-1}$$

The points on the scatterplot are labelled with the products $z_x z_y$. Points in the top right and bottom left quarters of the scatterplot have positive values of $z_x z_y$, whereas points in the other two quarters have negative products. The correlation coefficient, *r*, will be positive when points tend to be in the top right and bottom left quarters of the scatterplot and will be negative when most points are in the other two quarters. In the example above, $r = 0.877$.

Another common name for Pearson's *r* is the *product moment correlation coefficient*. Pearson's *r* is so widely used in statistics to describe the strength of relationships that it is often simply called *the* correlation coefficient.

7. An alternative equivalent formula for the correlation coefficient is:

$$r = \frac{s_{xy}}{s_x s_y}$$

where s_x and s_y are the standard deviations of the *x* and *y*, and

$$s_{xy} = \frac{\sum (x-\bar{x})(y-\bar{y})}{n-1}$$

is called the *covariance* between *x* and *y*.

Properties of Pearson's r

The value of Pearson's r indicates how strongly the two variables are associated with each other. A value of $r = 0$ suggests that the two variables are not strongly associated with each other (but, as described later, some types of relationship in which there is strong association may lead to zero, or near zero, values of r). Values close to +1 or −1 always indicate a strong relationship between the variables. No matter what the degree and nature of association between two variables, Pearson's r always lies in the range [−1, 1].

The examples which follow illustrate the interpretation of correlation.

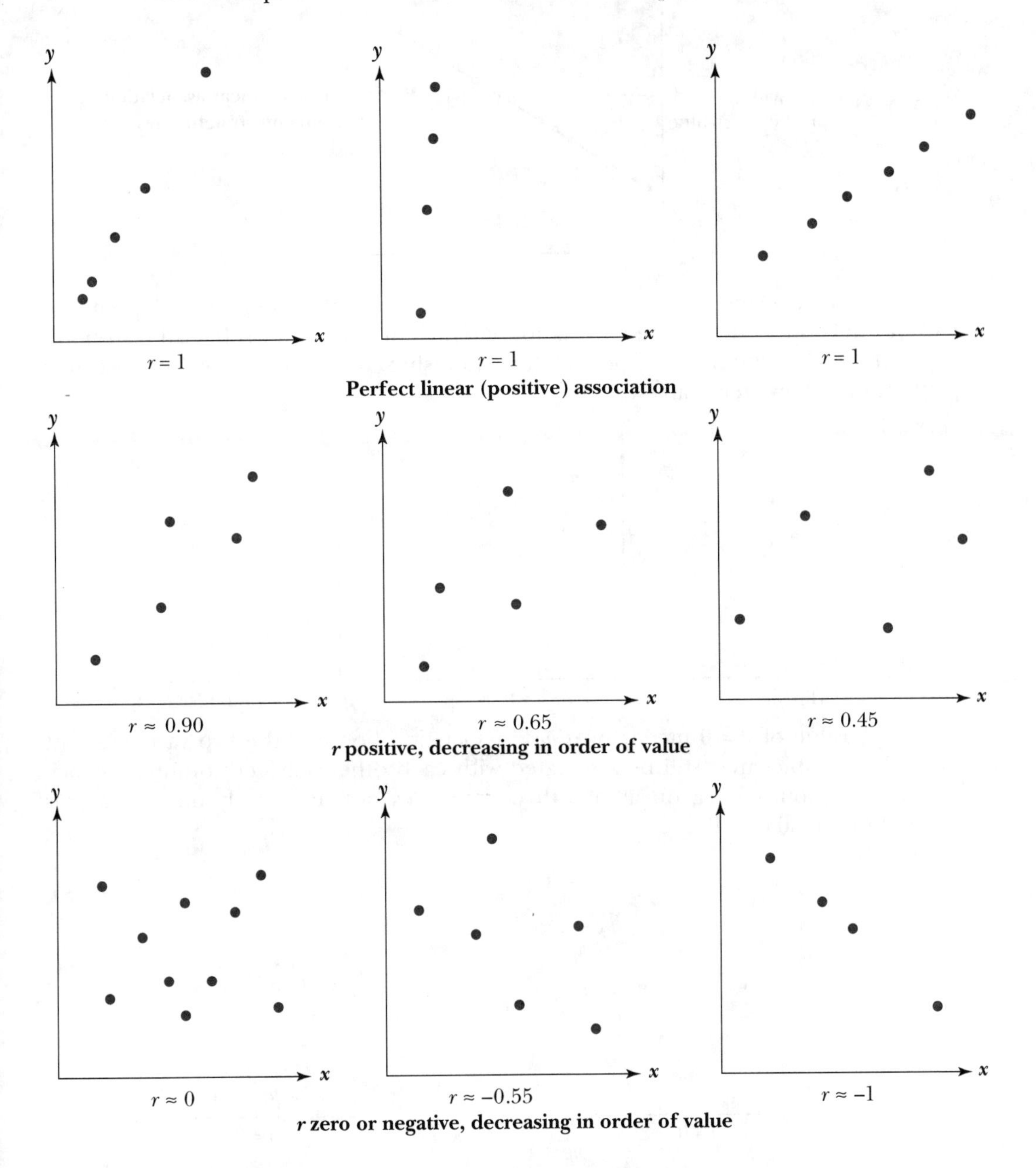

The correlation coefficient r is a measure of *linear* association, so that a value of $r = 1$ means there is 'perfect' positive linear association between the two variables; all the points on the scatterplot lie on a straight line with positive slope. A value of $r = -1$ means that there is perfect negative linear association and all the points lie on a straight line with negative slope. In either case, there is perfect predictability of one variable from the other based on the straight line passing through all the points.

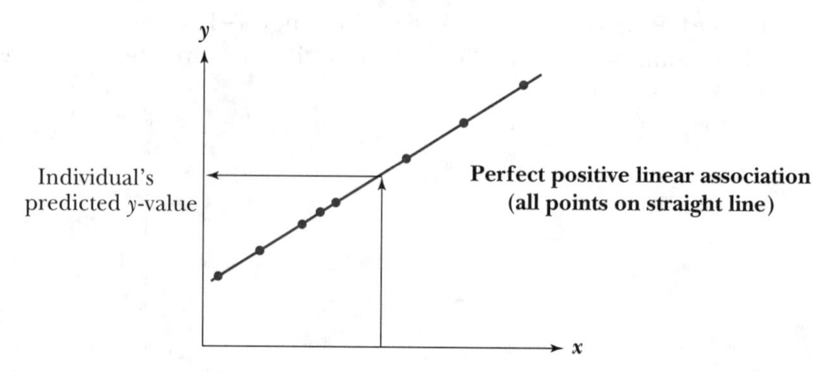

A *perfect nonlinear relationship* between two variables occurs when all points lie exactly on a curve. The correlation coefficient r does not reflect the strength of such a relationship. For such a relationship, the value of r will be less than 1.0 (and greater than −1.0).

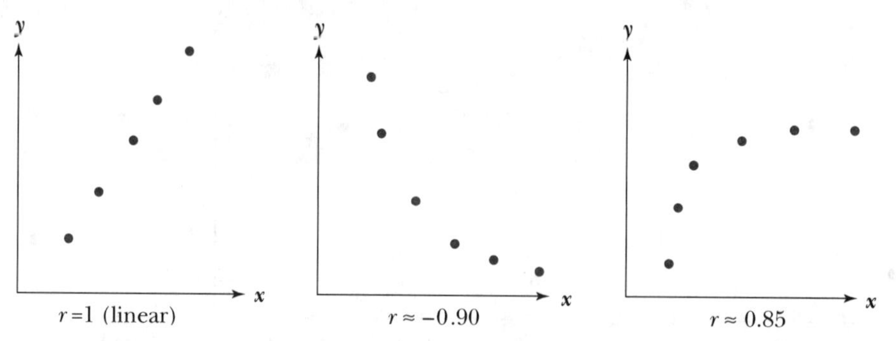

A value of $r = 0$ means no *linear* association between the two variables, but the variables may still be associated with each other. Perfect nonlinear association (points lying on a smooth curve) does not necessarily mean that r is close to 1.0 or −1.0.

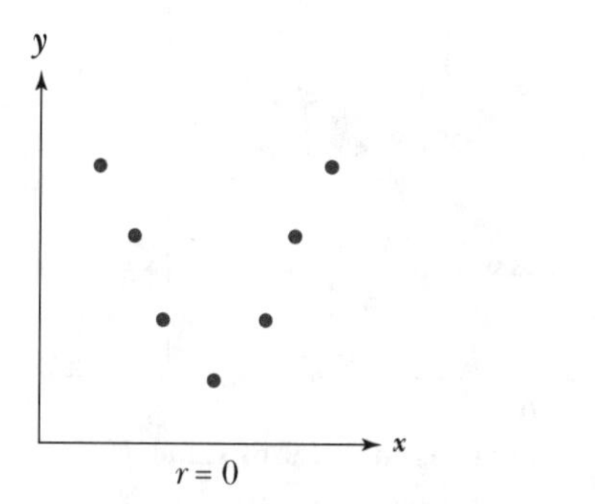

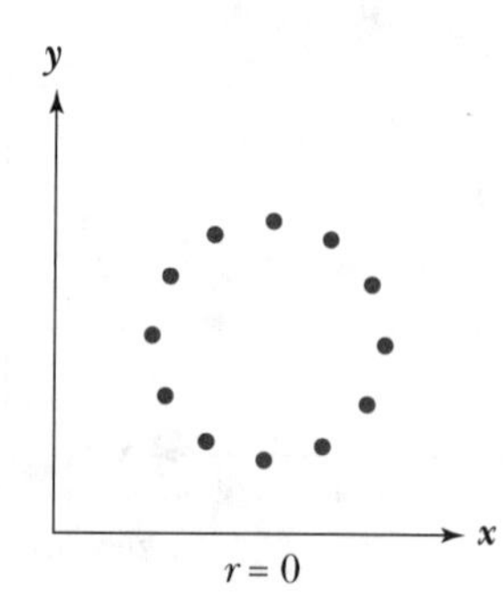

Perfect relationships (linear or nonlinear) rarely occur in practice. However, relationships are often nonlinear and care should be taken to draw a scatterplot of a data set before interpreting the value of Pearson's r.

As well as being a poor description of nonlinear association, r may be heavily influenced by single outliers (values that are far from the other points in the scatterplot). For bivariate data, a point may be a clear outlier in terms of the marginal distributions of x, or y, or both (in which case, it will have a strong influence on the perceived relationship). Of even greater interest is a point which is not an outlier in terms of *either* variable on its own, but whose position in the scatterplot is very atypical of the cloud of points. Such a point is distant from the one or more clusters of other points evident in the scatterplot; it is therefore not close to any line or smooth curve which reasonably describes the relationship for the remaining points.

Outliers can strongly influence the value of the correlation. This is illustrated by the example below where the addition of a single outlier causes a large change in the value of r.

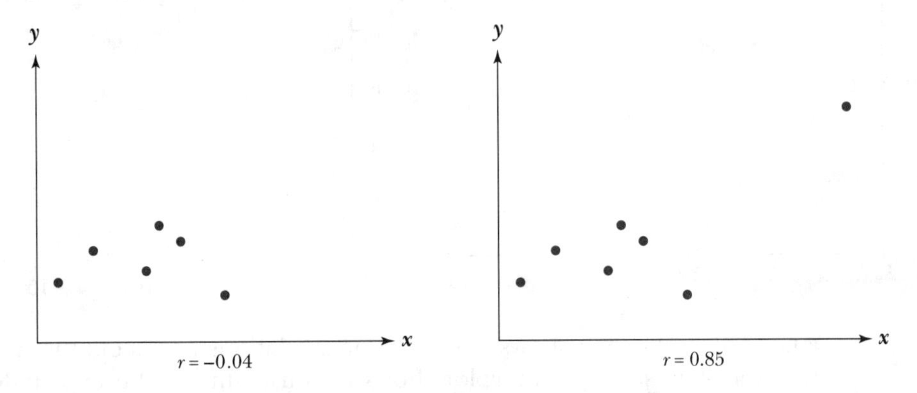

The examples we have seen so far indicate the need for caution in interpreting r. A scatterplot reveals what r does not. Anscombe uses four data sets to further emphasise the importance of the scatterplot.[8]

Data set 1		Data set 2		Data set 3		Data set 4	
Y	*X*	*Y*	*X*	*Y*	*X*	*Y*	*X*
8.04	10.00	9.14	10.00	7.46	10.00	6.58	8.00
6.95	8.00	8.14	8.00	6.77	8.00	5.76	8.00
7.58	13.00	8.74	13.00	12.74	13.00	7.71	8.00
8.81	9.00	8.77	9.00	7.11	9.00	8.84	8.00
8.33	11.00	9.26	11.00	7.81	11.00	8.47	8.00
9.96	14.00	8.10	14.00	8.84	14.00	7.04	8.00
7.24	6.00	6.13	6.00	6.08	6.00	5.25	8.00
4.26	4.00	3.10	4.00	5.39	4.00	12.50	19.00
10.84	12.00	9.13	12.00	8.15	12.00	5.56	8.00
4.82	7.00	7.26	7.00	6.42	7.00	7.91	8.00
5.68	5.00	4.74	5.00	5.73	5.00	6.89	8.00

8. F. J. Anscombe, 'Graphs in statistical analysis', *American Statistican*, 27, 1973, pp. 17–21.

For each of these four data sets, $\bar{x} = 9.0$, $s_x = 3.317$, $\bar{y} = 7.5$, $s_y = 2.032$ and $r = 0.816$. The 'basic' set of numerical summary statistics is therefore the same for all four data sets. That the data sets are, however, very different is clearly revealed by their scatterplots. The need to plot the data as part of the exploratory analysis of a relationship cannot be over-emphasised.

The scatterplot for data set 1 shows a positive linear association, whereas the scatterplot for data set 2 shows perfect nonlinear association.

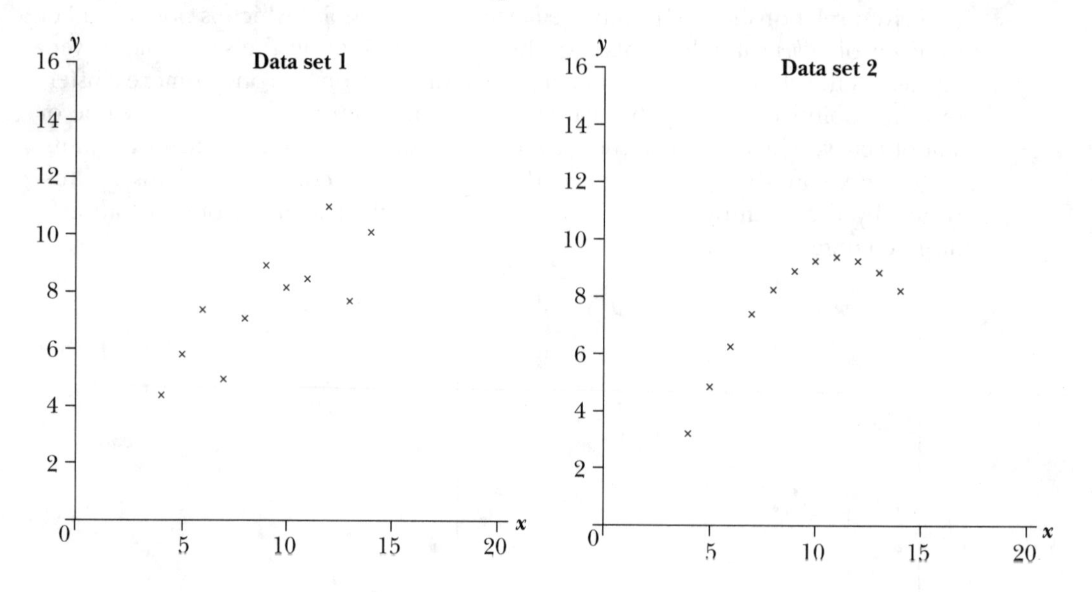

Data set 3's scatterplot shows a perfect linear relationship except for one outlier. However, data set 4's scatterplot shows no variability in the explanatory variable (and hence no evidence about the relationship) apart from that provided by one further x-value which is an outlier.

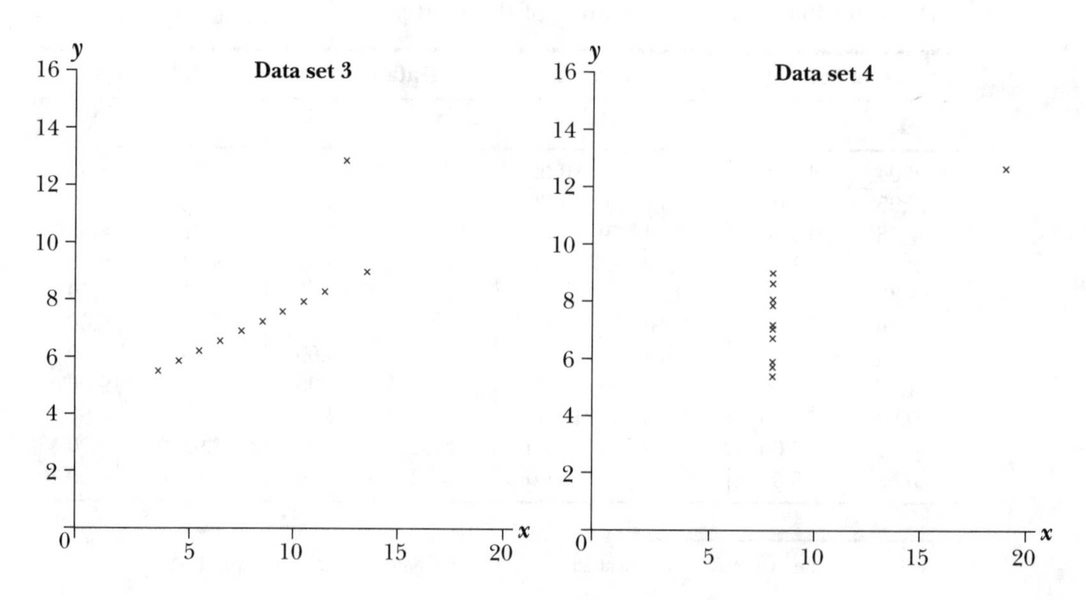

EXERCISE 4.3

1. If the correlation between average monthly temperature and monthly rainfall over 12 months, with temperature in degrees Fahrenheit and rainfall in inches, is 0.6, would the correlations be higher, lower or about the same when metric units (degrees Celsius and centimetres) are used for temperature and rainfall?
2. A research article reports that, in a group of subjects, the correlation coefficient between the weekly frequency of aerobic exercise and resting heart rate is –1.67. Is this reasonable?
3. Guess the correlation in the following data scatters. (With practice, you should be able to guess correctly the first significant digit.)

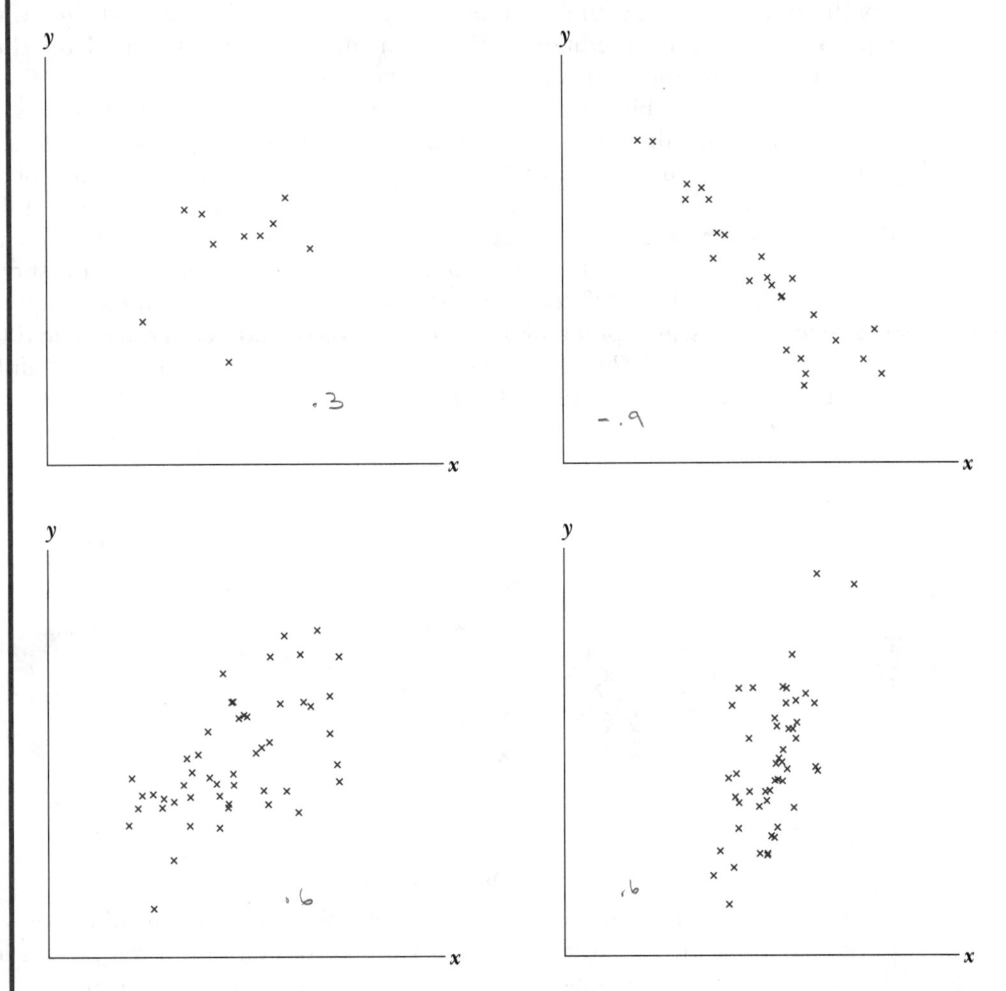

4. Consider the four scatterplots on page 128 that have the same numerical summaries, including the same correlation coefficient. Which of these is well-summarised by the numerical summaries?

4.4 FORM OF A RELATIONSHIP — LEAST SQUARES

A scatterplot is an excellent way to graphically show the relationship between the two variables in a bivariate data set, and Pearson's r provides a useful numerical description of the *strength* of this relationship. However, when the variables can be classified as an explanatory variable and a response, it is also useful to concisely describe the *form* of the relationship. The aim of the methods in this section is to find a trend line (or perhaps a curve) which describes the form of the relationship. As well as being descriptive, a trend line allows us to predict the response variable (y) from the explanatory variable (x).

Median trace

For time series, where the x values are chronologically ordered and often equispaced, running medians, running means and variants based on these provide useful methods of smoothing a relationship.

For more general bivariate relationships, these same 'running' methods can also be used, with the data arranged in increasing order of x. It is sometimes more convenient, however, just to group the data, based on the values of the x-variable (for example, lowest 10 x-values, next 10 x-values and so on) and to draw a *median trace* as a set of straight lines or a freehand curve through the medians for x and y, calculated for each group. The number of groups should be chosen (and adjusted if necessary) to produce a fairly smooth median trace.

The following scatterplot and table describe birth and death rates (per 1000) from 74 countries in 1976, sorted in order of their birth rates (and sorted alphabetically within each birth rate). How is death rate related to birth rate?

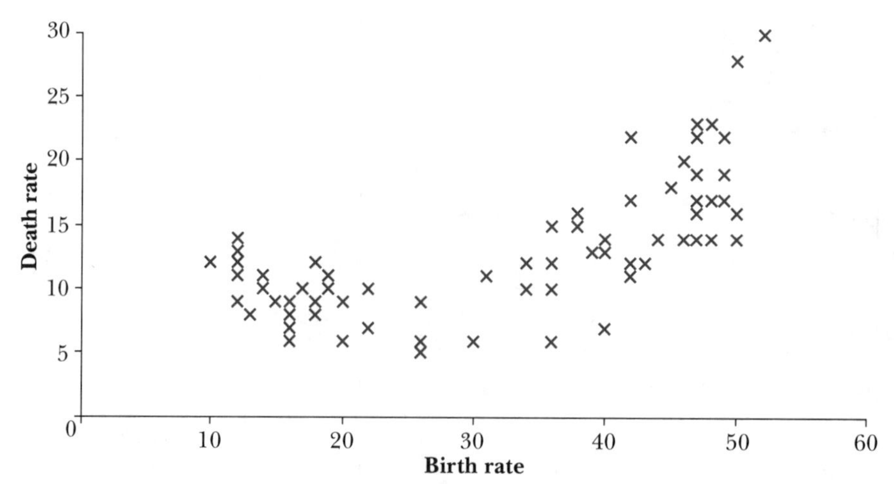

We will use a median trace to help describe the relationship. For the purposes of the median trace, the data will be grouped into seven groups, on the basis of their birth rates (the x-values). There is no unique method of grouping in cases such as this. The table on page 131 illustrates one such grouping that keeps different birth rates in separate groups and, subject to this constraint, makes the group sizes as equal as possible.

Country	Births/ 1000	Deaths/ 1000	Median births/1000	Median deaths/1000
Germany, Fed. Rep.	10	12		
Austria	12	13		
Belgium	12	12		
Germany, Dem. Rep.	12	14		
Sweden	12	11	12	11.5
Switzerland	12	9		
United Kingdom	12	12		
Netherlands	13	8		
France	14	11		
Italy	14	10		
United States	15	9		
Australia	16	8		
Canada	16	7		
Greece	16	9		
Japan	16	6	16.5	8.5
Bulgaria	17	10		
Hungary	18	12		
Spain	18	8		
USSR	18	9		
Yugoslavia	18	8		
Czechoslovakia	19	11		
Portugal	19	10		
Romania	19	10		
Cuba	20	6		
Poland	20	9		
Argentina	22	10	22	9
Chile	22	7		
Korea, Republic of	26	6		
Sri Lanka	26	9		
Taiwan	26	5		
Malaysia	30	6		
China	31	11		
Colombia	34	10		
Philippines	34	10		
Thailand	34	10		
Turkey	34	12		
Brazil	36	10	36	11
India	36	15		
South Africa	36	12		
Venezuela	36	6		
Burma	38	15		
Indonesia	38	16		
Egypt	39	13		
Guatemala	40	14		
Mexico	40	7		
Peru	40	13		
Cameroon	42	22	42	13
Ecuador	42	11		
Iran	42	12		
Vietnam	42	17		
Korea, Dem. Rep.	43	12		
Pakistan	44	14		

continued

Birth and death rates table *(cont'd)*

Country	Births/ 1000	Deaths/ 1000	Median births/1000	Median deaths/1000
Mozambique	45	18		
Zaire	45	18		
Ghana	46	14		
Nepal	46	20		
Angola	47	23	47	18
Bangladesh	47	19		
Madagascar	47	22		
Morocco	47	16		
Syria	47	14		
Tanzania	47	17		
Ethiopia	48	23		
Iraq	48	14		
Ivory Coast	48	23		
Rhodesia	48	14		
Uganda	48	17		
Nigeria	49	22	49	18
Saudi Arabia	49	19		
Sudan	49	17		
Algeria	50	16		
Kenya	50	14		
Upper Volta	50	28		
Afghanistan	52	30		

On the scatterplot below, the different groups of countries have been separated by vertical dashed lines in order to illustrate the calculations underlying the median trace. The median birth and death rates within each group of countries are shown in the table and a square has been drawn at these values on the scatterplot. The squares have been joined by straight lines to produce the median trace.

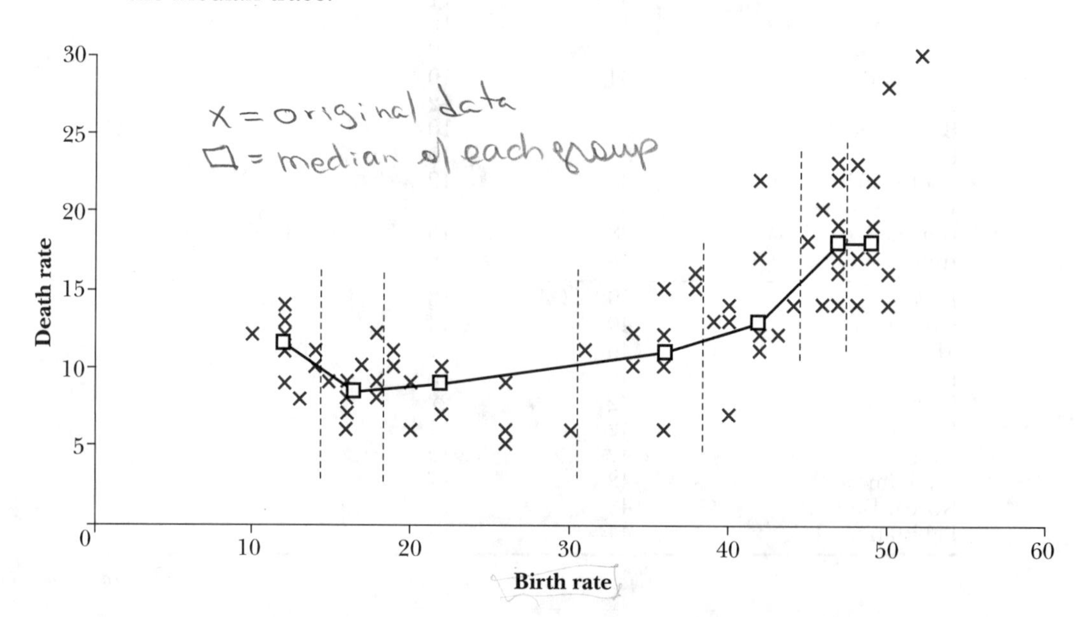

Such a procedure may produce a generally smooth relationship but does not produce a line or curve with simple functional form such as a straight line or quadratic. Although the median trace provides satisfactory predictions within the range of the data (that is, predictions of death rates for different countries whose birth rates are between 10 and 50), it is not clear how it might be used to extrapolate outside the range of the data (that is, to predict the death rate of a country with birth rate over 50 or under 100). A relationship that is described by a mathematical formula avoids this difficulty and is more easily generalised to situations where two or more explanatory variables affect a response.

Least squares

If supported by examination of a scatterplot, it is often useful to fit a simple mathematical curve to summarise the relationship. If appropriate, a straight line should be chosen since it is the simplest form of mathematical relationship. The *method of least squares* is a commonly used, objective method of fitting such a line. Before describing the details of this method, we will consider some general principles involved in choosing a good fitted line. Ideally, the line would pass through all the data values and would therefore also allow exact prediction to be made. In practice, however, it will not pass through all the points; some points will be above the line and others will be below it.

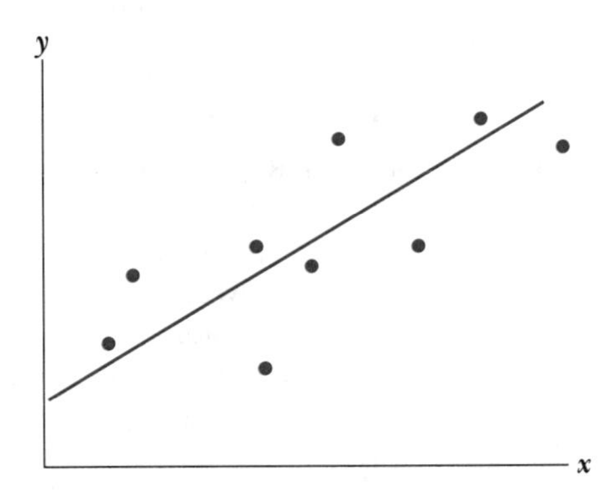

Consider how such a fitted line might be used to make *predictions* of the response variable y from the explanatory variable x. If the x-value for an individual is known, a prediction of the corresponding y-value can be read off the line in the following way:

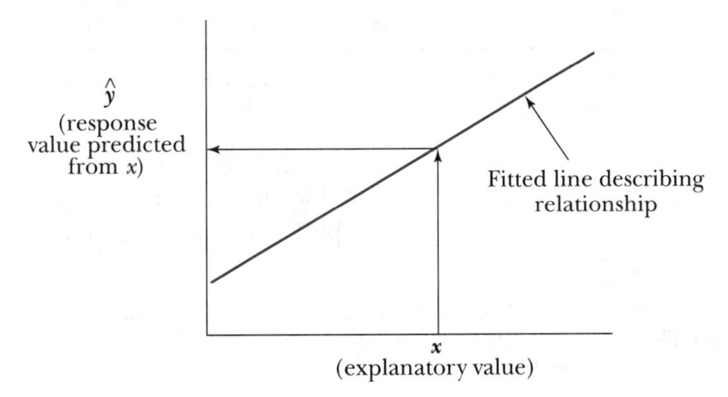

For example, consider the following table which describes the maximum January temperature (°F) from 1931 to 1960 and the latitude (°) of various cities in the USA.[9]

City	Max. Jan. temp. (°F)	Latitude (°)	City	Max. Jan. temp. (°F)	Latitude (°)
Mobile, Ala	61	30	Helena, Montana	29	46
Montgomery, Ala	59	32	Omaha, Nebraska	32	41
Juneau, Alaska	30	58	Concord, NH	32	43
Phoenix, Ariz.	64	33	Atlantic City, NJ	43	39
Litle Rock, Ark.	51	34	Albuquerque, NM	46	35
Los Angeles, Calif.	65	34	Albany, NY	31	42
San Francisco, Calif.	55	37	New York, NY	40	40
Denver, Col.	42	39	Charlotte, NC	51	35
New Haven, Conn.	37	41	Raleigh, NC	52	35
Wilmington, Del.	41	39	Bismarck, ND	20	46
Washington, DC	44	38	Cincinnati, Ohio	41	39
Jacksonville, Fla	67	38 *	Cleveland, Ohio	35	41
Key West, Fla	74	24	Oklahoma City, Okla	46	35
Miami, Fla	76	25	Portland, Ore.	44	45
Atlanta, Ga	52	33	Harrisburg, Pa	39	40
Honolulu, Hawaii	79	21	Philadelphia, Pa	40	39
Boise, Idaho	36	43	Charlestown, SC	61	32
Chicago, Ill.	33	41	Rapid City, SD	34	44
Indianapolis, Ind.	37	39	Nashville, Tenn.	49	36
Des Moines, Iowa	29	41	Amarillo, Tx.	50	35
Dubuque, Iowa	27	42	Galveston, Tx.	61	29
Wichita, Kansas	42	37	Houston, Tx	64	29
Louisville, Ky	44	38	Salt Lake City, Utah	37	40
New Orleans, La	64	29	Burlington, Vt	25	44
Portland, Maine	32	43	Norfolk, Va	50	36
Baltimore, Md	44	39	Seattle Tacoma, Wash.	44	47
Boston, Mass.	37	42	Spokane, Wash.	31	47
Detroit, Mich.	33	42	Madison, Wisc.	26	43
Sault Ste Marie, Mich.	23	46	Milwaukee, Wisc.	28	43
Minn St Paul, Minn.	22	44	Cheyenne, Wyoming	37	41
St Louis, Missouri	40	38	San Juan, Puerto Rico	81	18

Note: The latitude of Jacksonville was wrongly recorded in the published data, but we will not correct it in our initial analysis.

We are interested here in describing how temperature is related to latitude. In particular, how much warmer does it get (on average) for each degree of latitude we move south? A scatterplot helps to describe the relationship.

9. F. Mosteller and J. W. Tukey, *Data Analysis and Regression*, Addison-Wesley, Reading, Mass., 1977, pp. 73–74. (Original data from *The World Almanac and Book of Facts 1974*, Newspaper Enterprises Association, Mahwah, New Jersey, 1974, pp. 263, 704–705.)

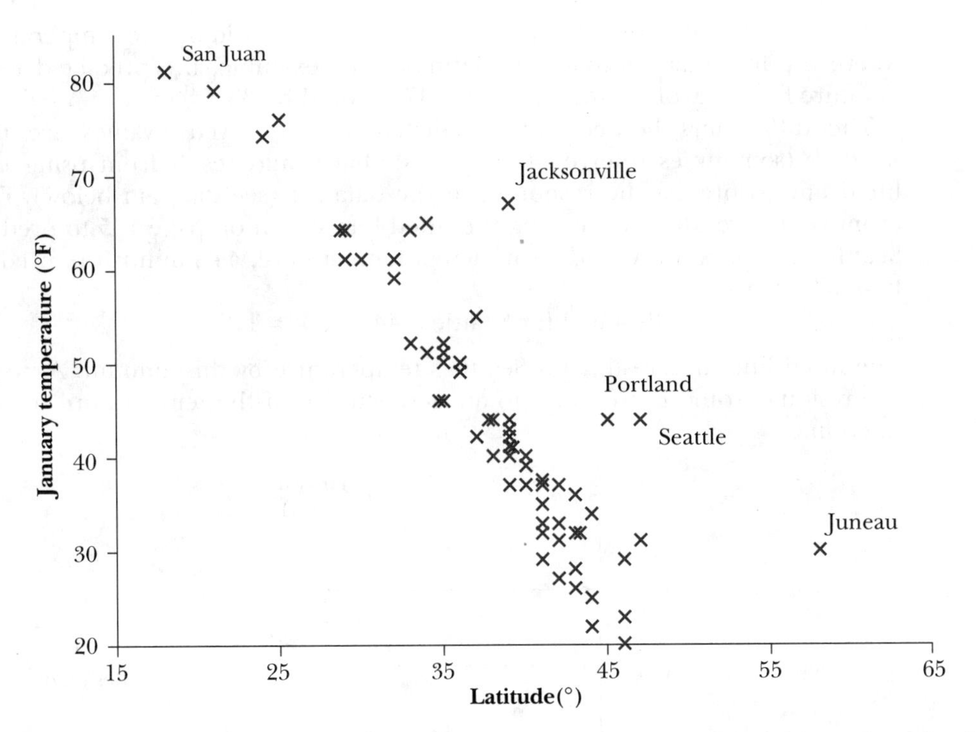

There are various outliers in this data set that we will investigate later in this section, but there is a clear trend for lower temperatures to be associated with higher latitudes.

Firstly, consider the use of the following straight line to represent the relationship.

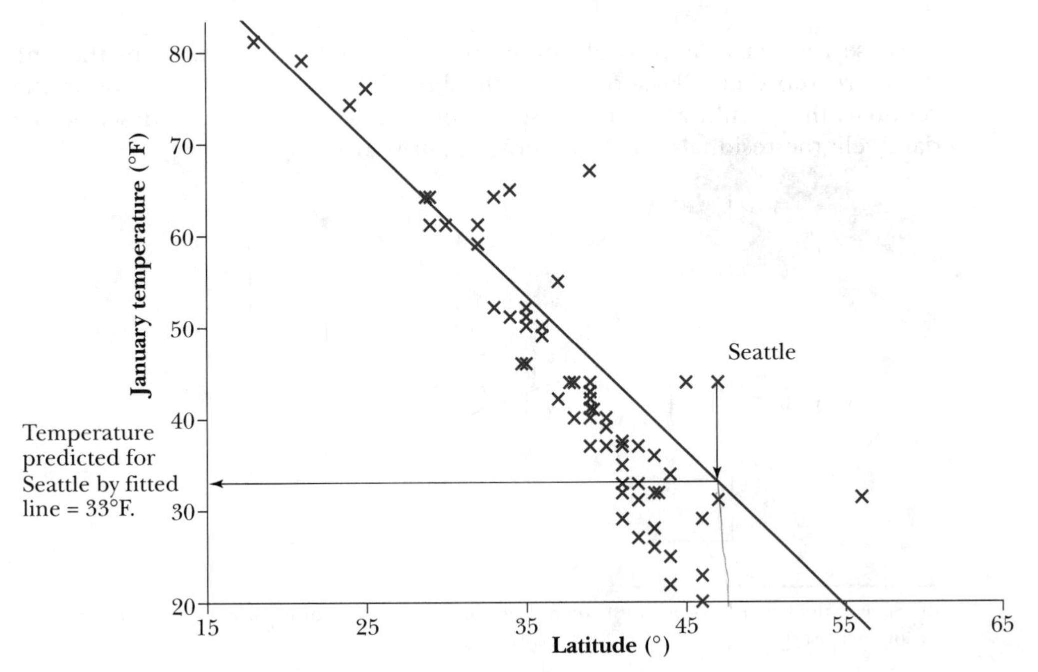

For each city, the predicted January temperature would be the temperature where the line reaches that city's latitude. For example, the predicted temperature for a city of Seattle's latitude (47°) would be 33°F.[10]

The differences between the predicted and observed y-values are the *residuals* (sometimes referred to as *errors*) that would result from using the fitted line to predict the responses in the data set (see diagram below). For example, the residual from using the fitted line shown on page 135 to predict Seattle's temperature would be its actual temperature, 44°, minus the prediction, 33°. Thus:

$$\text{Residual for Seattle} = 44° - 33° = 11°.$$

The fitted line underestimates Seattle's temperature by this amount. A negative residual would correspond to an overestimate of the temperature by the fitted line.

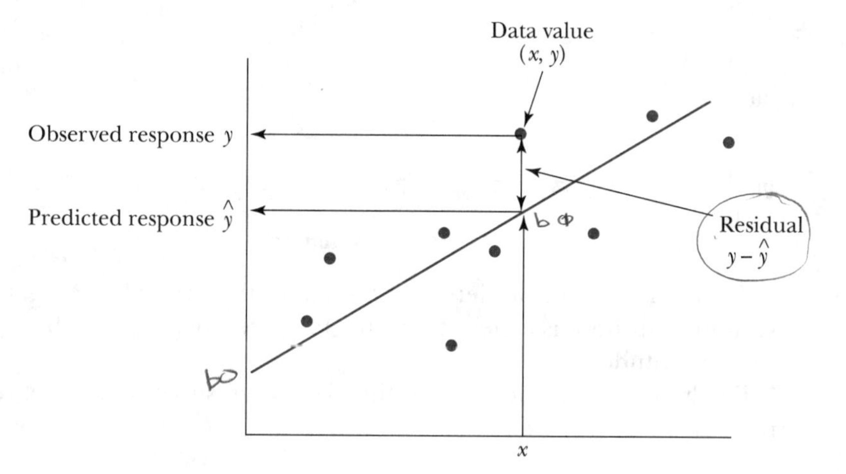

These residuals, or vertical deviations $(y - \hat{y})$, of the points from the line therefore represent 'lack of fit' of the line. The diagram below shows the residuals that would arise from using a fitted line that does not describe the data well; the residuals are, 'on average', fairly large.

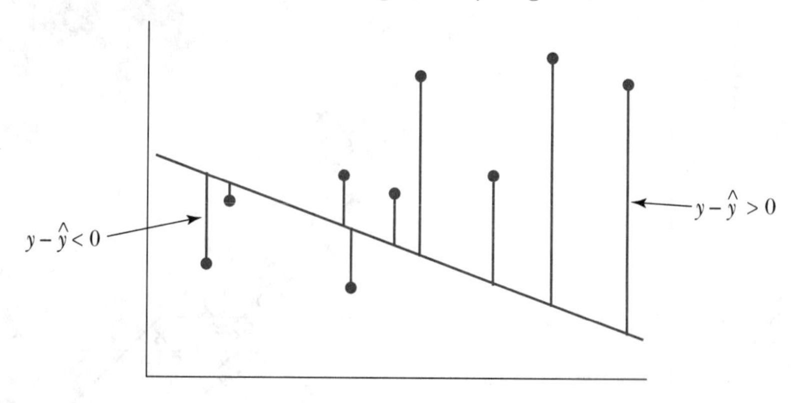

10. Similar 'predictions' (or *forecasts*) could be made of the temperatures for other cities not on the original list.

Good prediction requires these residuals to be as small as possible. If prediction is good, the line will closely fit the data. A better fit (smaller residuals) is given by the line below.

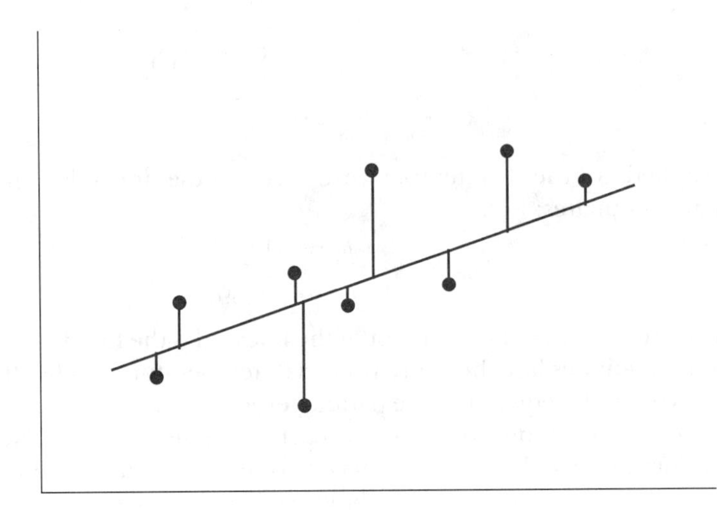

Balancing out the positive and negative residuals is desirable, but not enough. To obtain an objective method of fitting the line, we need to be more precise about how we combine the individual residuals to get a single measure of their size. The most useful overall measure is the sum of the squared residuals; that is:

$$\sum(y-\hat{y})^2 .$$

This measure is related to one of the measures of spread (the standard deviation) adopted in chapter 2. The *method of least squares* positions the fitted line to minimise the sum of the squared residuals.

To describe this mathematically, we denote the n observations (or cases) of the two variables x and y by the paired values $(x_1,y_1), (x_2,y_2), \ldots, (x_n, y_n)$. The general equation of a straight line is:

$$y = b_0 + b_1 x$$

where the constant b_0 represents the intercept and b_1 is the slope of the line. The residual for the ith observation, the point (x_i, y_i), is the observed y-value minus the y-value predicted by the line $y = b_0 + b_1 x$; that is:

$$y_i - (b_0 + b_1 x_i) = y_i - b_0 - b_1 x_i = y_i - \hat{y}_i$$

where $\hat{y}_i$ is the predicted value of y at $x = x_i$.

The method of least squares chooses the values of the intercept (b_0) and slope (b_1) to jointly minimise the sum of squared residuals; that is:

$$\sum_{i=1}^{n}(y_i - \hat{y}_i)^2 = \sum_{i=1}^{n}(y_i - b_0 - b_1 x_i)^2 .$$

Finding the values of b_0 and b_1 which minimise this quantity is a simple problem in multi- (in this case two-) variable calculus. Solving this problem mathematically[11] provides explicit formulae for the slope and intercept:

$$b_1 = \frac{\sum xy - \frac{1}{n}(\sum x)(\sum y)}{\sum x^2 - \frac{1}{n}(\sum x)^2}$$

and

$$b_0 = \bar{y} - b_1\bar{x}.$$

Applied to the US temperature data, these formulae give us the least squares estimates:

$$b_1 = -1.957$$

$$b_0 = 118.59.$$

The parameter b_0 is the intercept of the line and is the fitted response when $x = 0$. The least squares line therefore predicts a temperature of 118.59°F when the latitude is 0 (on the equator). The parameter b_1 is the slope of the line and describes how much y is predicted to rise for each increase of 1 in x. The temperature is therefore predicted to rise by –1.957°F for each increase in latitude of 1°; that is, it is predicted to drop by 1.957° for every 1° of latitude moved from the equator. The following diagram superimposes the least squares line on the scatterplot.

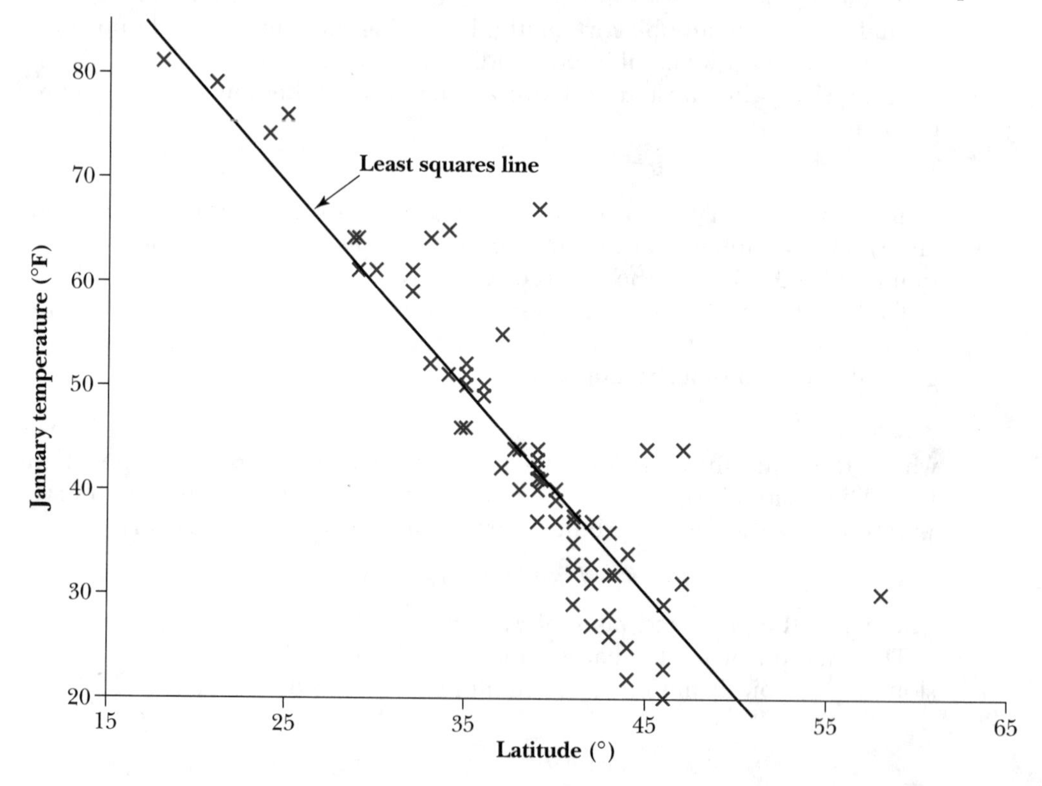

11. For a proof, see S. Weisman, *Applied Linear Regression*, John Wiley and Sons, New York, 1980, Appendix 1A.3.

The correlation coefficient has an interesting interpretation relating to least squares lines. The square of the correlation coefficient is called the *coefficient of determination.*[12] The coefficient of determination can also be written in the form:

$$r^2 = \frac{\text{Var}(\hat{y}_i)}{\text{Var}(y_i)} = 1 - \frac{\text{Var}(y_i - \hat{y}_i)}{\text{Var}(y_i)}.$$

The coefficient of variation therefore describes the proportion of the total variance of y_i that is explained by the least squares line, and is also 1 minus the proportion of the variance of y_i that is left unexplained.

We end this section by considering whether a straight line is an adequate description of the relationship between two variables, x and y.

Examining the fit of such a line can be based on examining the scatterplot of the data and superimposed line. For example, in the temperature example, the main problem is the four 'outliers' that are a considerable distance above the fitted line. These four cities have large positive residuals and therefore seem unusually warm in January for their latitude.

As with any other outliers in graphical displays, the data should be checked carefully to search for causes. Checking published raw data against an atlas shows that one of the latitudes (Jacksonville, marked with an asterisk in the table of data) was incorrectly recorded as 38° instead of 30°.

The other three 'outliers' (Seattle, Portland and Juneau) seem to have been correctly recorded, so we should look for other reasons why they should be so unusually warm for their latitudes. This is apparently caused by a warm ocean current flowing up the north-western coast of the United States and prevailing westerly winds in the winter raising the temperatures of these cities (which are all on the western coast).

What should we do with these three western coast cities? If they are retained in the data, the least squares line is likely to underestimate the January temperatures of other western coast cities (and correspondingly overestimate temperatures elsewhere). Another option is to delete these cities from the data set, noting that the resulting least squares line should not be used on the western coast of the United States. The best solution is to extend the model to allow it to explain temperatures in all of the United States, but this is beyond the scope of this book.

Outliers are highlighted if we plot residuals rather than the response. For example, the table on the next page describes the boiling point (°F) and barometric pressure (in.) of mercury at 17 locations in the Alps and in Scotland.[13]

The aim of the analysis is to find an equation that can be used to predict the barometric pressure (and, from this, the altitude) from the boiling point of water. The boiling point is therefore treated as the explanatory variable and the pressure is considered to be the response.[14]

12. In some books, it is denoted by R^2, but, in this book, we consistently use lower case letters to describe the numerical values of summary statistics.
13. J. D. Forbes, *Further Experiments and Remarks on the Measurement of Heights by the Boiling Point of Water,* trans. R. Soc. Edinburgh, 21, 1857, pp. 135–143.
14. If the aim of the analysis had been to *model* the relationship between pressure and boiling point, boiling point would have been treated as the response variable. However, for *predicting pressure,* the opposite classification is appropriate.

The scatterplot below the table describes the data and the least squares line is superimposed on the graph.

Case number	Boiling point (°F)	Pressure (in. Hg)
1	**194.5**	20.79
2	**194.3**	20.79
3	**197.9**	22.40
4	**198.4**	22.67
5	**199.4**	23.15
6	**199.9**	23.35
7	**200.9**	23.89
8	**201.1**	23.99
9	**201.4**	24.02
10	**201.3**	24.01
11	**203.6**	25.14
12	**204.6**	26.57
13	**209.5**	28.49
14	**208.6**	27.76
15	**210.7**	29.04
16	**211.9**	29.88
17	**212.2**	30.06

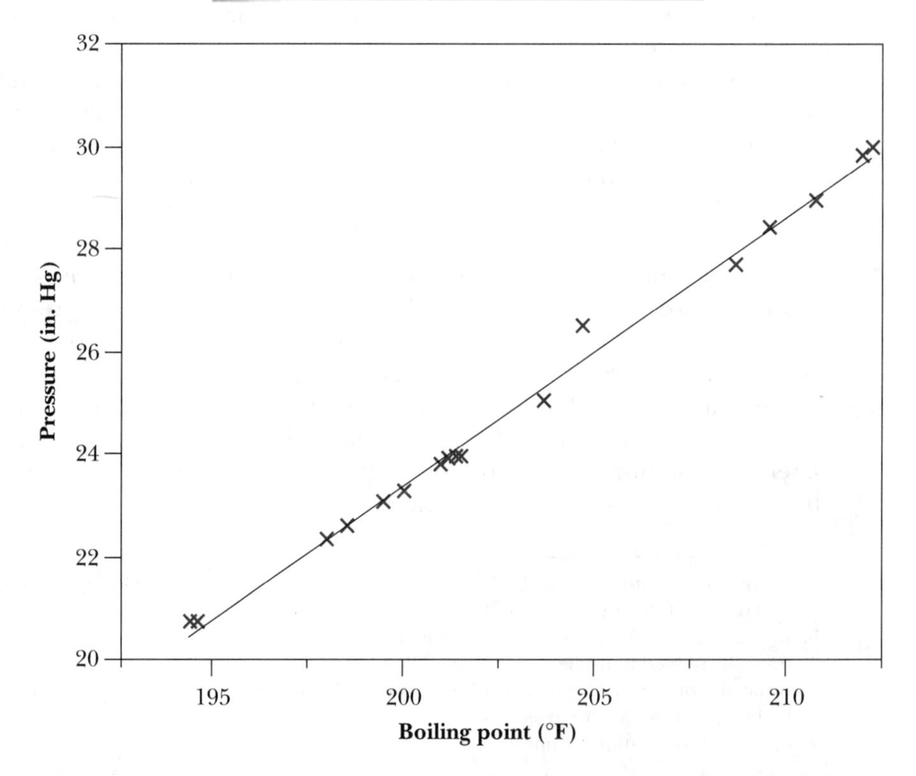

Plotting the residuals from the fitted least squares line results in the following plot:

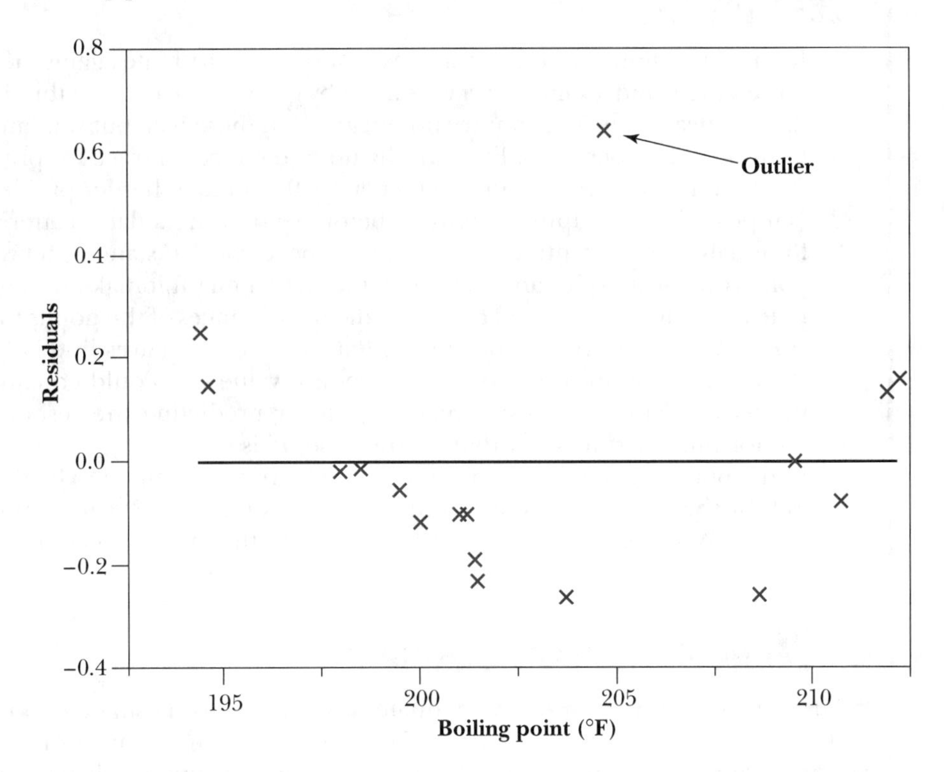

This plot highlights the single outlier in the data set. The observation has had a much higher pressure than would be expected for its boiling point. As the original data cannot be checked further and the value is so far from the trend in the rest of the data, the outlier would be dropped from the data set for further analysis; its existence would, however, be reported in any summary of the analysis.

A further feature is also clear from the residual plot. There seems to be curvature in the plot (at least after the outlier is removed). The evidence of curvature is slight in the original plot, and a straight line may allow adequate predictions from a practical perspective. However, the residual plot makes this curvature clear, and the relationship would be better modelled by a curve than a straight line, and this should allow more precise predictions.

In a similar way, if the latitude of Jacksonville is corrected in the scatterplot at the top of page 135, and the three outliers that we detected earlier (Portland, Seattle and Juneau) are removed, some curvature is evident in the relationship between January temperature and latitude.

A smooth curve fitted to the remaining data would be steeper at high latitudes than at low latitudes — the predicted difference in January temperature between cities with latitudes 15° and 25° would be less than the difference predicted between cities at 35° and 45°. Close to the equator, temperatures seem less dependent on latitude. Nonlinear relationships such as these will be described in the following section.

EXERCISE 4.4

1. In the subsection on median trace, a scatterplot of birth rate against death rate is presented, and a smooth curve is fitted by group medians. Use this fit to predict the death rate in a country not included in these data but with an annual birth rate of 45 per 1000. Explain why using the median trace for predictions for a country with a birth rate of 60 per 1000 is a much harder problem.
2. Is it possible to compute a residual before smoothing a data scatter?
3. Residuals are conventionally defined to be vertical distances between the points on a scatterplot and the fitted line. When might it make sense to compute residuals as horizontal or perpendicular distances of the points to the fit?
4. If the standard deviation of the residuals to a least squares line is $\frac{2}{3}$ of the standard deviation of the original response value, we would conclude that the least squares line is not very helpful for predicting the response. Use the formula of r^2 to show that, in this case, r^2 is 0.56.
5. In the boiling point data, atmospheric pressure data seemed to be described well by the least squares line, and yet the residual plot revealed an obvious outlier. Why was this outlier not obvious from the original scatterplot?

4.5 NONLINEAR RELATIONSHIPS

When a straight line is not an adequate description of the shape of a relationship, a curve may be fitted instead. There are two simple classes of nonlinear curves which are commonly used to describe relationships which are not linear.

Polynomials

The first of these classes is the class of polynomials, which includes linear and quadratic curves. When a straight line is inadequate to describe a relationship, a quadratic curve may do so. A quadratic curve is represented by the equation:

$$\hat{y}_i = b_0 + b_1 x_i + b_2 x_i^2 .$$

When using such a curve for prediction, the residuals can again be written as $(y_i - \hat{y}_i)$ and an overall measure of their size can be given by the residual sum of squares:

$$\sum (y_i - \hat{y}_i)^2 = \sum (y_i - b_0 - b_1 x_i - b_2 x_i^2)^2 .$$

As when fitting straight lines to data, we can choose the constants b_0, b_1 and b_2 to minimise this residual sum of squares.

The algebra and resulting equations for the constants are more complicated when fitting quadratic curves by least squares than when fitting straight lines, so we will not give details here. The methods are, however, conceptually similar. In practice, a computer would always be used to fit quadratic curves.

For example, after deleting the outlier in the barometric pressure example, the best-fitting quadratic curve is:

$$\hat{y}_i = 116.59 - 1.4165 x_i + 0.004752 x_i^2 .$$

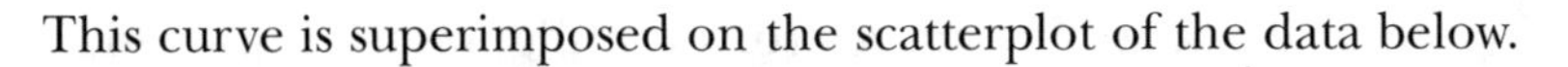

This curve is superimposed on the scatterplot of the data below.

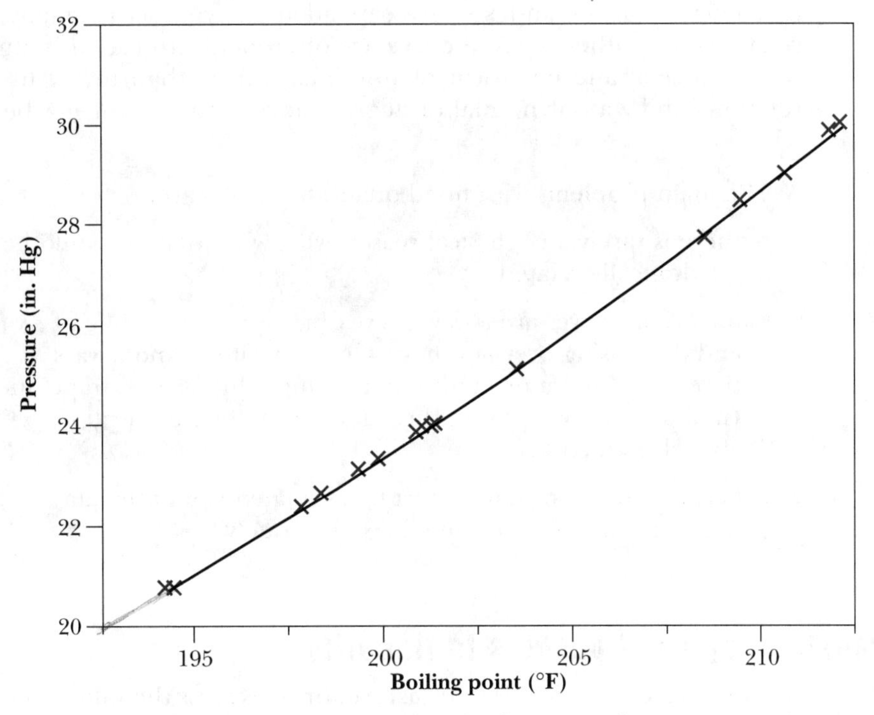

A scatterplot of the residuals from this fit does not show any pattern, so we would conclude that the quadratic curve seems to fit the data adequately.

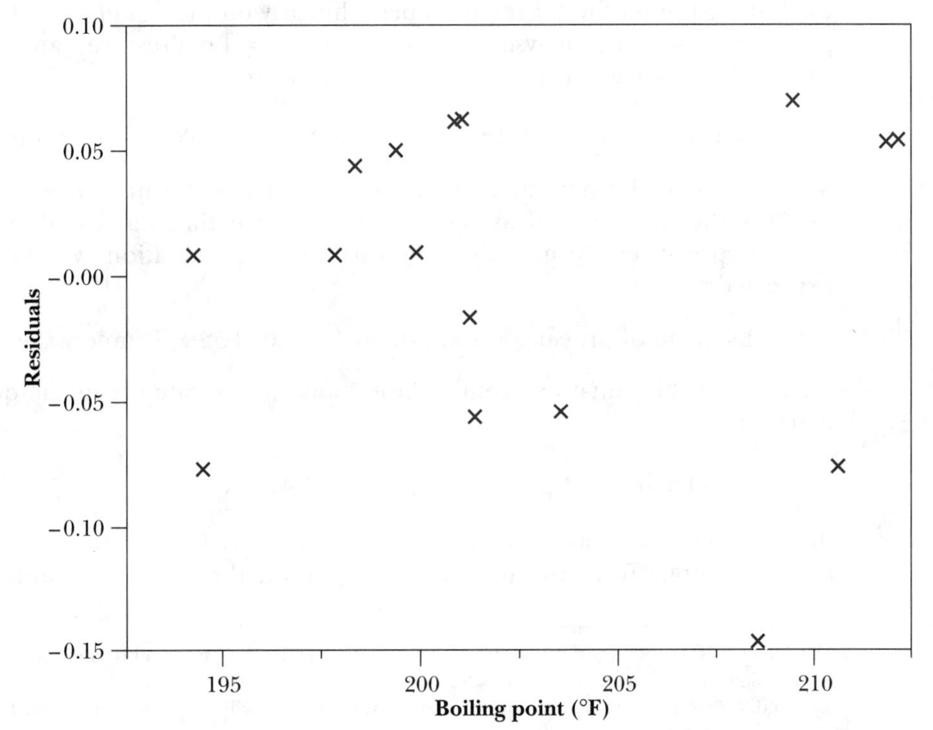

Although fitted quadratic curves (or even higher-order polynomials such as cubics, quartics and so on) can adequately describe many nonlinear relationships, other types of curve are often better to use. It is almost always better to seek another form of nonlinear curve than to try to describe a relationship by a polynomial of degree more than two (or maybe three).

The main problems with fitted quadratic curves are:

- there is rarely any physical reason why two variables should be quadratically related
- quadratic curves are always increasing over some range of x-values and decreasing over another range of x-values, and always have a maximum or minimum at some value of x. Extrapolation (in a positive or negative direction) may therefore lead to absurd predictions.
- because three constants must be estimated from the data rather than two, they can be determined less accurately.

Transforming relationships to linearity

An alternative approach is to transform or re-express the values of one or both variables in a bivariate relationship and then fit a straight line to the transformed variables.[15] For example, Forbes developed a theory suggesting that the boiling point of water might depend linearly on the *logarithm* of the atmospheric pressure. Re-analysing the data with y = Ln(Pressure) and fitting a straight line by least squares gives the relationship:

$$\text{Estimate of Ln(Pressure)}, \ \hat{y} = -0.9518 + 0.020\,52 \times \text{Temperature}.$$

Scatterplots of the data and of the residuals from the fit indicate that this line fits the data as well as the quadratic curve that was used earlier. To predict pressures from this relationship, the equation would be re-expressed as:

$$\text{Estimate of pressure} = \text{Exp}(-0.9518 + 0.020\,52 \times \text{Temperature}).$$

Although this equation seems a little more complicated than the quadratic equation:

$$\text{Estimate of pressure} = 116.59 - 1.4165\,x_i + 0.004\,752\,x_i^2$$

that was obtained earlier in this section, there is little difference between the predictions from the two equations within the range of boiling points

15. Sometimes, simple transformation of one or both variables does not readily linearise a relationship. In such circumstances, some nonlinear curve, other than a polynomial, is commonly used to describe the relationship between the variables. We will not pursue such cases further.

in the data. However, we would have more confidence in using the transformed model for predictions of pressures when the boiling point is more extreme (for example, higher than 220°F or lower than 190°F). Indeed, the fitted quadratic relationship suggests that, below the turning point of 149°F, pressure is not even an increasing function of boiling point. The difference between the two approaches is illustrated graphically below.

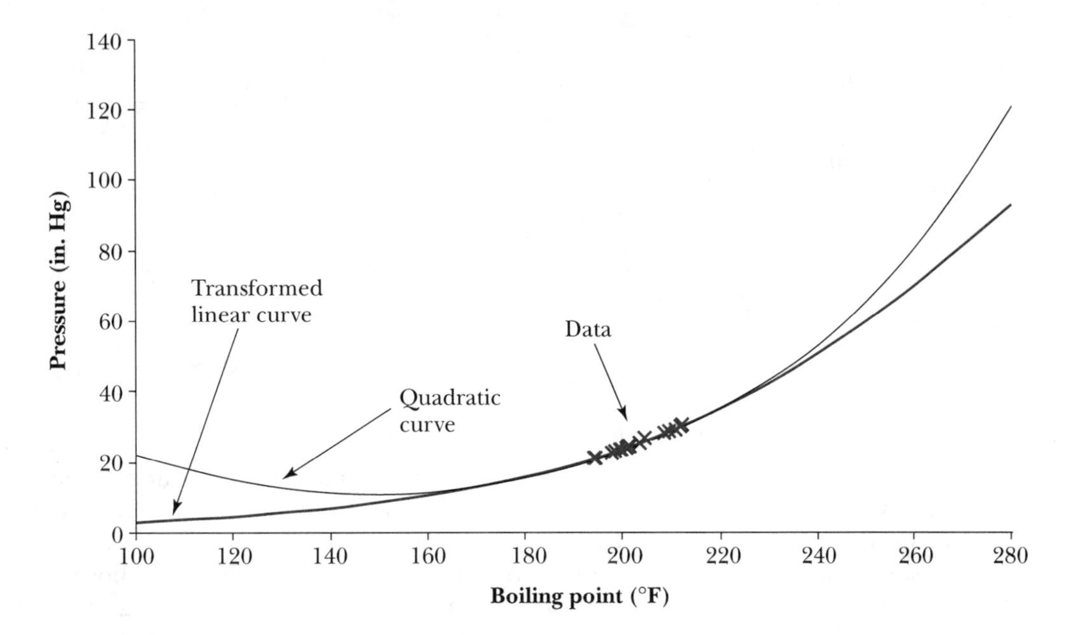

Of course, it would be very unwise to even consider extrapolation so far beyond the range of the data used to fit the relationship. But the comparison suggests that for some (but substantially less) extrapolation, a more reliable prediction is made from the transformed linear relationship.

Correlation and nonlinearity

A further complication with curved relationships is that the correlation coefficient is a misleading measure of association. As described in section 4.2, there can be a strong nonlinear relationship between two variables but the correlation coefficient between them may be close to zero. In order for the correlation coefficient to be a better description of the strength of the relationship, one or both of the variables should be transformed to linearise the relationship first.

For example, consider the mammal lifespan data presented in section 4.1. The correlation coefficient between lifespan and metabolic rate (O_2 consumption) is −0.47. However, as is clear from a scatterplot of the data (page 146), the relationship between the measurements is nonlinear, so Pearson's r would not adequately describe the strength of the relationship.

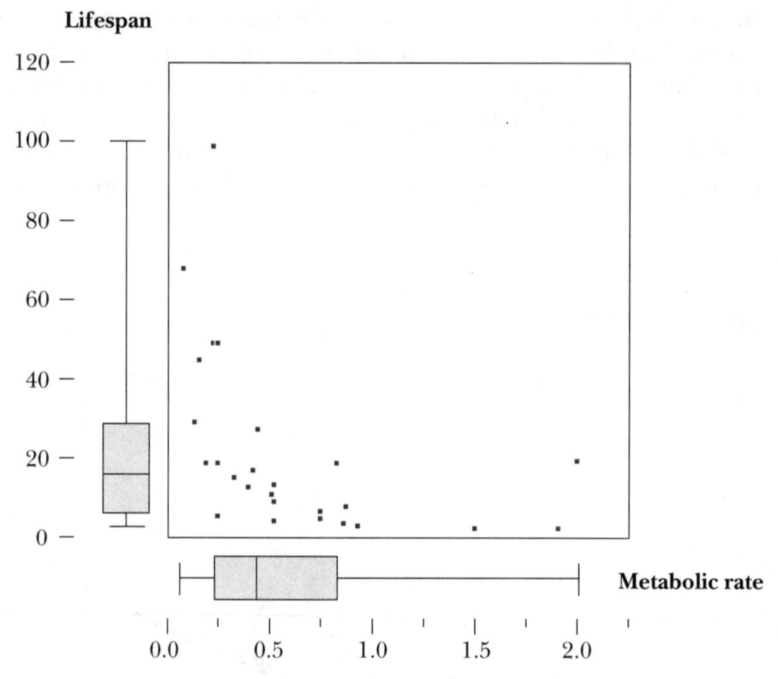

Separate transformation of each of the variables may be performed to make the distribution of sample values for that variable more symmetric. Although this is not guaranteed to linearise a relationship, it often does so in practice.

Since the marginal distributions of metabolic rate and lifespan above are both highly skew, a square root or logarithmic transformation is suggested to make each of these variables more symmetric. Logarithmic transformations of both variables are used in the scatterplot below.

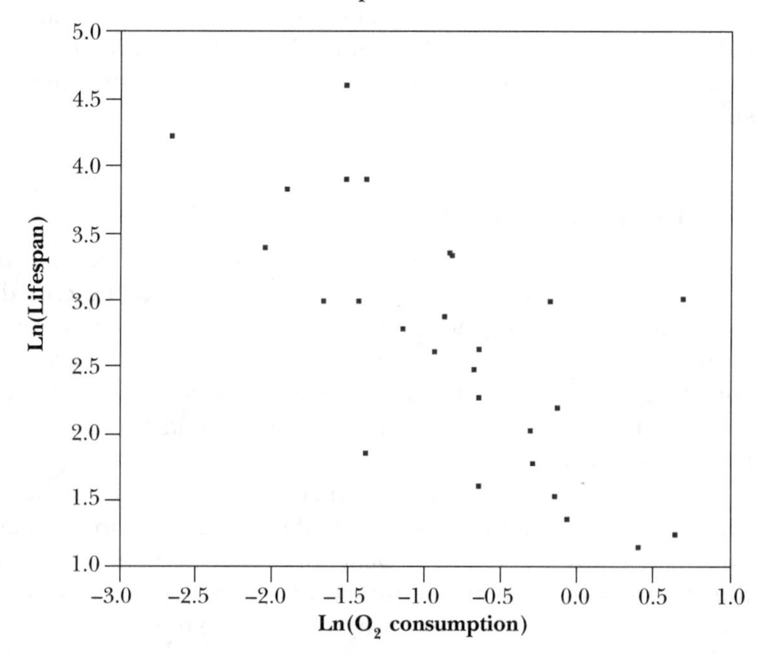

These transformations have not only made the marginal distributions more symmetric, but they have simultaneously linearised the relationship (at least approximately). The correlation coefficient between the transformed variables is −0.72 and this describes the strength of the relationship between lifespan and metabolic rate better than the correlation coefficient of the raw data, −0.47.

EXERCISE 4.5

1. In the subsection on polynomials, we fitted a quadratic curve to the boiling point data. Use this least squares quadratic fit to obtain the pressure that would correspond to a boiling point of 100°F. Does this extrapolation produce a reasonable result?
2. Consider a scatterplot in which the scatter of points has the shape of:
   (a) a square
   (b) a dagger pointing downwards
   (c) a large X.
   In each case, would the correlation be positive, negative or zero?
3. Is the relationship between hours of study on a statistics course and numerical score on the final exam likely to be linear? Explain why or why not.

4.6 MULTIVARIATE DATA

Many data sets contain more than two variables. Most techniques for analysing such data are beyond the scope of this book, but we will describe two graphical methods that are extensions of the scatterplot. Both are 'dynamic' methods that require interactive use of a computer for their effective application.

If there are three numerical variables in a data set, the simple scatterplot may be extended into a third dimension. This is, of course, impossible to display on paper but a computer can allow such a three-dimensional display to be rotated (using a mouse), giving an effective representation of the shape of the three-dimensional scatter of points.

A second plotting technique that is a useful tool for graphically exploring data sets is a *scatterplot matrix*. This simply consists of the scatterplots of all possible pairs of variables in the data set, arranged in an array. To illustrate the technique, we will present a scatterplot matrix of the multivariate data set listed on the next two pages.[16]

16. E. Anderson, 'The irises of the Gaspé Peninsula', *Bulletin of the American Iris Society*, 59, 1935, pp. 2–5.

Sepal length	Sepal width	Petal length	Petal width	Species	Sepal length	Sepal width	Petal length	Petal width	Species
5.1	3.5	1.4	0.2	*Setosa*	5.0	3.5	1.3	0.3	*Setosa*
4.9	3.0	1.4	0.2	*Setosa*	4.5	2.3	1.3	0.3	*Setosa*
4.7	3.2	1.3	0.2	*Setosa*	4.4	3.2	1.3	0.2	*Setosa*
4.6	3.1	1.5	0.2	*Setosa*	5.0	3.5	1.6	0.6	*Setosa*
5.0	3.6	1.4	0.2	*Setosa*	5.1	3.8	1.9	0.4	*Setosa*
5.4	3.9	1.7	0.4	*Setosa*	4.8	3.0	1.4	0.3	*Setosa*
4.6	3.4	1.4	0.3	*Setosa*	5.1	3.8	1.6	0.2	*Setosa*
5.0	3.4	1.5	0.2	*Setosa*	4.6	3.2	1.4	0.2	*Setosa*
4.4	2.9	1.4	0.2	*Setosa*	5.3	3.7	1.5	0.2	*Setosa*
4.9	3.1	1.5	0.1	*Setosa*	5.0	3.3	1.4	0.2	*Setosa*
5.4	3.7	1.5	0.2	*Setosa*	7.0	3.2	4.7	1.4	*Versicolor*
4.8	3.4	1.6	0.2	*Setosa*	6.4	3.2	4.5	1.5	*Versicolor*
4.8	3.0	1.4	0.1	*Setosa*	6.9	3.1	4.9	1.5	*Versicolor*
4.3	3.0	1.1	0.1	*Setosa*	5.5	2.3	4.0	1.3	*Versicolor*
5.8	4.0	1.2	0.2	*Setosa*	6.5	2.8	4.6	1.5	*Versicolor*
5.7	4.4	1.5	0.4	*Setosa*	5.7	2.8	4.5	1.3	*Versicolor*
5.4	3.9	1.3	0.4	*Setosa*	6.3	3.3	4.7	1.6	*Versicolor*
5.1	3.5	1.4	0.3	*Setosa*	4.9	2.4	3.3	1.0	*Versicolor*
5.7	3.8	1.7	0.3	*Setosa*	6.6	2.9	4.6	1.3	*Versicolor*
5.1	3.8	1.5	0.3	*Setosa*	5.2	2.7	3.9	1.4	*Versicolor*
5.4	3.4	1.7	0.2	*Setosa*	5.0	2.0	3.5	1.0	*Versicolor*
5.1	3.7	1.5	0.4	*Setosa*	5.9	3.0	4.2	1.5	*Versicolor*
4.6	3.6	1.0	0.2	*Setosa*	6.0	2.2	4.0	1.0	*Versicolor*
5.1	3.3	1.7	0.5	*Setosa*	6.1	2.9	4.7	1.4	*Versicolor*
4.8	3.4	1.9	0.2	*Setosa*	5.6	2.9	3.6	1.3	*Versicolor*
5.0	3.0	1.6	0.2	*Setosa*	6.7	3.1	4.4	1.4	*Versicolor*
5.0	3.4	1.6	0.4	*Setosa*	5.6	3.0	4.5	1.5	*Versicolor*
5.2	3.5	1.5	0.2	*Setosa*	5.8	2.7	4.1	1.0	*Versicolor*
5.2	3.4	1.4	0.2	*Setosa*	6.2	2.2	4.5	1.5	*Versicolor*
4.7	3.2	1.6	0.2	*Setosa*	5.6	2.5	3.9	1.1	*Versicolor*
4.8	3.1	1.6	0.2	*Setosa*	5.9	3.2	4.8	1.8	*Versicolor*
5.4	3.4	1.5	0.4	*Setosa*	6.1	2.8	4.0	1.3	*Versicolor*
5.2	4.1	1.5	0.1	*Setosa*	6.3	2.5	4.9	1.5	*Versicolor*
5.5	4.2	1.4	0.2	*Setosa*	6.1	2.8	4.7	1.2	*Versicolor*
4.9	3.1	1.5	0.2	*Setosa*	6.4	2.9	4.3	1.3	*Versicolor*
5.0	3.2	1.2	0.2	*Setosa*	6.6	3.0	4.4	1.4	*Versicolor*
5.5	3.5	1.3	0.2	*Setosa*	6.8	2.8	4.8	1.4	*Versicolor*
4.9	3.6	1.4	0.1	*Setosa*	6.7	3.0	5.0	1.7	*Versicolor*
4.4	3.0	1.3	0.2	*Setosa*	6.0	2.9	4.5	1.5	*Versicolor*
5.1	3.4	1.5	0.2	*Setosa*	5.7	2.6	3.5	1.0	*Versicolor*

Sepal length	Sepal width	Petal length	Petal width	Species	Sepal length	Sepal width	Petal length	Petal width	Species
5.5	2.4	3.8	1.1	*Versicolor*	6.4	3.2	5.3	2.3	*Virginica*
5.5	2.4	3.7	1.0	*Versicolor*	6.5	3.0	5.5	1.8	*Virginica*
5.8	2.7	3.9	1.2	*Versicolor*	7.7	3.8	6.7	2.2	*Virginica*
6.0	2.7	5.1	1.6	*Versicolor*	7.7	2.6	6.9	2.3	*Virginica*
5.4	3.0	4.5	1.5	*Versicolor*	6.0	2.2	5.0	1.5	*Virginica*
6.0	3.4	4.5	1.6	*Versicolor*	6.9	3.2	5.7	2.3	*Virginica*
6.7	3.1	4.7	1.5	*Versicolor*	5.6	2.8	4.9	2.0	*Virginica*
6.3	2.3	4.4	1.3	*Versicolor*	7.7	2.8	6.7	2.0	*Virginica*
5.6	3.0	4.1	1.3	*Versicolor*	6.3	2.7	4.9	1.8	*Virginica*
5.5	2.5	4.0	1.3	*Versicolor*	6.7	3.3	5.7	2.1	*Virginica*
5.5	2.6	4.4	1.2	*Versicolor*	7.2	3.2	6.0	1.8	*Virginica*
6.1	3.0	4.6	1.4	*Versicolor*	6.2	2.8	4.8	1.8	*Virginica*
5.8	2.6	4.0	1.2	*Versicolor*	6.1	3.0	4.9	1.8	*Virginica*
5.0	2.3	3.3	1.0	*Versicolor*	6.4	2.8	5.6	2.1	*Virginica*
5.6	2.7	4.2	1.3	*Versicolor*	7.2	3.0	5.8	1.6	*Virginica*
5.7	3.0	4.2	1.2	*Versicolor*	7.4	2.8	6.1	1.9	*Virginica*
5.7	2.9	4.2	1.3	*Versicolor*	7.9	3.8	6.4	2.0	*Virginica*
6.2	2.9	4.3	1.3	*Versicolor*	6.4	2.8	5.6	2.2	*Virginica*
5.1	2.5	3.0	1.1	*Versicolor*	6.3	2.8	5.1	1.5	*Virginica*
5.7	2.8	4.1	1.3	*Versicolor*	6.1	2.6	5.6	1.4	*Virginica*
6.3	3.3	6.0	2.5	*Virginica*	7.7	3.0	6.1	2.3	*Virginica*
5.8	2.7	5.1	1.9	*Virginica*	6.3	3.4	5.6	2.4	*Virginica*
7.1	3.0	5.9	2.1	*Virginica*	6.4	3.1	5.5	1.8	*Virginica*
6.3	2.9	5.6	1.8	*Virginica*	6.0	3.0	4.8	1.8	*Virginica*
6.5	3.0	5.8	2.2	*Virginica*	6.9	3.1	5.4	2.1	*Virginica*
7.6	3.0	6.6	2.1	*Virginica*	6.7	3.1	5.6	2.4	*Virginica*
4.9	2.5	4.5	1.7	*Virginica*	6.9	3.1	5.1	2.3	*Virginica*
7.3	2.9	6.3	1.8	*Virginica*	5.8	2.7	5.1	1.9	*Virginica*
6.7	2.5	5.8	1.8	*Virginica*	6.8	3.2	5.9	2.3	*Virginica*
7.2	3.6	6.1	2.5	*Virginica*	6.7	3.3	5.7	2.5	*Virginica*
6.5	3.2	5.1	2.0	*Virginica*	6.7	3.0	5.2	2.3	*Virginica*
6.4	2.7	5.3	1.9	*Virginica*	6.3	2.5	5.0	1.9	*Virginica*
6.8	3.0	5.5	2.1	*Virginica*	6.5	3.0	5.2	2.0	*Virginica*
5.7	2.5	5.0	2.0	*Virginica*	6.2	3.4	5.4	2.3	*Virginica*
5.8	2.8	5.1	2.4	*Virginica*	5.9	3.0	5.1	1.8	*Virginica*

The table describes the sepal width and length and the petal width and length from a sample of irises. The irises belong to three different species which are also distinguished in the data. The following scatterplot matrix shows all pairs of scatterplots of the four numerical measurements. Different symbols have been used to distinguish the three species of iris. The differences between the species are displayed clearly in the scatterplot matrix.

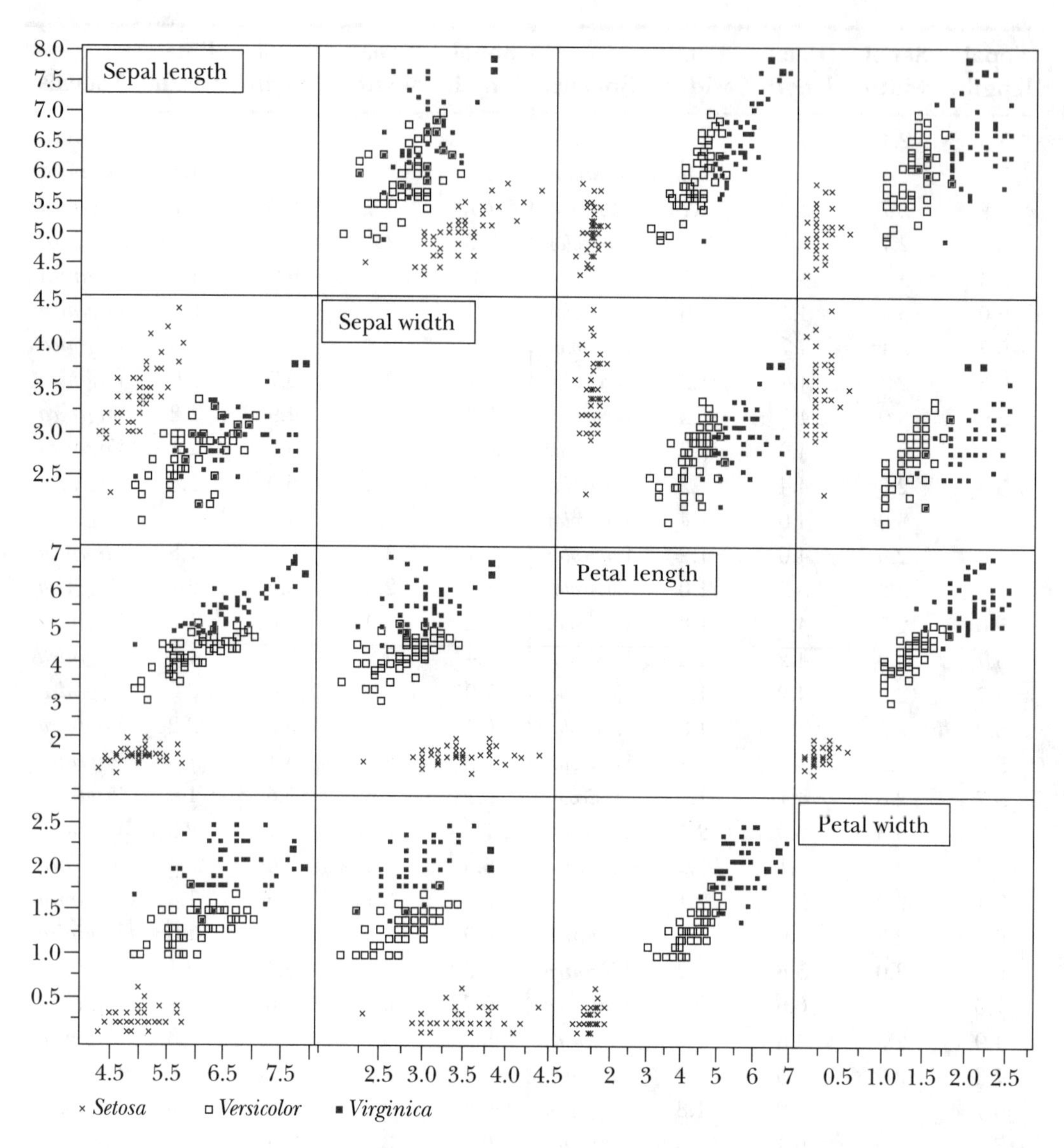

Further information about the relationship between the variables can be obtained from a scatterplot matrix if it is displayed by a computer program that can link together the crosses on all the scatterplots belonging to the same individual. By moving the computer's mouse over some of the points with the mouse button held down (*brushing* the points), the linked points in all scatterplots are highlighted. This can show, for example, whether a feature such as an outlier or cluster of points that is apparent in one scatterplot is caused by (or at least linked to) values that are also unusual in the other scatterplots.

Brushing in a scatterplot matrix, like rotation of three-dimensional scatterplots, is a dynamic procedure that cannot be easily illustrated in a textbook. However, the following example provides some indication of what the technique can achieve.

Let us re-examine the data set which related maximum January temperatures in various US cities to their latitudes. The scatterplot matrix below shows these two variables and a third variable, the longitude of the city (degrees west of Greenwich). For reasons that will become apparent later, the third variable used is the negative of the longitude rather than the longitude itself. We have also omitted the lower left quarter of the scatterplot matrix since the scatterplots there are mirror images of those in the top right quarter.

The least squares line describing the relationship between temperature and latitude has been drawn on the scatterplot of these two variables. Several points are well below the fitted line; these are cities which are more than 5°F colder than predicted by the least squares line for their latitude. To examine them, a computer mouse is used to brush over these particular points, and the computer's response is to immediately highlight the corresponding points in the other two scatterplots. In the diagram below, the highlighted points are displayed as crosses.

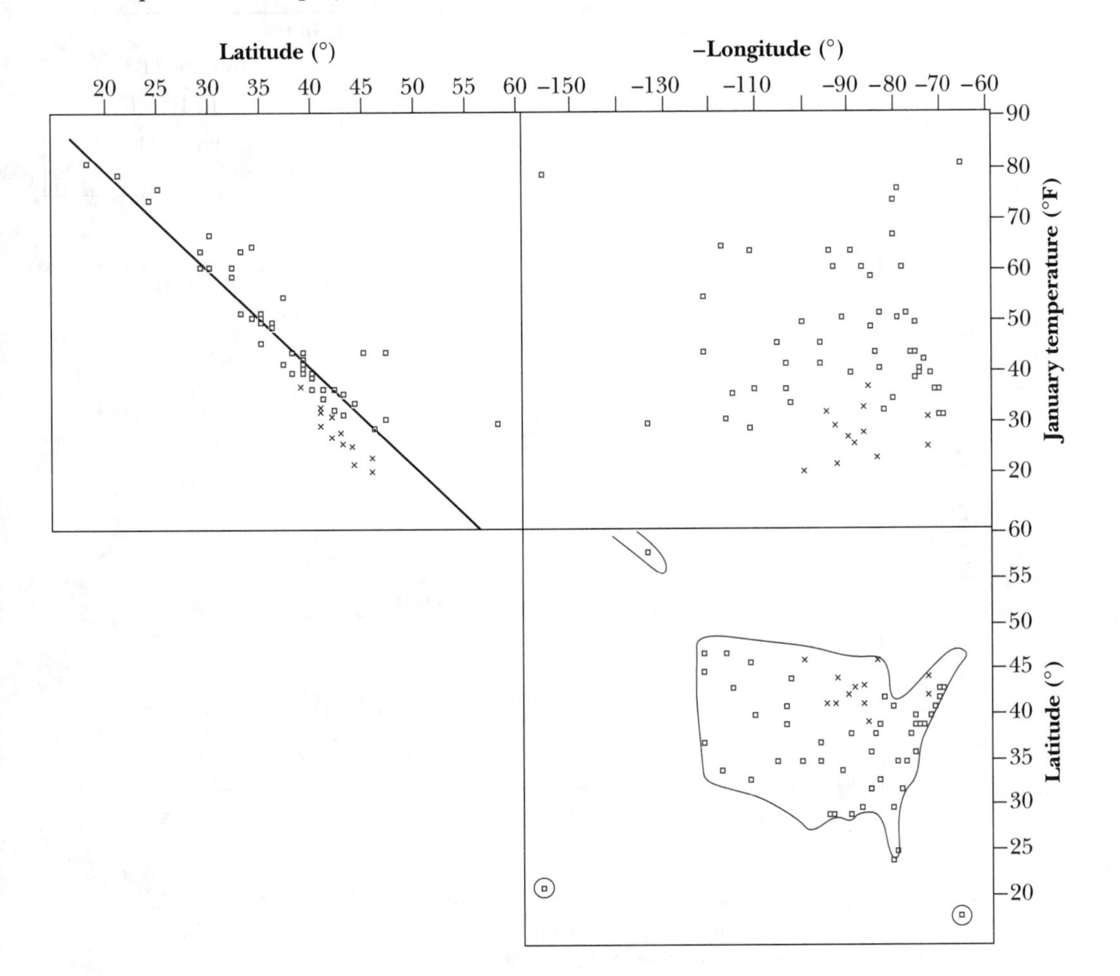

The scatterplot on the bottom right on page 151 displays the latitude of the cities on the vertical axis and the negative of the longitude on the horizontal axis. This is equivalent to showing the positions of all the cities on a map of the United States. The highlighted crosses therefore show the geographical location of these unusually cold cities. The cities are unusually cold because they are so far inland and have an extreme continental climate — very cold in the winter and hot in the summer.

In the scatterplot matrix of the iris data, information about the iris species was coded using a different plotting symbol for each species. Different plotting symbols can similarly be used in a scatterplot (or scatterplot matrix) to represent the value of a third numerical variable. For example, in the scatterplot of maximum January temperatures and latitudes below, the altitudes of the cities have been represented with different plotting symbols. While many high altitude cities are relatively cold for their latitudes (below the least squares line), there are exceptions which could be further investigated.[17]

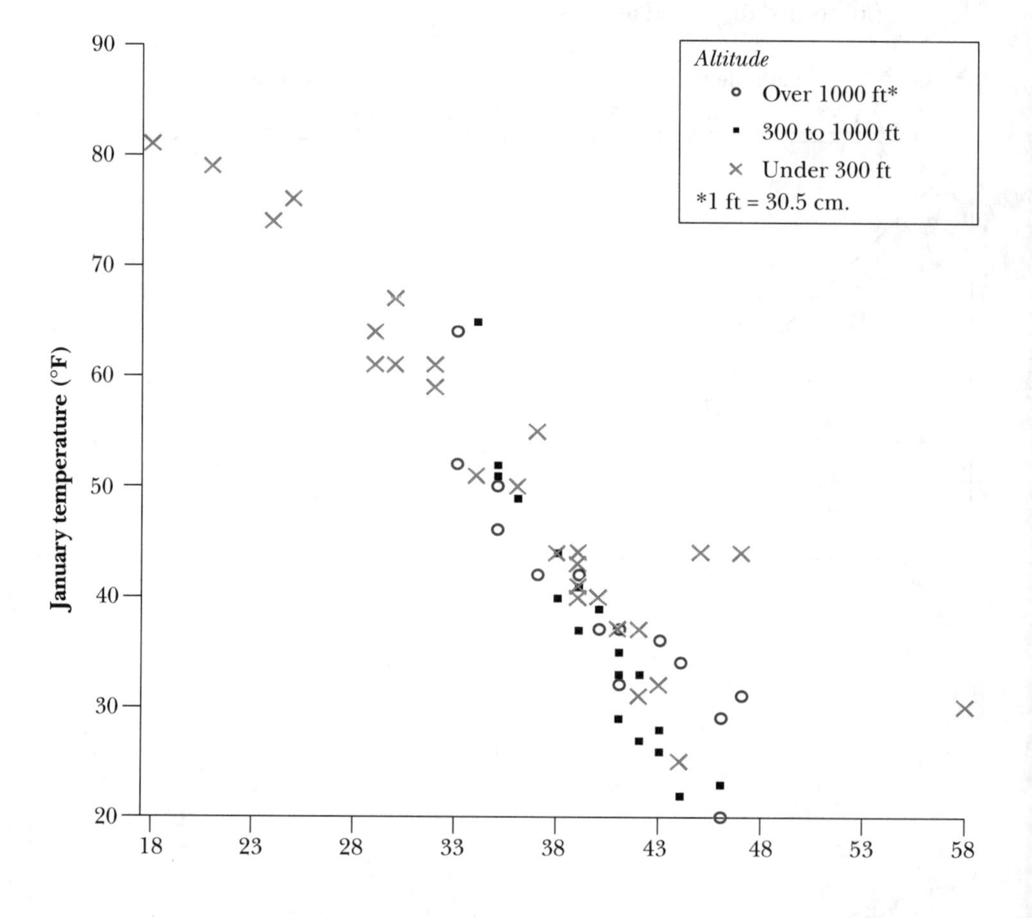

17. Related methods have been proposed to represent the values of two or more additional variables with different shapes and sizes of plotting symbol (including faces with different characteristics!), but it is rarely useful to display more than one extra numerical or grouping variable in this way.

EXERCISE 4.6

1. The scatterplot below is a component of the scatterplot matrix of the US January temperatures data, flipped vertically.

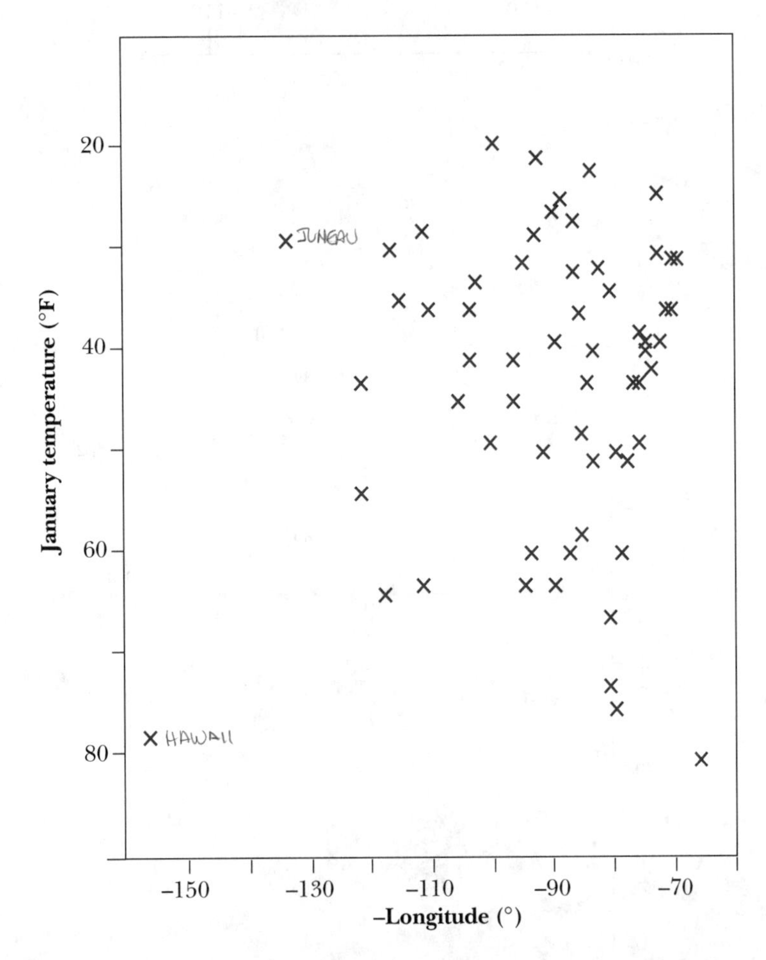

The positions of the cities on this scatterplot are a distortion of their positions on a map. For example, Hawaii is the city on the bottom left and Juneau, Alaska, is on the top left. The reason for this is that January temperatures and latitudes are quite strongly related, so that cities with high latitudes tend to have low January temperatures, and vice versa. However, the relationship between latitude and temperature is not exact, so some cities are higher, or lower, in this diagram than they would appear on an accurate map.

This distortion of the map can be interpreted in terms of cities that are unusually warm or cold for their latitudes. Describe how this plot has moved some groups of cities from their geographic positions. Consideration of the highlighted points (plotted as X instead of O in the scatterplot matrix) should help.

2. The January temperatures could have been plotted as numbers located at the scatterplot position indicated by latitude and (–longitude) — in other words, as temperatures on a map.

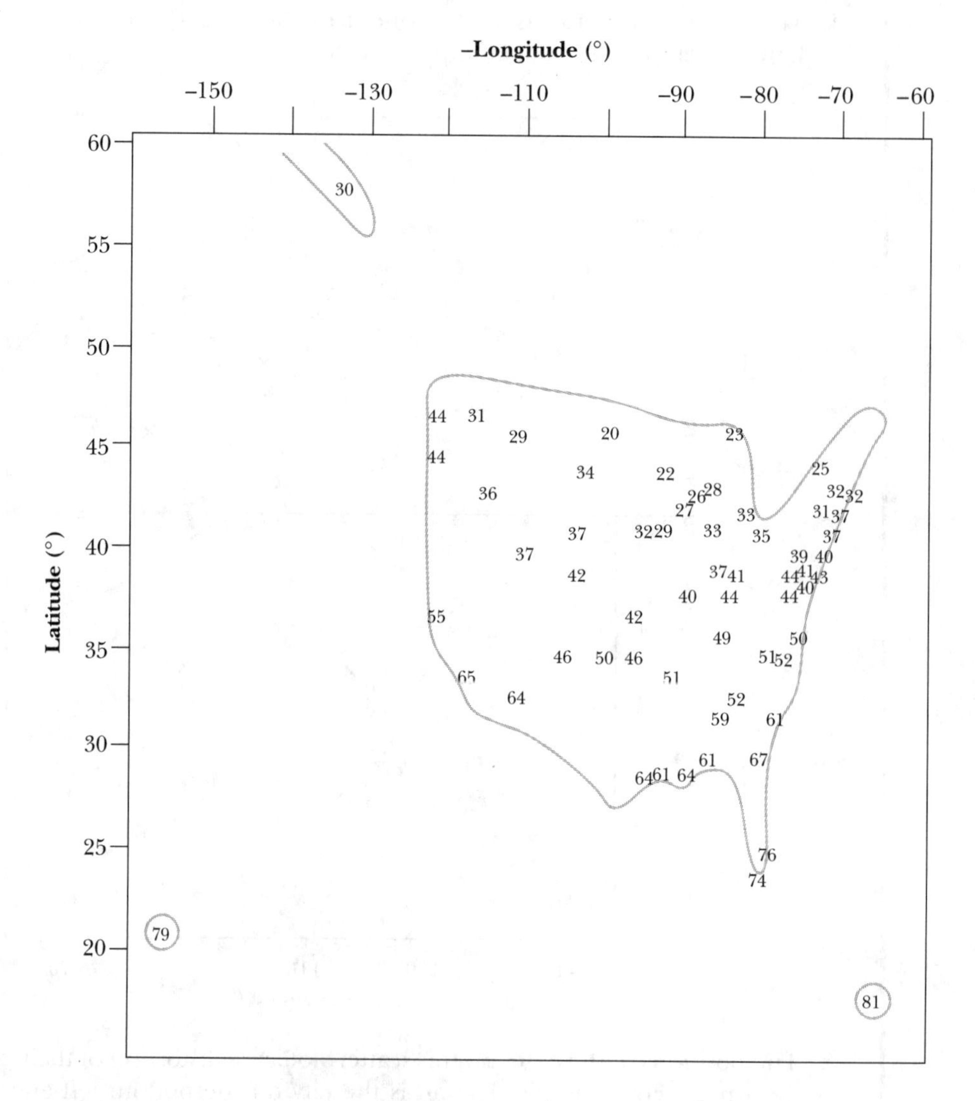

Consider the contiguous part of the United States (omitting Alaska, Hawaii, and Puerto Rico) in the map above. Imagine a smooth surface suspended over this page whose height at the cities' positions is indicated by the temperature — perhaps one millimetre above the page for each Fahrenheit degree. What shape would the smooth surface have? In particular, if water were poured onto the centre of the map, where would it drain?

3. Examine the matrix plot of the iris data set. Describe the features of the *Setosa* variety that distinguish it from the other two. Do the same thing for the distinguishing features of *Versicolor* and *Virginica* varieties.

4.7 CAUSATION AND EXPERIMENTAL DESIGN

Correlation measures *association*, but is often wrongly regarded as measuring *causation*. For example, consider the following table which shows the number of visits made to national parks in Victoria[18] and the number of aluminium cans sold in Australia[19] between 1978 and 1989.

Year	Victorian national park visits ('000)	Aluminium beverage cans sold ('000 000)
1978	3400	917
1979	4160	1290
1980	4366	1196
1981	5056	1360
1982	6491	1466
1983	7000	1393
1984	6349	1596
1985	7731	1577
1986	7858	1827
1987	8115	2037
1988	8354	2321
1989	8815	2523

A scatterplot of the data shows a strong association between the number of cans sold and the number of visits to national parks in Victoria — they are strongly correlated. However, very few people would claim that changes to the number of beverage cans sold were the *cause* of the increase in park visits (or vice versa). Variation in both measurements is caused by other factors that are not displayed in the scatterplot — in particular, the passage of time.

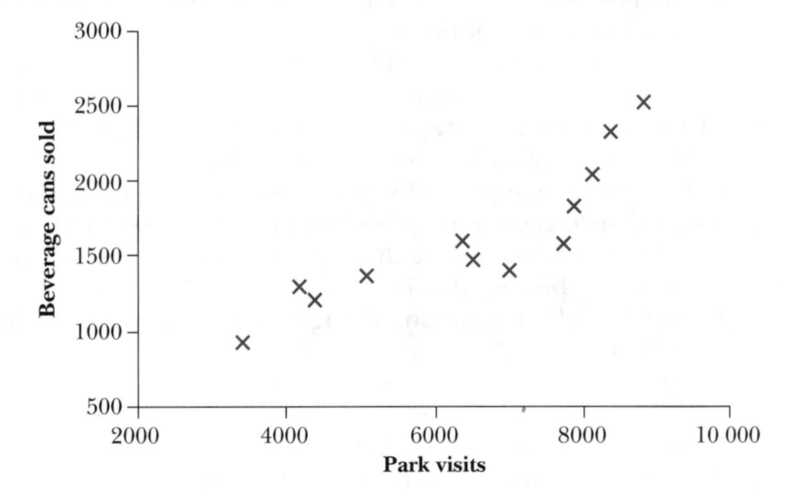

18. Australian Bureau of Statistics, *Striking a Balance*, AGPS, Canberra, 1992, p. 228 and Industry Commission, *Recycling, Volume 2: Recycling of Products*, AGPS, Canberra, 1991. Permission also from Department of Natural Resources and Environment, Victoria.
19. ibid.

In other examples, however, interpreting the association between two variables as an indication of a causal relationship is less obviously unreasonable. For example, an association between incidence of lung diseases and cigarette smoking has been observed; should this relationship be interpreted as a causal one? The issue was hotly debated for many years.

When the data collector has no control over the values of either of the variables, but can simply record their values, the data are called *observational* and the data themselves cannot establish that any relationship between them is causal. Any such implication must come from other knowledge about the nature of the two variables. However, if the data collector can control the values of one variable, such as the temperature at which a chemical reaction is conducted, then it is possible to infer a causal relationship between this variable and a second response variable, such as the concentration of a particular chemical produced in the reaction, provided other aspects of the experiment are well designed. Such data are called *experimental*. We will compare observational and experimental data further in chapter 7.

EXERCISE 4.7

1. From your general knowledge, which of the following pairs of variables would probably be positively associated?
   (a) a country's population; and the average number of years schooling attained by the population over age 25, for each of the last 25 years
   (b) the ratio of women to men registered at a university, during each year from 1980 to 2000; and the proportion of women in the graduating class in each year
   (c) the amount of 'calorie-reduced' foods in a person's diet; and the per cent overweight of the person
   (d) the concentration of fertiliser solution a farmer spreads on his wheatfield; and the yield of wheat, over several different experimental plots
2. Of the variables that are positively correlated in question 1, which correlations are attributable to a direct causal link?
3. Use the example suggested by the variables below to explain that an association between two variables need not reflect a causal link.
   X: number of cars passing through a certain intersection in each of the 24 hours of the day (that is, 24 numbers)
   Y: number of buses passing though a certain intersection in each hour of a 24 hour day (24 numbers)
4. Use the example suggested below to explain how association can be used to suggest causal mechanisms.
   X: presence of mould on a Petrie dish
   Y: reduced bacterial growth on a Petrie dish
   (*Hint*: Penicillin occurs in certain moulds.)

Problems

Although most problems in this chapter can be completed using a scientific calculator with regression functions and graph paper, you are encouraged to use a computer to answer the problems. Your instructor will advise you on the availability of a statistical program.

1. At the 1992 Summer Olympic Games in Barcelona, Spain, and the 1994 Winter Olympic Games in Lillehammer, Norway, the media spent a lot of time discussing the number of medals won by each country's athletes. The implication was that the comparison was of some importance. However, larger countries would be *expected* to win more medals than smaller countries, simply because of their larger populations. The table below shows information about all the medal-winning countries.

Rank	Country	Medals summer 1992	Medals winter 1994	Total medals	Population
1	Unified Team*	112	34	146	231.5
2	United States	108	13	122	260.7
3	Germany	82	24	106	81.1
4	China	54	3	57	1190.4
5	Italy	19	20	39	58.1
6	France	29	5	34	57.8
7	Norway	7	26	33	4.3
8	Canada	18	13	31	28.1
9	Cuba	31	0	31	11.1
10	Hungary	30	0	30	10.3
11	South Korea	29	0	29	45.1
12	Australia	27	1	28	18.1
13	Japan	22	5	27	125.1
14	Britain	20	2	22	58.1
15	Spain	22	0	22	39.3
16	Netherlands	16	4	20	15.4
17	Poland	19	0	19	38.7
18	Romania	18	0	18	23.2
19	Bulgaria	16	0	16	8.8
20	North Korea	9	6	15	23.1
21	Sweden	12	3	15	8.8
22	Austria	2	9	11	8.0
23	Finland	5	6	11	5.1
24	New Zealand	10	0	10	3.4
25	Switzerland	1	9	10	7.0
26	Kenya	8	0	8	28.2

continued

Medal-winning countries table *(cont'd)*

27	Czechoslovakia	7	0	7	10.4
28	Denmark	6	0	6	5.2
29	Turkey	6	0	6	62.1
30	Indonesia	5	0	5	200.4
31	Slovenia	2	3	5	2.0
32	Jamaica	4	0	4	2.6
33	Nigeria	4	0	4	98.1
34	Belgium	3	0	3	10.1
35	Brazil	3	0	3	158.7
36	Croatia	3	0	3	4.7
37	Ethiopia	3	0	3	58.7
38	Iran	3	0	3	65.6
39	Latvia	3	0	3	2.7
40	Morocco	3	0	3	28.6
41	Algeria	2	0	2	27.9
42	Estonia	2	0	2	1.6
43	Greece	2	0	2	10.6
44	Ireland	2	0	2	3.5
45	Israel	2	0	2	5.1
46	Lithuania	2	0	2	3.8
47	Mongolia	2	0	2	2.4
48	Namibia	2	0	2	1.6
49	South Africa	2	0	2	43.9
50	Argentina	1	0	1	33.9
51	Bahamas	1	0	1	0.3
52	Colombia	1	0	1	35.6
53	Ghana	1	0	1	17.2
54	Malaysia	1	0	1	19.3
55	Mexico	1	0	1	92.2
56	Pakistan	1	0	1	121.9
57	Peru	1	0	1	23.7
58	Philippines	1	0	1	69.8
59	Qatar	1	0	1	0.5
60	Suriname	1	0	1	0.4
61	Taiwan	1	0	1	21.3
62	Thailand	1	0	1	59.5

* The Russian Federation plus various other countries that were formerly part of the USSR
Source: R. Famighetti (ed.), *The World Almanac and Book of Facts 1995*, Funk and Wagnalls, Mahwah, New Jersey, 1995, p. 858. Reprinted with permission. Copyright © 1994. K-III Reference Corporation. All rights reserved.

(a) Draw a scatterplot of the total medals against population.

(b) Fit a straight line by least squares to this scatterplot. Examine the residuals and comment on the fit of the line.

(c) Is there a subset of countries for which the line fit in (b) seems to be a very poor fit? Can you give a reason for this? Eliminate those countries and refit the least squares line. Re-examine the residual plot and comment.

(d) If the Russian Federation (population 149.6 million) had competed separately from the other countries in the United Team, how many medals would the least squares line predict that they would have won?

(e) Do you think the prediction in (d) would be high or low? (*Hint*: Look at the size of the residual for the United Team.)

2. In the recent Olympics, some viewers, especially from the smaller countries, felt that the number of medals should be standardised to account for the very wide range of populations, and that a per capita number of medals for a country was a fairer comparison. Others felt that this was unfair to the countries with large populations — that having twice as many people did not lead to twice as many medals. If standardisation is performed adequately, there should be no systematic relationship between the adjusted medal count and population. Try two or three different adjustments, such as:

$$\text{Index} = \frac{\text{Medal count}}{\text{Population}} \quad \textit{or} \quad \text{Index} = \frac{\text{Medal count}}{\sqrt{\text{Population}}}$$

and examine their effectiveness with a scatterplot and correlation coefficient. Write a report on your findings.

3. The research expenditure data graphed in section 4.2 shows a positive association between expenditure from the two sectors, but there is some question about its linearity. The five countries for which business expends more than 1.5% of GDP (Switzerland, Japan, Sweden, Germany and USA) seem to have less expenditure in the public sector than one would estimate from the linear relationship suggested by the other countries. For this problem, we will call this shortfall 'underfunding'.

(a) Predict the underfunding by the public sector in the five countries mentioned.

(b) Could the underfunding vanish with a quadratic fit? If your computer software allows it, try the quadratic fit, and examine the residual plot.

4. Find a linearising transformation for the relationship between cycles to failure and amplitude of stress for the stress loads data presented in section 4.1. (*Hint*: Consider logarithm and reciprocal transformations.) Once you have a linearising transformation, fit a least squares line and re-express this relationship between the variables in their original units.

5. Examine the scatterplot of the aircraft emissions data in section 4.2. While there is a hint of a strong linear relationship between the two variables, there are several apparently exceptional points. Identify these exceptional points. Fit a least squares line to the data with and without the exceptional points. Comment on the different fits.

6. Using the Olympics Medals data, examine the ratio of winter to summer games medals — call it the 'winter index' — as an index of that country's relative interest in winter sports. Omit from consideration any countries with fewer than 4 medals.

(a) Is the total number of medals related to the winter index?

(b) Does the winter index reflect latitude? Select 10 countries from your list, look up their latitude and compare with the winter index.

7. A study of the relationship between the final *mark* and *hours* of study in a certain course results in a relationship:

$$\text{Mark} = 2 + \frac{\text{Hours} - 50}{25} + \text{Residuals}.$$

The standard deviation of the residuals is 0.5.

(a) Draw a freehand diagram, approximately to scale, of the above line. Invent 10 observations which fit roughly with the description given above and draw them on your diagram.

(b) Indicate on your diagram how you would use the diagram to predict the final mark for a person studying only 25 hours.

(c) Use the scatter of points around the line to give a rough estimate of the accuracy of your prediction in part (b). (You should assume that the least squares line is accurately known, although the residual for the person being predicted is, of course, unknown.)

8. In a stand of Douglas Fir trees, the following data were recorded, relating the tree height (feet) to the tree volume (cubic feet). Here is a graph of the data:

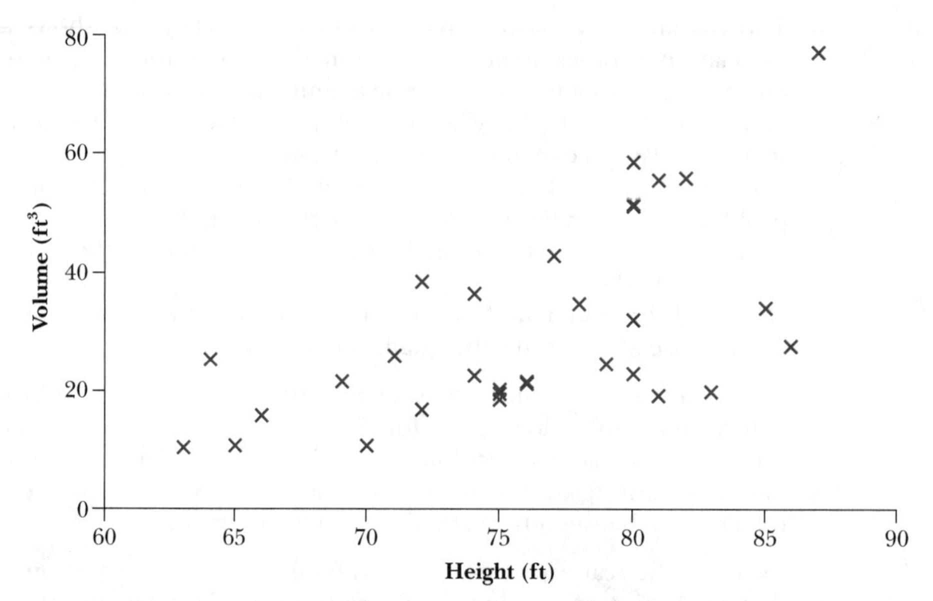

Source: MINITAB

The regression of volume on height produces the least squares equation:

$$\text{Volume} = -87.1 + 1.54 \text{ height}.$$

The regression of height on volume produces the least squares equation:

$$\text{Height} = 69.0 + 0.232 \text{ volume}.$$

(a) Which least squares line would be useful for judging the volume of a particular tree based on its height? Draw both lines on the scatterplot.

(b) Guess the value of the standard deviation of the residuals, using only the graph and the least squares line you have drawn. Remember that approximately 95% of the residuals will be within two standard deviations of their mean (zero). Therefore, a band between two standard deviations above and below the least squares line should enclose approximately 95% of the residuals. What does this tell you about the precision of the prediction of volume from height?

(c) Guess the value of the correlation coefficient of *height* and *volume.*

(d) If tree height were measured in metres instead of feet, would the correlation coefficient increase, decrease or stay the same? Explain.

9. A taxi company wishes to analyse the petrol consumption of its fleet of taxis, after one month's experience. Data are compiled for 100 taxis on the number of kilometres travelled in the month, *K*, and the litres of petrol consumed in that same month, *L*. The owner of the company suspects that a few drivers have been inflating their petrol bills and charging the company for too much petrol. All the cars are the same make and model and yet there seems to be quite a bit of variation in the petrol purchased. How would you analyse the data to assist the owner of the taxi company in assessing the situation? Be as specific as possible in your recommendations.

10. Your boss gives you the following data on sales and advertising expenditure for 10 different regions in which your company operates:

Region	**1**	**2**	**3**	**4**	**5**	**6**	**7**	**8**	**9**	**10**
Advertising ($)	100	100	200	200	500	500	900	900	900	900
Sales ('000) ($)	10	5	25	25	25	20	35	25	25	40

(a) Summarise the data in a way that will be useful to your boss, who has never taken a statistics course.

(b) What additional information should you ask for from your boss to do a good job in (a)?

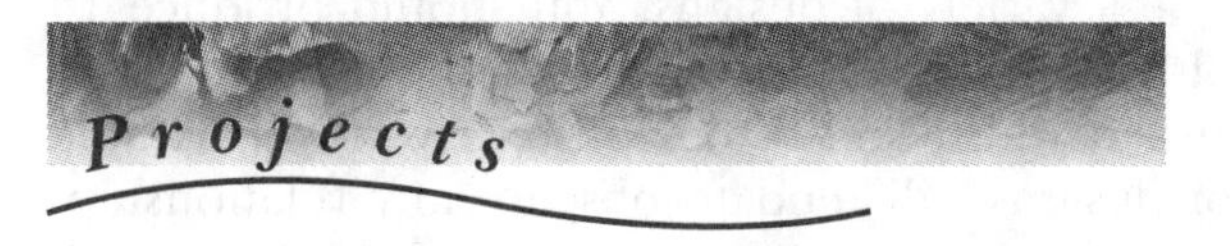

Projects

1. **Paper planes — width and point of balance**
The object of this project is to determine whether the width of a paper aeroplane is related to the distance that it will fly. In chapter 2, a project was described in which the effect of paper size on distance flown was examined, for a fixed design of paper aeroplane. In this project, you will use the same size of paper for each aeroplane, but will build your aeroplanes from a variety of designs. While you are free to design your own aeroplanes, here is some guidance for the ways in which you can vary your design.

- The first fold usually halves the paper to determine the centre axis of the aeroplane; if the paper is not square, you may choose either the long or short axis of the piece of paper for the direction of flight.

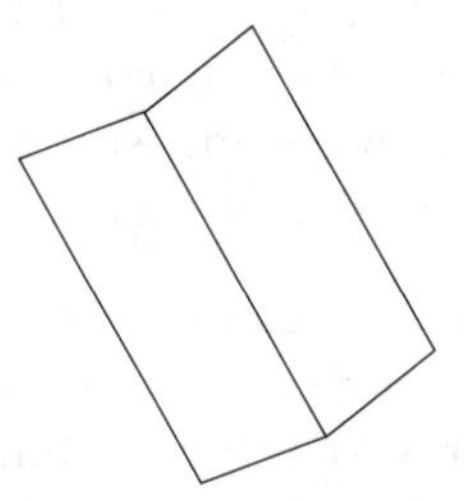

- The next fold is usually one that forms the nose of the aeroplane. You have the option of making a very sharp point or a broader point. Also, you can fold the point back for extra nose weight if you think this will help.

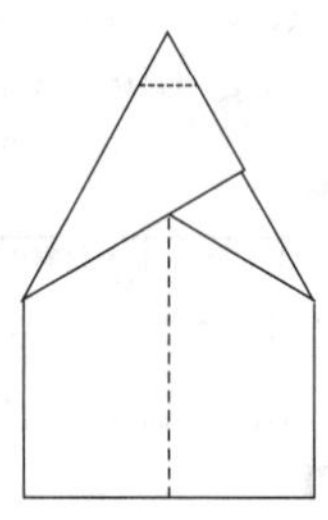

- You can use other folds to form the fuselage. The width of the aeroplane and flight characteristics can also be affected by the depth of the fuselage.

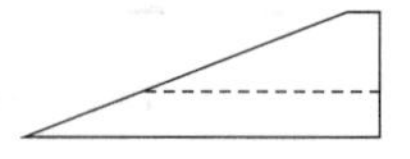

- Another device that is sometimes used is to fold the edge of the wings upward.
- You may also try to add trim tabs to your aeroplanes — small upward folds at the back of the wing.

It is important that there be a variety of designs. You should produce 10 or more designs whose width varies from about 5 cm to about 20 cm, so if there is a relationship between width and distance flown you will be able to detect it. The variety of designs will tend to obscure any relationship, but if the relationship does show through the variety of designs you will have shown that it is a very 'robust' effect. Be sure to identify your aeroplanes with a code to avoid confusion.

It would be a good idea to check that each of the aeroplanes actually 'flies'. A definition of 'fly' that will do for this project is that the aeroplane moves forward — an unfolded piece of paper wafting to the ground would be declared not to have 'flown'; nor would a compressed ball of paper or similar deviations from the normal paper aeroplane!

Using a standard launch procedure, launch each of the models several times until its average distance flown is well established. The definition of distance flown is to be the same as was used in the paper aeroplane project of chapter 2. Make a note of any anomalies in the flying behaviour, recording the aeroplane identification and the nature of the anomaly.

Now construct a table, and in each row record the identification code, the width and the average flight distance for one of your models. Answer the following questions based on your data:

(a) Is there any apparent dependence of flight distance on the width of the paper aeroplane? (Draw a scatterplot to examine this first, and then use any summary techniques you think will be helpful.) If there seems to be a dependence, does the relationship appear linear over the entire range of widths you used? If not, did you notice any anomalies of the flying behaviour that are related to width?

(b) An important factor in the flying behaviour of paper aeroplanes is the position of the centre of gravity. The nearer it is to the front of the aeroplane, the faster the aeroplane will head down, but if it is too far back, the aeroplane will not 'fly'. By balancing each aeroplane on a pencil or the edge of a ruler, determine the position of the centre of gravity. Express this position as a distance from the front, expressed as a proportion of the total length. Call this the 'balance index'. Is the flying distance related to this balance index? You should judge this by first constructing a scatterplot, and then using any other summary techniques you think would be helpful.

(c) Report your findings very briefly. For each part above, the scatterplot, some numerical summary, and your verbal conclusion would suffice. *Note:* Keep your data set. It can be used again in connection with other projects in later chapters.

2. **Weather data**

Among the world's major cities, average annual precipitation varies from 1.1 inches per year in Cairo to 95 inches per year in Singapore. Similarly, the range of temperatures (over a 30 year period) varies from only 31°F in Singapore (66°F to 97°F) to 133°F in Montreal (–35°F to 97°F). Is there a relationship between low precipitation and high temperature range? Here are the data from a few cities:

City	Max. temp (°F)	Min. temp (°F)	Range	Precipitation (inches)*
Auckland, New Zealand	33	90	57	49.1
Bombay, India	46	110	64	71.2
Capetown, South Africa	28	103	75	20.0
Dublin, Ireland	8	86	78	29.7
Jerusalem, Israel	26	107	81	19.7
Madrid, Spain	14	102	86	16.5
Nairobi, Kenya	41	87	46	37.7
Rome, Italy	20	104	84	29.5
Singapore	66	97	31	95.0
Toronto, Canada	–26	105	131	32.2

* 1 inch = 2.54 cm. 32°F = 0°C.

Source: R. Famighetti (ed.), op. cit., 1995, p. 184. Every fifth city in the list was selected.

(a) Examine the relationships between precipitation and the minimum temperatures, between precipitation and the maximum temperatures, and between the precipitation and the range of temperatures. Comment on the nature of these relationships.

(b) From the above data, and using least squares lines, estimate the average annual precipitation in Montreal and the range of temperatures in Cairo.

(c) Compare your estimates in (b) with the actual values: 40.8 inches for Montreal and 83°F for Cairo. Are your estimation errors within the range you expected, based on the tabulated data? Explain.

(d) Record the minimum and maximum daily temperatures for your city during a two week period. Also record the amount of precipitation for each 24 hour period. These statistics are usually available from your local daily newspaper. Is there a relationship similar in nature to the one you found from the international data?

3. **Lifespan and metabolic rate**

Refer to the mammal lifespan data in section 4.1 and the scatter diagram for these data in the subsection 'Clusters of Points' of section 4.2. There does seem to be a relationship between the lifespan and metabolic rate, but it clearly is not linear. In this project, you will investigate the form of the relationship between the two variables.

(a) Is it possible to transform the data so that the scatter is approximately linear? If so, compute the correlation coefficient between the two transformed variables. (*Hint*: Compute the product of the variables — is it approximately constant? How can the relationship $x \times y$ = constant be rewritten as a linear relationship between a transformation of x and a transformation of y?)

(b) Use the following approach to drawing a smooth curve through the points in the original scatterplot. Sort the mammals into increasing order of their O_2 intake. Then use the techniques from chapter 3 to smooth the reordered column of lifespans as though they were a time series. Plot these 'smoothed' lifespans against O_2 intake.

(c) Discuss the differences between the two approaches described in parts (a) and (c) above.

4. **Iris species**

Consider the iris data from section 4.5. Describe the relationship between petal length and sepal length for each of the three iris species separately, and also for the combined data set. Discuss what this tells you about the dangers of ignoring a grouping variable, such as the iris species.

5. **Vehicle production**

The data set on page 165 describes motor vehicle production by country in 1993. Study the relationship between the number of commercial and passenger vehicles. Summarise your findings.

Country	Passenger vehicles manufactured	Commercial vehicles manufactured
Argentina	286 964	55 386
Australia	348 509	131 272
Austria	40 777	3 942
Belgium	347 427	56 528
Brazil	1 102 119	288 142
Canada	1 349 081	888 652
China	221 390	1 088 610
CIS*	1 207 500	599 500
Czech Rep./Slovakia	210 100	23 950
France	2 836 280	319 437
Germany	3 753 341	237 309
Hungary	10 600	5 110
India	199 571	172 059
Italy	1 117 009	150 186
Japan	8 497 094	2 730 451
South Korea	1 592 669	457 389
Malaysia	115 000	0
Mexico	835 079	245 065
Netherlands	80 246	18 946
Poland	266 000	38 000
Spain	1 505 949	261 691
Sweden	279 002	58 384
Taiwan	277 900	105 000
United Kingdom	1 375 524	193 410
United States	5 981 048	4 883 157
Yugoslavia	7 359	954

*CIS is the Commonwealth of Independent States — essentially a selection of States from the former USSR.

Source: R. Famighetti (ed.), op. cit., 1995, p. 208.

CHAPTER 5

Exploring categorical data

There is no excellent beauty that hath not some strangeness in the proportion.
Francis Bacon (1561–1628)

5.1 CATEGORICAL DATA

Most of the data sets used to illustrate the ideas of chapters 2–4 consisted of batches of numbers. We described various graphical and numerical summaries of such batches that highlight important characteristics of the data. For example, the age of each employee of a large corporation might be recorded, giving a batch of numerical data

36, 23, 51, 54, 29, 26, ...

which could be explored through the use of stem-and-leaf plots, histograms, box plots or other graphical displays suggested in chapter 2.

In this chapter, we will explore data sets in which the individual values are not numbers but categories. If, for example, we record the religion of each employee instead of age, we get a batch of categorical data of the form:

Christian, Muslim, Christian, Christian, Christian, Hindu, Atheist, ...

Note that some of the categories will usually appear more than once in the batch. (Different employees may have the same religion.) There is usually a small number of possible categories that are decided on before data collection begins.[1]

EXERCISE 5.1

1. A beachcomber wants to record some information about her collection of items picked up from the beach. Suggest three categorical variables and three numerical variables that she could use to describe the objects. Among the collection are stones, shells, driftwood, dried seaweed and a miscellany of manufactured things.
2. A questionnaire on religious affiliation includes the question: 'With which major religious group are you affiliated?' The respondent has a number of options to choose from:
   (a) Christian
   (b) Muslim
   (c) Hindu
   (d) Buddhist
   (e) other
   (f) none.
   The responses are coded numerically (Christian = 1, Muslim = 2, Hindu = 3, Buddhist = 4, other = 5, none = 6) before the data are analysed by computer. An analyst extracts from a computer output the fact that the average of the religion codes is 1.5 and the standard deviation is 0.3. Is this useful information?

---

1. A catch-all category, 'Other', is often added in case some category has been inadvertently omitted.

5.2 DISPLAYING CATEGORICAL DATA

Different methods are needed to display categorical data than those described in chapter 2 for numerical data. The main aim is to display and compare the frequencies of the distinct categories in the batch. The basic summary of the batch of values is therefore a frequency table containing the number of times each category occurs. For example, the following table summarises the causes of bushfires in parks in the State of Victoria in 1989–90.[2]

Cause	Number of fires (frequency)	Relative frequency (proportion)
Deliberately lit	26	0.382
Lightning	22	0.324
Burn-off escape	5	0.074
Other	13	0.191
Unknown	2	0.029
Total	**68**	**1.000**

There are two common displays that are used to show such frequencies and relative frequencies: pie charts and bar charts.

A *pie chart* of the data partitions a circle into segments, each of which represents a category and has an area proportional to the frequency of the corresponding category. Conventionally, the categories are arranged clockwise, starting at '12 o'clock'.

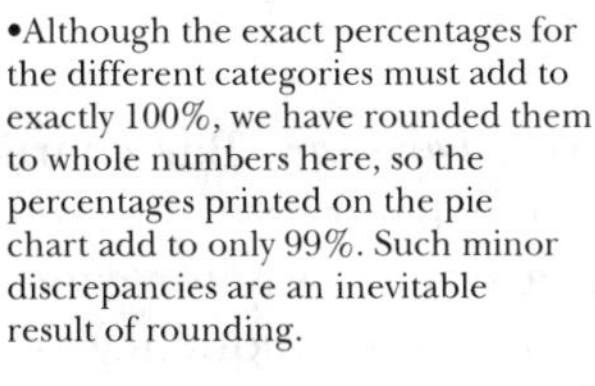
•Although the exact percentages for the different categories must add to exactly 100%, we have rounded them to whole numbers here, so the percentages printed on the pie chart add to only 99%. Such minor discrepancies are an inevitable result of rounding.

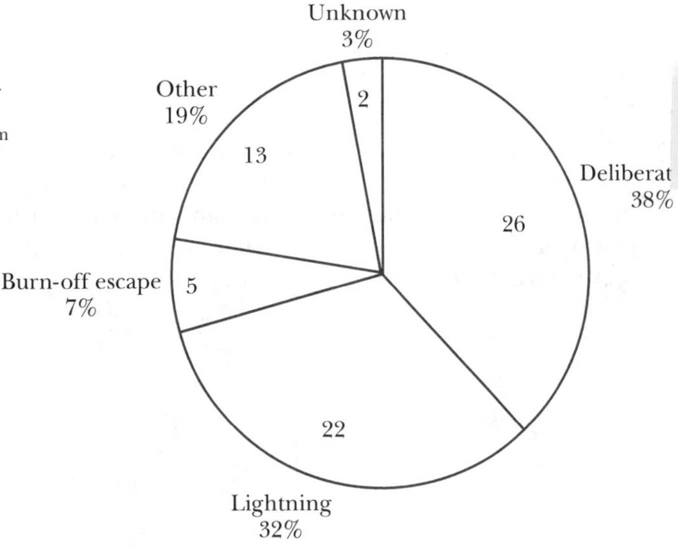

2. Australian Bureau of Statistics, *Striking a Balance*, AGPS, Canberra, 1992. (Commonwealth of Australia copyright reproduced by permission.) Note that, as with most other summaries of data, some information is lost by summarising the raw data in a frequency table. In this case the time order of the causes of the 68 fires has been lost.

Anyone carrying out exploratory data analysis should normally use a computer to draw pie charts (and to produce most of the other graphical and numerical summaries described in this book) but, if it must be drawn by hand, a table like the following one can be used to systematically perform the calculations.

Cause	Frequency	Cumulative frequency	Angle (°) Cum. freq. $\times \frac{360}{68}$
Deliberately lit	26	26	137.6°
Lightning	22	48 (= 22 + 26)	254.1°
Burn-off escape	5	53 (= 48 + 5)	280.6°
Other	13	66 (= 53 + 13)	349.4°
Unknown	2	68 (= 66 + 2)	360°
Total	**68**		

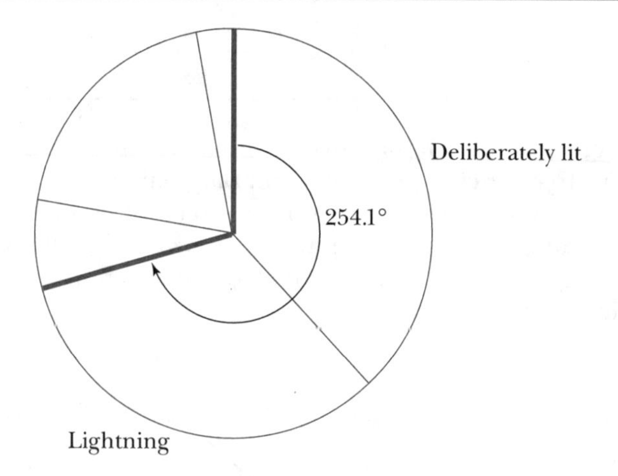

The second common display for categorical data is a *bar chart.* Bar charts have already been described in section 2.5 for displaying batches of discrete numerical values. They are drawn and interpreted in a similar way for categorical data, the height of each bar being proportional to the frequency (or relative frequency) for that category.

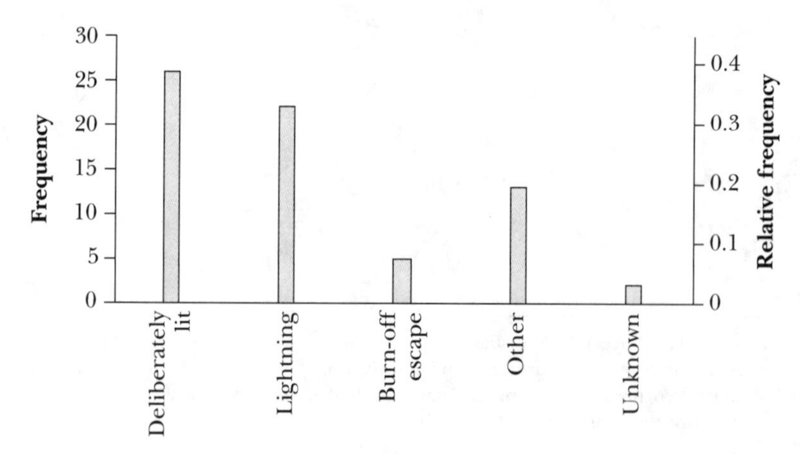

The vertical axis on the bar chart may either be labelled with the frequencies of the categories or their relative frequencies or, as on page 170, with both.

Pie charts and bar charts have different relative merits. Pie charts show more clearly the proportions corresponding to the various categories. For example, it is clear that about three-quarters of the fires were caused by lightning, deliberate lighting or burn-off escape, and about one-quarter had other or unknown causes. This is less clear from the bar chart. It is necessary to go beyond visual comparisons and study the relative frequencies shown on the right-hand vertical axis in order to read the same information from a bar chart.

On the other hand, bar charts allow a more precise comparison of the different categories. For example, it is more immediately apparent in the bar chart that more fires were deliberately lit than caused by lightning.

Frequency tables, pie charts and bar charts can be enhanced in a variety of ways by adhering to various conventions and by following principles of data presentation which will be outlined in chapter 6.[3]

EXERCISE 5.2

World numbers for adherents of all religions in 1993 were estimated as follows.[4]

Religious group	Number of adherents ('000)
Christians	1 869 751
Muslims	1 014 372
Non-religious	912 874
Hindus	751 360
Buddhists	334 002
Atheists	241 852
Chinese folk	140 956
New religionists	123 765
Tribal religionists	99 736
Other	87 286
Total	**5 575 954**

(a) Find the relative frequencies of all categories.
(b) Display the table in a pie chart.
(c) Display the table in a bar chart.
(d) Comment on the relative merits of the pie and bar charts.

3. Note that some software programs can also produce misleading variants of pie charts and bar charts! These are also discussed in chapter 6.
4. *Encyclopaedia Britannica Book of the Year,* 1994. Reprinted with permission, Encyclopaedia Britannica Inc, Chicago.

5.3 COMPARING BATCHES OF CATEGORICAL VALUES

As with numerical data, two or more batches of categorical data cannot be efficiently compared with a single-variable summary for each batch. In particular, it is difficult to compare multiple pie charts. However, two or more bar charts can be combined to effectively compare different batches. There are two main variants. One uses frequency and the other uses relative frequency to scale and label the vertical axis.

The following table describes the responses of a group of adolescent girls to a question about whether they would prefer to gain weight, lose weight or maintain their present weight. The responses were split into two batches depending on the race of the respondent.[5]

	Reducers	Maintainers	Gainers	
White	352	152	31	535
African-American	47	28	24	99
	399	180	55	634

In the first of the following two bar charts, the vertical axis is frequency. It is clear from this bar chart that there are far more White adolescents than African-American adolescents in the study. The display also effectively shows the proportions of reducers, maintainers and gainers within each batch, but does not make it easy to compare the two batches.

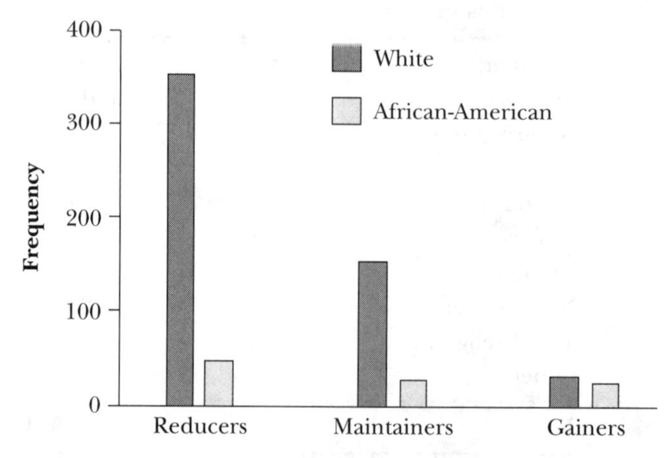

The second bar chart (on the top of page 173) uses the relative frequency within each batch as its vertical scale. These relative frequencies are shown in the table below.

	Reducers	Maintainers	Gainers	
White	0.658	0.284	0.058	1.0
African-American	0.475	0.283	0.242	1.0

5. J. S. Gross, Weight Modification and Eating Disorders in Adolescent Boys and Girls, Doctoral Dissertation, University of Vermont. In D. C. Howell, *Statistical Methods for Psychology*, Duxbury Press, Belmont, CA., 1992, p. 154.

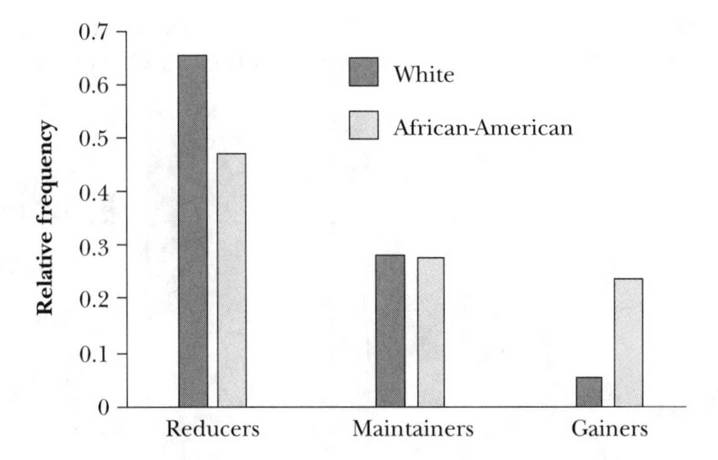

Standardising the bar heights within each batch in this way more effectively allows the types of responses within the batches to be compared. It is much more obvious now that the proportion of African-Americans who would prefer to gain weight is far higher than the corresponding proportion of Whites. This display loses information about the numbers of Whites and African-Americans in the study. Such a loss would usually be regarded as acceptable, since the lost information is of subsidiary importance in the interpretation of these data.

Further variants of the two basic bar charts can be obtained by stacking the bars on top of each other to produce displays that are called *stacked bar charts.*

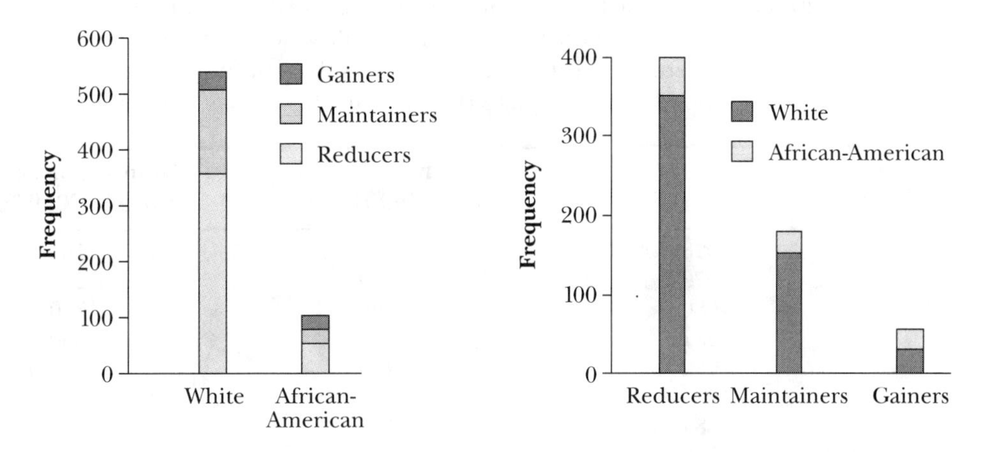

The stacked bar chart on the left, above, highlights the difference in the numbers of Whites and African-Americans in the study, shown by the total heights of the stacks. This information is, however, of little interest here. The stacked bar chart on the right similarly highlights the differences between the total numbers with each attitude to weight loss, a feature of more interest. In both graphs, however, information about the differences in the other categorisation is somewhat obscured.

Stacking the bars in the relative frequency bar chart is only meaningful when the stacking is done by race, as in the bar chart below.

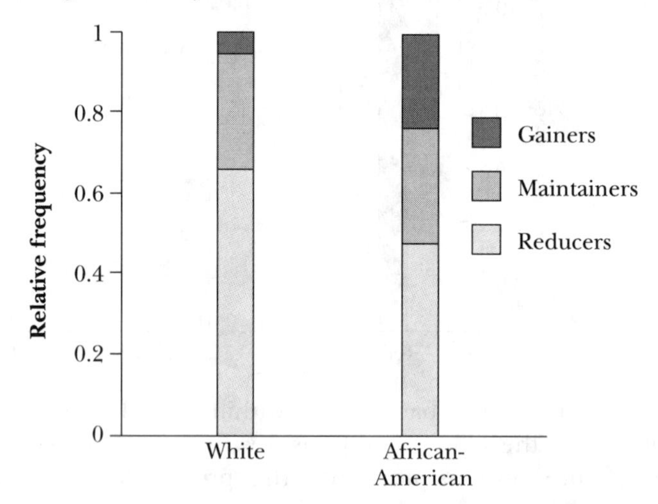

In this example, however, there is little to be gained from stacking the bar charts. Although the standardised stacked bar chart of relative frequencies still effectively displays the difference between the proportions of gainers and reducers in the races, it is now harder to compare the proportions of maintainers because these data are displayed in the middle of each stacked bar.

Stacked bar charts do allow effective comparison when there are large numbers of batches, especially when there are only two (or at most three) categories within each batch. For example, the table below describes the recovery of used aluminium beverage cans in Australia from 1978 to 1989.[6] Note that this table is not in the form of a conventional frequency table since each can is not classified as either sold or returned; returned cans are a subset of those which were sold. The table at the top of the next page is a standard frequency table of the data.

Year	Number of cans sold ('000 000)	Number of cans returned ('000 000)
1978	917	165
1979	1290	297
1980	1196	550
1981	1360	680
1982	1466	733
1983	1393	752
1984	1596	816
1985	1577	820
1986	1827	950
1987	2037	1100
1988	2321	1300
1989	2523	1566

6. Industry Commission, *Recycling, Volume 2: Recycling of Products*, AGPS, Canberra, 1991. Permission also from Comalco Limited.

Year	Number of cans returned ('000 000)	Number of cans unreturned ('000 000)	Number of cans sold ('000 000)
1978	165	752	917
1979	297	993	1290
1980	550	646	1196
1981	680	680	1360
1982	733	733	1466
1983	752	641	1393
1984	816	780	1596
1985	820	757	1577
1986	950	877	1827
1987	1100	937	2037
1988	1300	1021	2321
1989	1566	957	2523

Treating each year's data as a separate batch, these data may be effectively displayed with either of the two types of stacked bar chart, as shown below.

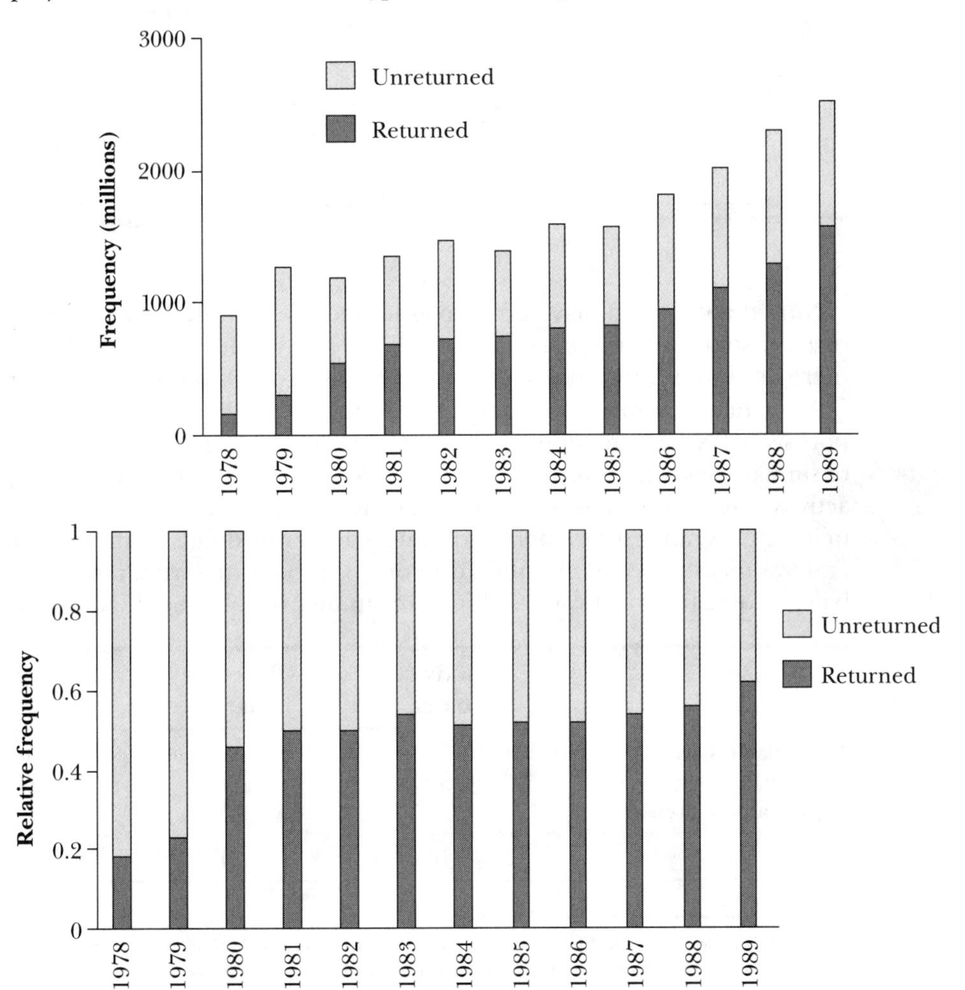

The first of these stacked bar charts shows clearly the trend in both the total number of aluminium cans sold (the total heights of the stacks) and the number recycled (the shaded heights). The standardised version shows more clearly the trends in the proportion of cans returned. An alternative form of the unstandardised display, called a *dot chart,* is shown below.

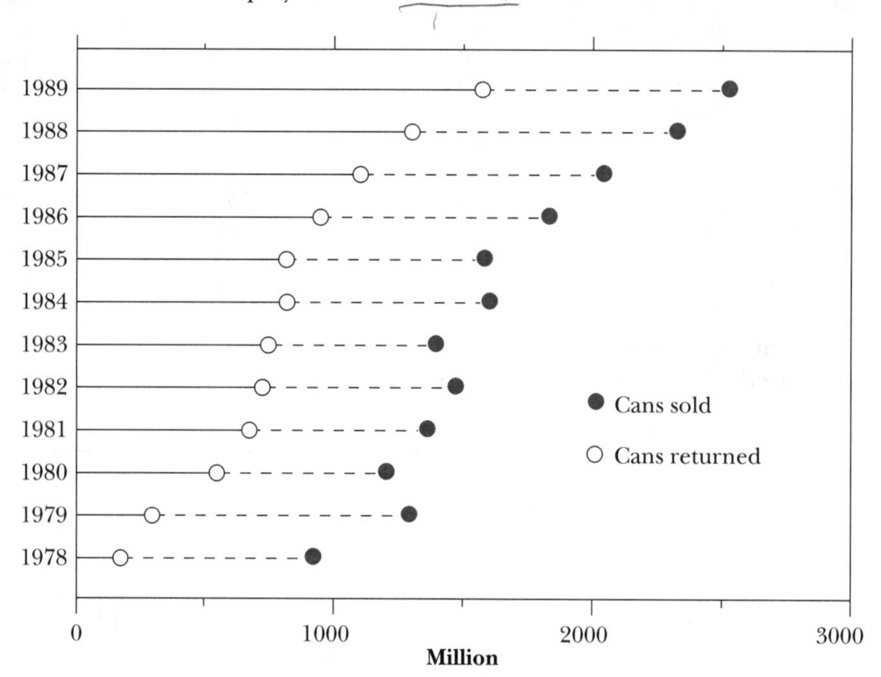

Note that the data may also be considered as two time series of the numbers of cans sold and numbers of cans returned. As illustrated by this example, there are often different (and complementary) ways to represent a data set.

A further example is provided by the following data which relate to osteoporosis (a loss of bone minerals) in the elderly. Three groups of elderly subjects of similar ages were given different programs; the first group was given physical activity, the second was given physical therapy and the third group was given neither program. After a period of time, each individual's change in bone mineral was measured and classified into one of three categories, resulting in three batches of categorical values which are summarised by the table below.[7]

	Activity batch	Therapy batch	No treatment batch
Appreciable loss	15	22	38
Little change	30	32	15
Appreciable increase	25	16	7
Total	**70**	**70**	**60**

7. R. R. Johnston and G. K. Bhattacharyya, *Statistics: Principles and Methods,* John Wiley and Sons, New York, 1987, p. 525. Reprinted by permission of John Wiley and Sons Inc.

The following standardised stacked bar chart highlights the differences between the effects of the three treatments.

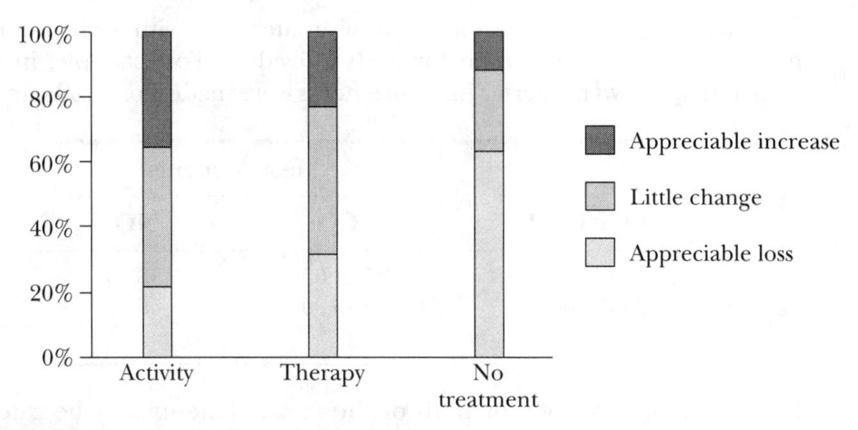

EXERCISE 5.3

1. The table below presents information about a sample of households in Great Britain in 1975–76 who had moved during the preceding 12 months and who had kept the same head of household during this period.[8]

Nature of employment of head of household	Main reason for moving				
	Housing	*Environment*	*Job*	*Personal*	*Other*
Manual (skilled and unskilled)	221	83	99	86	118
Nonmanual (professional, managerial, etc.)	133	66	78	49	111

   (a) Draw an unstacked bar chart of the counts to allow comparison of the two groups of workers. Draw a similar unstacked bar chart of the relative frequencies with the various reasons for moving. Which bar chart do you find more useful and what does it tell you?

   (b) Draw an appropriate stacked bar chart of the relative frequencies. Is it more or less useful than the bar charts you drew in question 1?

2. Compare the dot chart of the aluminium can returns in this section with the bar chart preceding it. Which chart would be better for this particular data set?
3. The dot plot relating to the recycling of aluminium cans clearly shows trends in both the total number of cans sold and the number of recycled cans. Suggest a simple modification of the dot plot that would highlight the trend in the number of unreturned cans. Draw it and describe the trend.

8. *General Household Survey*, Office of Population Censuses and Surveys, HMSO, London, 1976.

5.4 Bivariate categorical data

In chapter 4, we saw examples of bivariate data with observations on two numerical variables recorded for each individual. For example, in the aircraft emission data, two numerical measurements were made from each aircraft; that is:

	Measurements	
Individual	**CO**	**NO_x**
B747-400	32.68	48.45
B747-200	126.90	56.89
...	...	...

In other situations, one or both of the measurements may be categorical. In sections 2.6, 2.7 and 5.3, we saw examples of this in which one of the values was categorical and was used to split the data into batches. For example, in section 2.7 the percentage turnout and winning party in each electorate at the 1996 national election in New Zealand were presented.

	Measurements	
Electorate	**Party**	**Turnout (%)**
1	National	87.42
2	Labour	87.95
3	Labour	88.08
...	...	...

Similarly, the raw osteoporosis data in section 5.3 are of the form:

	Measurements	
Individual	**Change in bone mineral**	**Treatment**
1	Appreciable loss	Therapy
2	Little change	Activity
3	Appreciable loss	No treatment
...	...	...

When one measurement is categorical and can be meaningfully considered to split the individuals into batches, we can use the methods described in sections 2.6 and 2.7 (when the other variable is numerical) or in section 5.3 (when the other variable is categorical) to display the data. Using a categorical variable to split the data into batches is appropriate when the categorical variable can be considered to affect the other variable. The relationship is said to be *causal* and we are usually interested in how the explanatory variable affects the response variable, but not in the proportions of individuals in the explanatory variable categories. Our analysis is said to be *conditional* on the values of the explanatory variable.

Sometimes, however, neither of the variables can be naturally considered to split the individuals into batches. For example, the following table describes the heights of 205 married couples, each couple being classified according to the heights of both spouses.[9,10] Tabulated count data of this form are often called a *contingency table.*

		Wife			
		Tall	**Medium**	**Short**	**Total**
	Short	12	25	9	46
Husband	**Medium**	20	51	28	99
	Tall	18	28	14	60
	Total	50	104	51	205

These data are in a similar format to the osteoporosis data at the start of the previous section. However, neither of the classifications (by height of husband or by height of wife) naturally splits the data into batches; the two classifications have equal status.

The standard bar chart of batches of single categorical measurements can be extended to a three-dimensional bar chart of such bivariate data (see below), but they are very hard to read. They also disobey some of the principles of good graphical display that will be covered in chapter 6.

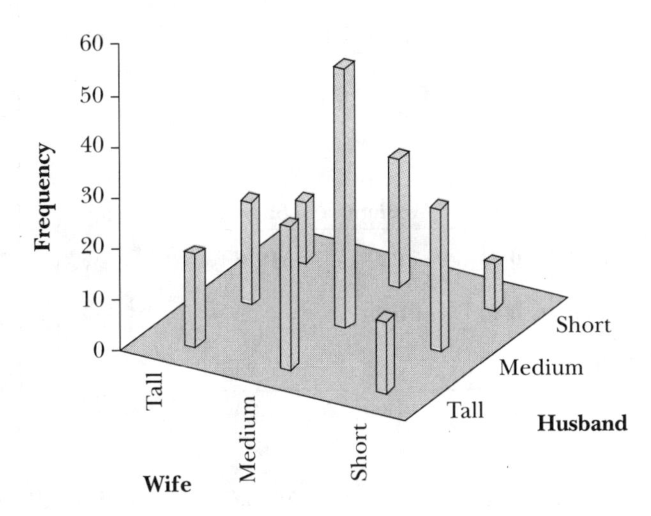

9. S. E. Fienberg, *The Analysis of Cross-classified Categorical Data*, MIT Press, Cambridge, MA, 1977, p. 26.
10. Ungrouped data would have allowed the use of methods discussed in chapter 3 to analyse the relationship between heights of husbands and wives.

Although neither of the two variables naturally splits the data into batches, insight may be obtained with standardised stacked bar charts.

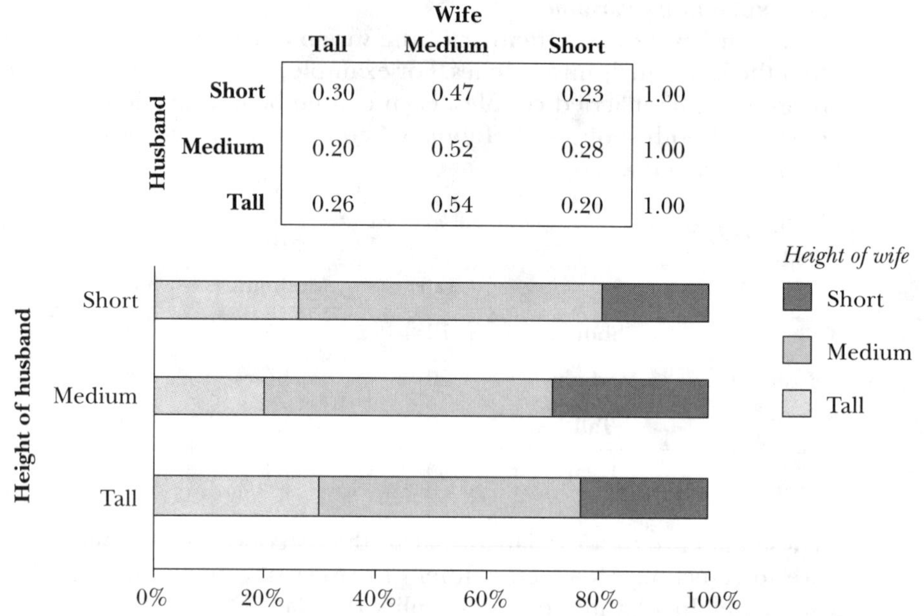

Husband	Wife: Tall	Wife: Medium	Wife: Short	
Short	0.30	0.47	0.23	1.00
Medium	0.20	0.52	0.28	1.00
Tall	0.26	0.54	0.20	1.00

The stacked bar chart above uses the height of the husbands to split the data into batches. The complementary bar chart below uses the height of the wives to form three batches.

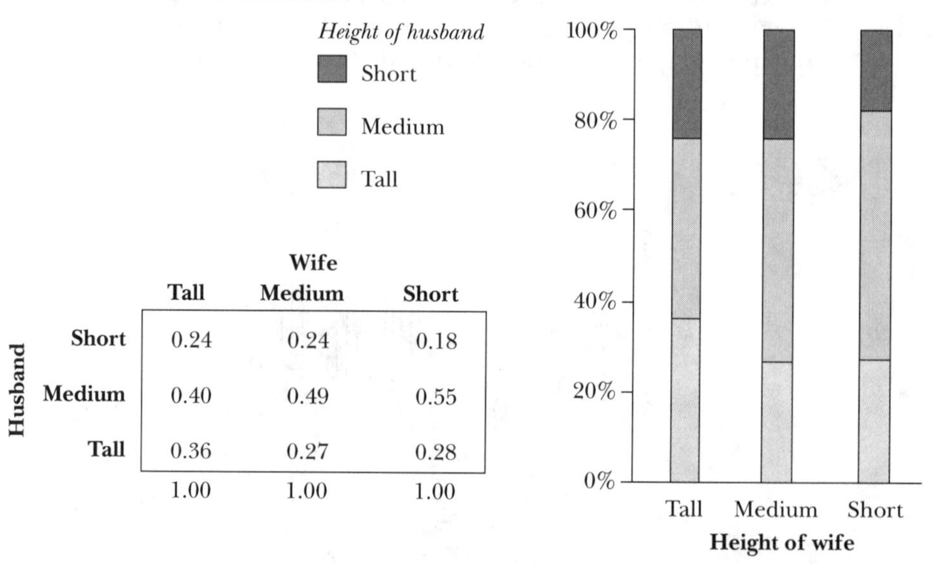

Husband	Wife: Tall	Wife: Medium	Wife: Short
Short	0.24	0.24	0.18
Medium	0.40	0.49	0.55
Tall	0.36	0.27	0.28
	1.00	1.00	1.00

When one categorical variable can be considered to affect the other (a causal relationship), it is also possible to analyse the data conditionally on the value of one variable. This is equivalent to treating the 'explanatory' variable as splitting the data into batches and using the methods of the previous section to display the data.

The final type of data that will be considered in this section arises when a numerical and a categorical measurement are made from each individual in a study, and the relationship is causal with the numerical variable being the explanatory variable. For example, a study was conducted in Warsaw to determine the proportions of girls who had started menstruating at different ages. A total of 3898 girls of various ages between 8 and 19 were asked whether they had started menstruating.[11] Age (the numerical explanatory variable) is known to affect whether the girls have started menstruating (the categorical response variable). The raw data were of the form:

	Measurements	
Girl	**Age (to nearest month)**	**Menstruating?**
1	12 yrs, 9 months	No
2	14 yrs, 2 months	Yes
3	16 yrs, 0 months	Yes
4	11 yrs, 8 months	No
...	...	...

The published data grouped the girls into age groups (usually of 3 months). We will treat the age of each girl in an age group as being the exact middle age of that group. The data are summarised in the following frequency table.

Age class (to nearest month)	Middle age, x	Number menstruating	Number not menstruating	Total	Proportion menstruating
8 yrs, 6 mths–9 yrs, 11 mths	9.21	0	376	376	0.0000
9 yrs, 12 mths–10 yrs, 5 mths	10.21	0	200	200	0.0000
10 yrs, 6 mths–10 yrs, 8 mths	10.58	0	93	93	0.0000
10 yrs, 9 mths–10 yrs, 11 mths	10.83	2	118	120	0.0167
10 yrs, 12 mths–11 yrs, 2 mths	11.08	2	88	90	0.0222
11 yrs, 3 mths–11 yrs, 5 mths	11.33	5	63	68	0.0735
11 yrs, 6 mths–11 yrs, 8 mths	11.58	10	95	105	0.0952
11 yrs, 9 mths–11 yrs, 11 mths	11.83	17	94	111	0.1532
11 yrs, 12 mths–12 yrs, 2 mths	12.08	16	84	100	0.1600
12 yrs, 3 mths–12 yrs, 5 mths	12.33	29	64	93	0.3118
12 yrs, 6 mths–12 yrs, 8 mths	12.58	39	61	100	0.3900
12 yrs, 9 mths–12 yrs, 11 mths	12.83	51	57	108	0.4722
12 yrs, 12 mths–13 yrs, 2 mths	13.08	47	52	99	0.4747
13 yrs, 3 mths–13 yrs, 5 mths	13.33	67	39	106	0.6321
13 yrs, 6 mths–13 yrs, 8 mths	13.58	81	24	105	0.7714
13 yrs, 9 mths–13 yrs, 11 mths	13.83	88	29	117	0.7521
13 yrs, 12 mths–14 yrs, 2 mths	14.08	79	19	98	0.8061
14 yrs, 3 mths–14 yrs, 5 mths	14.33	90	7	97	0.9278
14 yrs, 6 mths–14 yrs, 8 mths	14.58	113	7	120	0.9417
14 yrs, 9 mths–14 yrs, 11 mths	14.83	95	7	102	0.9314
14 yrs, 12 mths–15 yrs, 2 mths	15.08	117	5	122	0.9590
15 yrs, 3 mths–15 yrs, 5 mths	15.33	107	4	111	0.9640
15 yrs, 6 mths–15 yrs, 8 mths	15.58	92	2	94	0.9787
15 yrs, 9 mths–15 yrs, 11 mths	15.83	112	2	114	0.9825
15 yrs, 12 mths–19 yrs, 3 mths	17.58	1049	0	1049	1.0000

11. J. K. Lindsey, *Modelling Frequency and Count Data*, Oxford University Press, New York, 1995, p. 85.

Using the categorical variable to split the data into batches is not helpful since menstruation is the response variable in the experiment.[12] We initially use the 25 different age groups to split the data into 25 batches and draw a standardised stacked bar chart of the response for each of the 25 batches as described in section 5.3. Each stacked bar chart has been drawn on the diagram below at the centre of the corresponding age group. The heights of the dark bars show the proportions of girls menstruating at each age.

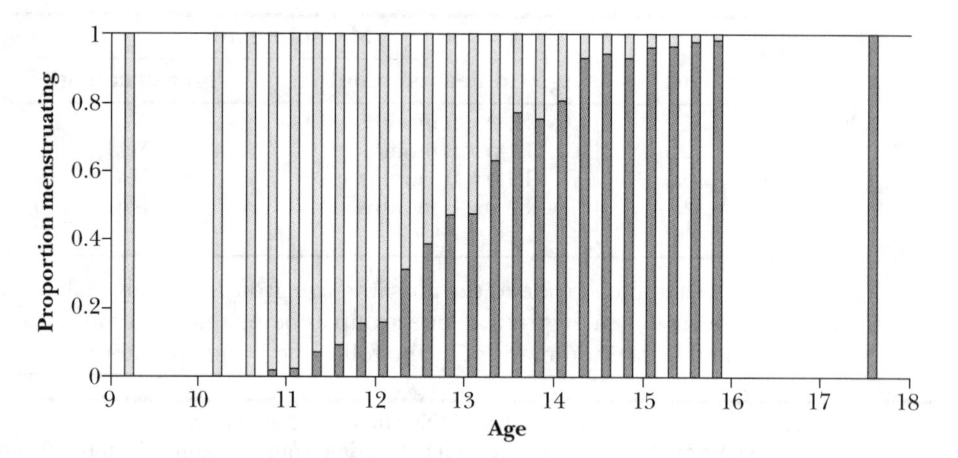

We may extend the exploratory analysis of data such as these by seeking to describe the relationship between dose and response by an appropriate line that may be superimposed on the diagram to summarise the relationship. Such a line should be close to the observed proportions of girls menstruating at each age but should smooth the data. The line will also allow us to estimate the proportions menstruating at other ages, both within the range of ages used in the experiment (interpolation) and at more extreme ages (extrapolation).[13]

---

12. The diagram below shows histograms of the ages of the girls, split into groups by whether or not the girls are menstruating. The histograms show the age distribution of the girls used for the study more clearly than the proportions menstruating at each age.

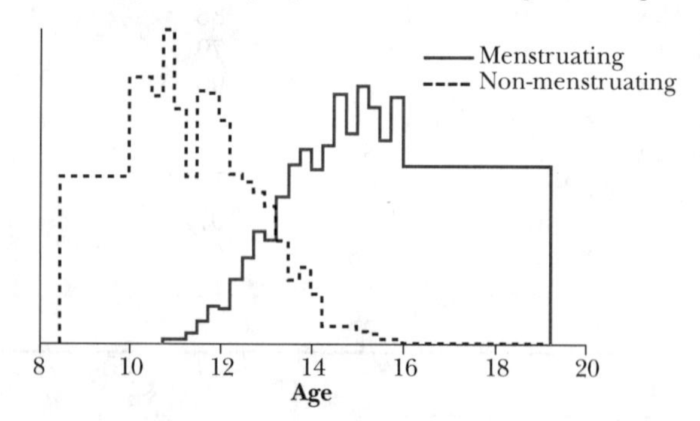

13. Extrapolation should, however, be done with great care, as cautioned in chapter 4.

It would be a simple matter to fit a straight line by eye to these data. For example, the line shown below might be seen as a reasonable attempt at predicting the 'smooth' relationship.

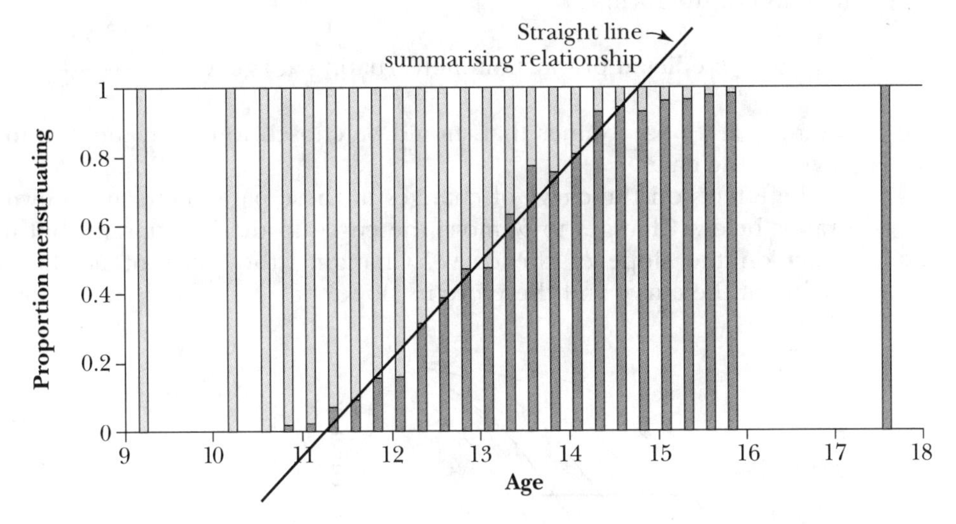

This straight line allows reasonable predictions of the proportion of girls menstruating between ages 11.5 and 14.5. From the slope of the line, we can state that approximately 25% of girls start menstruating each year.

However, for ages greater than 15 or less than 11, this straight line would suggest that more than 100% or less than 0% menstruate respectively. Both of these outcomes are clearly impossible. Modelling the relationship between the proportion menstruating and age with a straight line is clearly not a sensible representation of the relationship over the full range of ages in the study.

For this reason, it is necessary to choose a line which is not straight. Various forms of curves could be used, but statisticians generally prefer a sigmoidal curve such as that in the graph below.

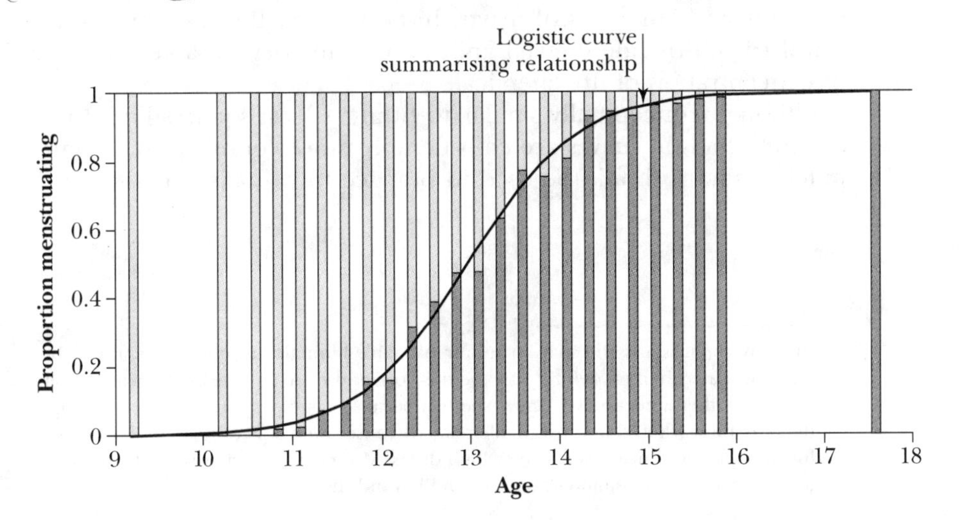

A sigmoidal curve is usually chosen so that it tends towards 0 as the explanatory variable (age) decreases and approaches 1 at very high ages.[14] A commonly chosen mathematical form for a curve with these properties is a logistic function of the form:

$$\text{Predicted proportion menstruating at age } x = \frac{e^{b_0 + b_1 x}}{1 + e^{b_0 + b_1 x}}$$

where b_0 and b_1 are values that should be chosen to obtain the best fit of the curve to the data.

The effects on the curve of changes in these parameters are shown in the graphs below. Changes to b_0 move the curve to the left or right but have no effect on the slope of the curve. Changing the value of b_1 changes the scaling of the curve and therefore its slope.

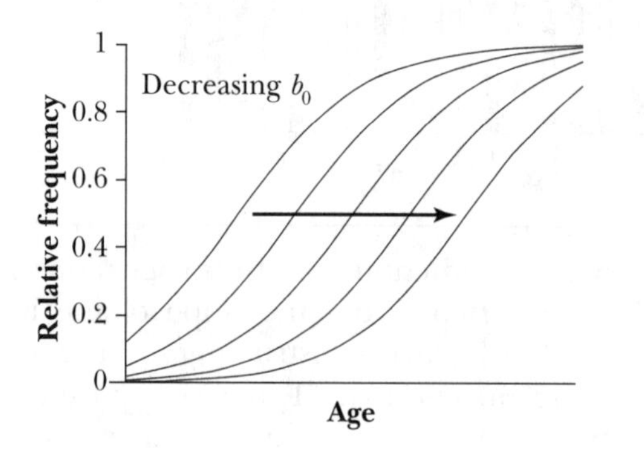

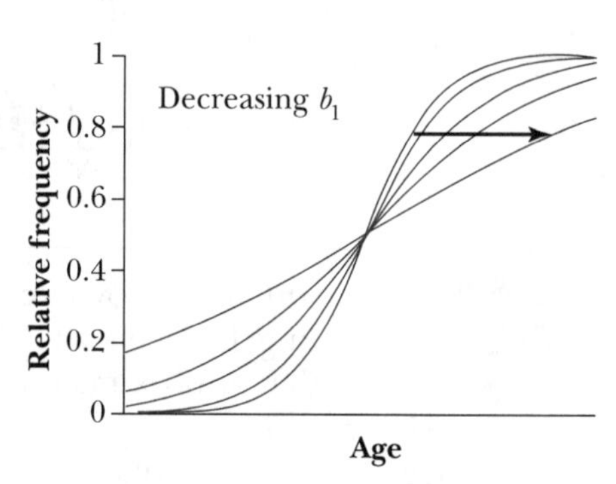

In this respect, the logistic parameters b_0 and b_1 are similar to the intercept and slope of the straight lines that were fitted by least squares in chapter 4; changing the least squares estimate b_1 affects the slope of the least squares line, whereas changing the intercept b_0 shifts the line.

A very important application of models of this kind is in the testing of chemical products for toxicity. This may either be tests to determine the kill rates of pesticides at different doses or tests to examine the incidence of side effects from high doses of drugs. In both cases, the response would be categorical (death or side effects) and the explanatory variable would be the dose or concentration of the chemical.

Although conceptually similar to fitting a least squares line, the mechanics of fitting logistic curves are considerably more complicated. A statistical computer program should be used in practice to fit logistic curves to this type of data.

14. In many applications, this may be an unreasonable limitation. For example, it may be inappropriate when modelling how the dose of an insecticide (the explanatory variable) affects whether a beetle dies (the response) because some proportion of beetles might die of natural (that is, not insecticide-related) causes even at zero dose, and some might live (due to resistance) even at very high doses. More complex curves can be fitted to allow for these possibilities, but they are beyond the scope of this book.

EXERCISE 5.4

1. The following data relate to educational attainment of Americans in 1988. Values are millions of Americans in the various categories.[15]

	Age group					
Education	**25–34**	**35–44**	**45–54**	**55–64**	**≥65**	**Total**
Did not complete high school	5.9	4.8	5.2	7.0	13.2	36.1
Completed high school	17.9	13.2	9.9	8.6	9.4	58.9
College, 1–3 years	9.1	7.3	3.7	2.8	2.9	25.8
College, ≥ 4 years	10.2	9.3	5.0	3.2	3.0	30.8
Total	43.0	34.7	23.8	21.6	28.5	151.6

   (a) How would percentages be computed that would compare the mix of educational attainment across the age groups?
   (b) The total population is less than the known population in 1988 of approximately 225 000 000, so there are about 75 000 000 people missing from this table. Why is this?
   (c) Assuming that the mix of education in the age groups reflects the educational patterns in the United States at corresponding times in the past, has the educational pattern changed over the last half a century?
   (d) Is the group attaining only high school or less education older than the group with at least some college education?

2. Refer to the stacked bar chart for the menstruation age data. If possible, read off the age at which 90% of the girls in the study have begun to menstruate. If it is not possible, use the graph to guess the answer for the 3898 girls in the study.

5.5 CAUSAL ASSOCIATION IN CATEGORICAL RELATIONSHIPS

The table on page 186 describes the relationship between socioeconomic status and smoking habits in a group of volunteer male US federal employees. Each individual was classified into one of three categories of socioeconomic status and was also asked whether they were current smokers, former smokers or had never smoked.[16]

---

15. Mark S. Hoffman (ed.), *The World Almanac and Book of Facts 1990*, Pharos Books, New York, 1990, p. 179.
16. D. S. Moore and G. P. McCabe, *Introduction to the Practice of Statistics*, 2nd ed., Freeman, New York, 1989, p. 605. Used with permission.

Smoking	Socioeconomic status			Total
	High	**Middle**	**Low**	
Current	51	22	43	116
Former	92	21	28	141
Never	68	9	22	99
Total	211	52	93	356

We are interested in whether socioeconomic status is related to smoking. The following stacked bar chart describes the relationship.

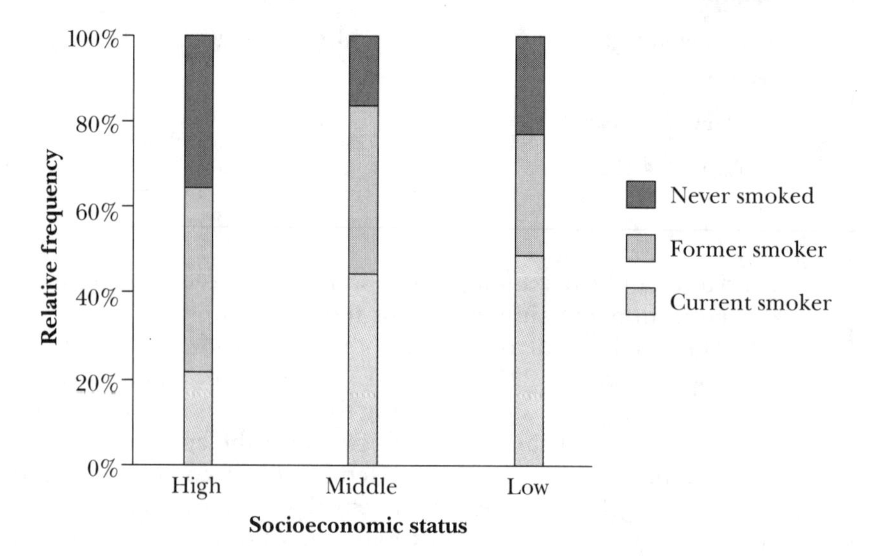

Clearly, there is a larger proportion of current smokers in the middle and low socioeconomic groups than in the high socioeconomic group. A smaller proportion of those in the middle socioeconomic group have never smoked than in either the high or low socioeconomic groups.

We should, however, be extremely cautious about reporting such findings because some of the differences are small and the sample size that the table is based on is also fairly modest. In chapter 9 we will describe ways to assess whether such differences are worth reporting.

Although there seems to be a relationship between the variables, is it causal? It is, perhaps, unlikely that the smoking habits of employees would affect their socioeconomic status, but does an employee's socioeconomic status affect that person's smoking habits? It is difficult and possibly dangerous to conclude causation from association (as has also been discussed in chapter 4).

In section 4.8, we discovered that an apparent relationship between two numerical variables may be caused by a third random variable that affects the other two. Such lurking variables may exist with similar effect on categorical measurements.

The following table describes renewals of subscriptions to the magazine *American History Illustrated* in early 1979.[17]

Month	Renewed	Not renewed	Total	Renewal rate
January	23 545	22 410	45 955	0.512
February	5 869	3 288	9 157	0.641

The magazine was pleased to note an increase in the overall renewal rate from 51.2% to 64.1%.

However, there is a third categorical variable that is strongly related to both the month and whether the subscriptions were renewed; the third variable is the type of subscription. The following frequency table breaks the subscriptions down into the subscription categories within each month.

Month	Category	Renewed	Not renewed	Renewal rate
January	Gift	2 918	676	0.812
	Previous renewal	14 488	3 876	0.789
	Direct mail	1 783	1 203	0.597
	Subscription service	4 343	16 519	0.208
	Catalogue agent	13	136	0.087
February	Gift	704	180	0.796
	Previous renewal	3 907	1 233	0.760
	Direct mail	1 134	1 090	0.510
	Subscription service	122	742	0.141
	Catalogue agent	2	43	0.044

For each separate subscription category, the renewal rate actually dropped from January to February! This is an example of an effect often generically referred to as *Simpson's Paradox.* As with other 'paradoxes' there is no real contradiction; it just takes a bit more thought to understand why your initial intuition is wrong.

The subscription category is strongly related to both renewal and the month. Some categories have much higher renewal rates than others and there are different proportions of subscriptions in the different categories in the two months. The main reason for the overall increase in renewal rate is that the proportion of subscriptions in the subscription service category has dropped greatly from January to February. This category has a relatively low renewal rate in both months.

Simpson's Paradox is an extreme case of an effect that happens very often in studies where the investigator has no control over the values of the explanatory variable (*observational* studies). Whenever groups are compared in an

17. C. H. Wagner, 'Simpson's paradox in real life', *The American Statistician*, 36, 1982, p. 47.

observational study in order to assess how a grouping variable affects a response, a lurking variable (such as the subscription category above) may affect the comparison. We can never be sure that some unmeasured lurking variable has not affected the relationship between the two variables being studied. We must combine general knowledge with statistical methods to make wise judgements from comparative, observational studies. As with numerical variables, we must be careful that an observed relationship between two categorical variables is really a causal one and not just a result of their relationships to a further 'lurking' variable.[18]

EXERCISE 5.5

1. From your general knowledge, which of the following pairs of categorical variables would likely have the strongest association? Rank them from strongest to weakest.
   (a) sex and readership of a major women's magazine
   (b) sex and occupation
   (c) occupation and favourite spectator sport
   (d) occupation and favourite colour
2. Of the pairs of variables that are likely to be related in question 1, which associations are attributable to a direct causal link?
3. In the age and education data in question 1 of section 5.4, is the relationship causal for an individual?

18. In chapter 7, we will discuss the effect of lurking variables further, and describe how data may be collected to avoid (or minimise) this problem.

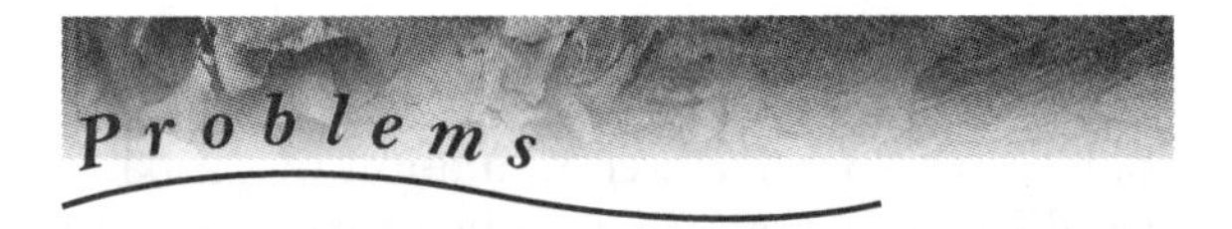

1. The following table lists the numbers of adherents of various religions by geographical area, in 1993 (figures are in millions).

	Africa	Asia	Europe	L. Am.	N. Am.	Oceania	USSR*	World
Christians	341.2	300.4	409.7	443.0	241.1	22.7	111.6	1869.8
Muslims	284.8	668.3	13.6	1.4	3.3	0.1	42.8	1014.3
Non-religious	2.6	721.1	57.5	18.4	24.7	3.6	84.9	912.9
Hindus	1.6	746.5	0.7	0.9	1.3	0.4	0.0	751.4
Buddhists	0.0	332.1	0.3	0.6	0.6	0.0	0.4	334.0
Atheists	0.3	167.2	16.7	3.3	1.3	0.5	52.4	241.8
Other	72.6	356.1	3.4	7.4	9.7	0.3	2.6	451.8
Total	703.1	3291.7	501.9	475.0	282.0	27.6	294.7	5576.0

*Russia and other countries in the former USSR

Source: *Encyclopaedia Britannica Book of the Year*, 1994. Reprinted with permission of Encyclopaedia Britannica Inc., Chicago.

(a) The table above does not make it easy to compare the distribution across religions. Describe which percentages would facilitate this comparison.

(b) In comparing the distribution of religions across geographical regions, what graphical method should you use? (Here is a list of choices: multiple bar chart, stacked bar chart, dot chart, pie chart. Also consider whether the absolute or relative frequencies should be used.)

(c) If your primary interest were comparison of the mix of religions in the USSR with those in other countries, how would this affect your graphical display in (b)?

(d) How would you graphically display the geographical distribution of Christians and Muslims in a way that facilitates the comparison?

(e) What graphical display would be best for displaying the geographical distribution of Christians?

(f) Write a report on what you have learned from your analysis about the geographical spread of religions.

2. An asthma unit attached to a hospital emergency ward routinely records the smoking status (nonsmoker, former smoker, current smoker) of each patient, and also the usual level of exercise undertaken by the patient (active, moderate, sedentary). It also keeps records of the number of visits per year. A researcher wants to summarise the data for the 745 patients who have been on file for at least 12 months. The aim is to examine whether smokers have more asthma visits than nonsmokers, after taking exercise activity into account.

(a) How many variables are involved in this study, based on the above description, and which variables are categorical?
(b) A table showing the number of cases for each combination of smoking status and exercise level would show whether these two variables were associated, but would not tell anything about the association of these two variables with asthma. What tabular display would show how these categorical variables relate to asthma?
(c) How would the display in (b) be portrayed graphically?

3. The following data show religion and government types for countries with a population over 10 million in 1981.

Country	Religion[a]	Government[b]	Country	Religion	Government
Afganistan	Is.	Soc. Rep.	Korea, South	Bu., Ch., Co., others	1-party Rep.
Algeria	Is.	Soc. Rep.	Malaysia	Is.	Parl. Mon.
Argentina	Ch.	Fed. Rep.	Mexico	Ch.	Fed. Rep.
Australia	Ch.	Fed. Parl. Mon.	Morocco	Is.	Mon.
Bangladesh	Is.	Rep.	Nepal	Hi., Bu.	Const. Mon.
Brazil	Ch.	Rep.	Netherlands	Ch.	Parl. Mon.
Burma	Bu.	Soc. Rep.	Nigeria	Is., Ch.	Fed. Rep.
Canada	Ch.	Fed. Parl. Mon.	Pakistan	Is.	Rep.
Chile	Ch.	Rep.	Peru	Ch.	Rep.
China	Co., Bu., Ta., others	Com. Rep.	Philippines	Ch.	Rep.
Colombia	Ch.	Rep.	Poland	Ch.	Soc. Rep.
Czechoslovakia	Ch.	Soc. Rep.	Romania	Ch.	Soc. Rep.
Egypt	Is.	Rep.	Spain	Ch.	Parl. Mon.
France	Ch.	Rep.	Sri Lanka	Bu.	Rep.
Germany	Ch.	Fed. Rep.	Sudan	Is.	1-party Rep.
Ghana	Ch., Is., others	Rep.	Taiwan	Co., Bu., Ta.	Rep.
Hungary	Ch.	Rep.	Thailand	Bu.	Parl. Mon.
India	Hi.	Parl. Rep.	Turkey	Is.	Rep.
Indonesia	Is.	Rep.	Uganda	Ch., Is.	Rep.
Iran	Is.	Isl. Rep.	United Kingdom	Ch.	Parl. Mon.
Iraq	Is.	Soc. Rep.	USA	Ch.	Fed. Rep.
Italy	Ch.	Rep.	Venezuela	Ch.	Rep.
Japan	Bu., Sh.[c]	Parl. Mon.	Vietnam	Bu.	Soc. Rep.
Kenya	Ch., others	1-party Rep.	Yugoslavia	Ch.	Soc. Rep.
Korea, North	Bu., Co., others	Com. Rep.	Zaire	Ch., others	1-party Rep.

a. Majority religion. If relative percentages not known, or if no majority religion, then all major religions are listed. Bu. = Buddhism, Ch. = Christianity, Co. = Confucianism, Hi. = Hinduism, Is. = Islam, Sh. = Shintoism, Ta. = Taoism.

b. Abbrebiations used are: Com. = Communist, Const. = Constitutional, Fed. = Federal, Isl. = Islamic, Mon. = Monarchy, Parl. = Parliamentary, Rep. = Republic, Soc. = Socialist.

c. The majority adhere to both Shintoism and Buddhism.

Source: Adapted from *Reader's Digest Book of Facts*, Reader's Digest Association, London, 1986, pp. 148–165.

(a) Summarise the distribution of the primary religions in these countries.
(b) Summarise the distribution of the government types.
(c) Is there an association between the two? Construct appropriate tables and/or charts to examine the relationship between primary religion and government type.

(*Hint*: You should combine some categories to reduce the compexity of the summaries.)

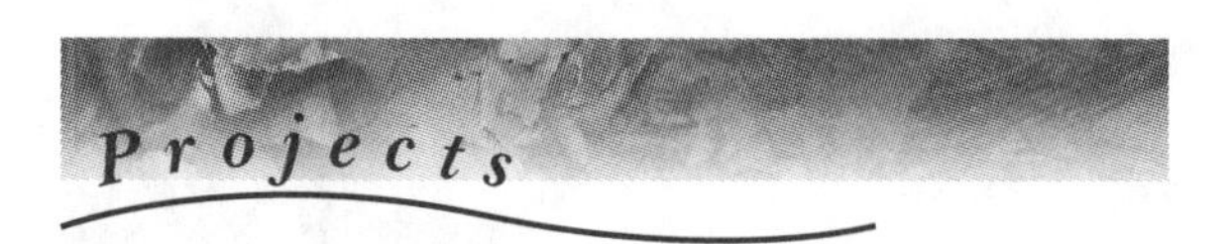

1. **Religions**
   Using computer software, construct the various plots suggested in problem 1, parts (b) to (e). Write a short paragraph summarising the most remarkable features of each graph.

2. **Housing survey**
   For a residential neighbourhood near you, record the following variables from approximately 30 houses:
   (a) type of facing (stucco, vinyl siding, brick, ... or whatever categories are familiar to you)
   (b) quality of landscaping (elaborate, ordinary, minimal)
   (c) presence of a privacy screen (trees, hedge or fence) blocking view of house from street.

   Explore the relationships among these variables, using suitable graphs. Summarise your findings in words.

3. **Car occupancy**
   In this project, you will examine car occupancy and factors that affect it. Firstly, select three locations that you expect to have different driver characteristics. For example, you might select the entrance to a car park at your college, the entry to a supermarket car park and an intersection near the business centre of your town.

   Collect information about the number of occupants and the manufacturer of 50 cars passing at each location. To simplify your analysis of the data, classify the number of occupants as *one*, *two* or *more than two*, and classify their countries of origin into two groups (for example *Japan* and *Other*). Explore whether country of origin or location have any effect on car occupancy.

CHAPTER 6

Principles of data presentation

Truth is stranger than fiction, but not so popular.

Anonymous

6.1 General principles

In earlier chapters, various graphical displays were presented to help us extract useful information from data. Tabular displays, such as tables of means or frequency tables, may similarly summarise important aspects of a data set. Appropriate displays highlight features in the data that are difficult to detect by examining tables of the 'raw data'. This use of graphics and tables is an important and integral part of *data exploration.* We have concentrated on a few general purpose types of display, each of which can be easily and quickly applied to a common data structure. These displays can be used for a wide variety of data.

Data exploration is essentially an introspective activity; the intention is to glean information from the data, rather than to communicate that information to others. Some standard data summaries, such as stem-and-leaf plots and standard deviations, are not well understood by non-statisticians and are therefore not widely used when reporting the features in data.[1]

This chapter discusses issues relating to the use of graphical displays for effective *data presentation,* a stage that follows exploration and analysis of the data. Data presentation is concerned with how the information you have gleaned from the data is conveyed to others, in company reports, newspaper articles, research reports and so on. Although this chapter is mainly concerned with the preparation of data displays for data presentation, the underlying principles are also relevant to the production of graphics during data exploration and should also be kept in mind at that stage of the analysis process.

Many of the types of graphical display that were described in earlier chapters can also be used for data presentation. When presenting such displays to others we need to be especially careful to ensure that they clearly and faithfully display the information that we have extracted from the data. Both the graphics and supporting textual material must be carefully designed for effective communication.[2]

There are three ways to present information about data to others: tabular displays, graphical displays and textual descriptions.

Tabular displays

When data are collected, they are usually recorded in a table. Such tables of raw data are usually of little direct interest to others unless the readers need to further analyse the same data. We usually analyse the data further ourselves and summarise them before presentation.

---

1. However, as the population becomes increasingly statistically literate, especially through changes in the school syllabus, a wider array of methods will become available for communicating information about data to others.
2. In practice, data exploration and data presentation are not conducted strictly in that order. How the data will be presented later is often considered during exploration, and presentation of results often highlights features that merit further exploration.

However, tabular displays are often effective summaries for very simple data sets. For example, the following table describes the percentages of a population in groups *A*, *B* and *C* as effectively and concisely as any graphical display.

Group	Number	Percentage
A	362	46%
B	253	32%
C	170	22%
Total	785	100%

Sometimes, a lot of detailed information needs to be conveyed to the reader. Tabular displays can then be a concise way to present the information. As with other types of data display, care must be taken to present tabular information effectively. Large tabular displays can present a lot of detail, but the 'broad picture' is often obscured by this detail; it is hard to see the wood for the trees. Graphical displays allow the eye to pick up such 'summary information' from data more readily.

Graphical displays

Graphical displays represent features of the data by means of spatial patterns on paper (or on a computer screen). They are especially useful ways to show the features of large data sets since the human eye can readily detect, interpret and retain patterns.

Simple examples of graphical displays are bar charts, pie charts, histograms, box plots and scatterplots. With more complex data, different specialised graphical displays can often be devised to effectively show other aspects of the data. Examples of such graphical displays will be shown later in this chapter. Graphical displays will provide the majority of our illustrations of the principles of effective data presentation in this chapter.

Textual descriptions

Paragraphs of text are rarely adequate descriptions of a set of data on their own. Graphical and tabular displays invariably convey information much more clearly and in a much more immediate and memorable way. However, graphics and tables must be integrated into a report, and there is definitely a place for text to describe the source and background of the data and to summarise the notable features of the display. Text should be integrated into the graphics or table if possible, perhaps with annotations identifying outliers or other features.

The annotations on the 'atomic table' graph on page 196, preferably together with a brief textual commentary, serve well to highlight the periodicity of properties of chemical elements.[3]

3. E. R. Tufte, *The Visual Display of Quantitative Information*, Graphic Press, Cheshire, Connecticut, 1983, p. 105.

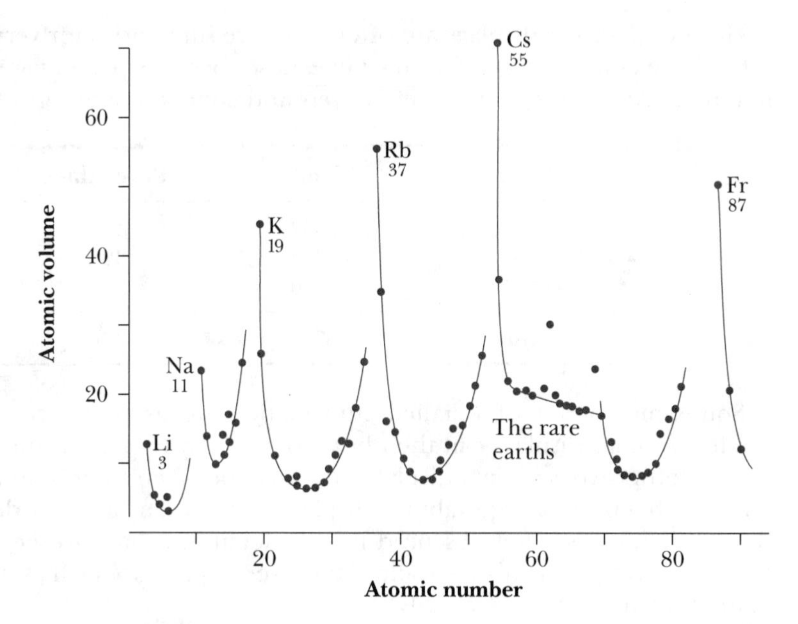

Text should be used to summarise and interpret information in tables and graphs, but not to simply repeat in words information that has already been clearly presented in another form. Such repetition tends to obscure rather than inform.

Excellence

What is it that makes a statistical display of data excellent? There is no better discussion of this than in Tufte's book *The Visual Display of Quantitative Information.*[4] You are encouraged to examine the many examples of the art of data display which are shown in that text. To quote Tufte:

> *Excellence in statistical graphics consists of complex ideas communicated with clarity, precision and efficiency.*

Even simple ideas need to be presented with that same clarity, precision and efficiency. Any statistical graphic should show the data efficiently and truthfully, should not distort the information in the data and should be closely integrated with numerical and verbal descriptions of the data.

Examples of good and bad graphics, taken from Tufte and elsewhere, will be used in this chapter to illustrate the principles of graphic excellence. The same principles that lead us to producing effective graphical displays of data will also be guides to critically assessing the effectiveness and truthfulness of graphical displays that have been produced by others.

4. ibid., p. 13.

EXERCISE 6.1

1. Which of the following displays are appropriate for data presentation?
   (a) stem-and-leaf plot
   (b) pie chart
   (c) bar chart
   (d) box plot
2. In general, when would you use a tabular display in preference to a graphical display?
3. It is an important skill to summarise the aspects of what you have learned from a set of data that will be of interest to your reader. A general reader is not interested in technical details or the steps in your analysis, only in the 'message' that the data contain. In section 5.3, several plots and tables were given of the weight preference data. Write a simple sentence to summarise the relationship between *race* and *weight preference.*
4. Many publications insist that graphical and tabular displays are accompanied by a title and short paragraph describing the notable features of the display. This allows a browser to understand the message in the display without reading the full article. In sections 4.4 and 4.6, the relationship between January temperature and latitude in various US cities was investigated. Annotate the scatterplot below and add a title and brief caption that would allow the reader to understand what you have learned from the data. (The latitude of Jacksonville has been corrected in the display; this error in the raw data should not be commented on.)

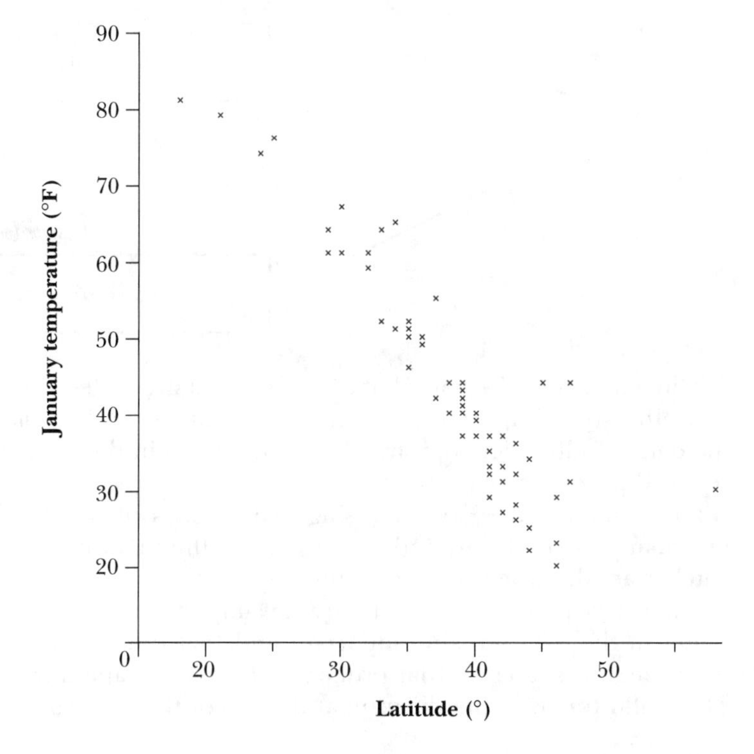

6.2 'SIGNAL' AND 'NOISE'

The aim of data presentation is to convey the information you gleaned from exploratory data analysis clearly to the reader. You should therefore be clear about the message you want to convey before producing graphics for presentation to others, and you should ensure that this message faithfully and unambiguously describes the data.

Many projects and reports are filled with bar charts, pie charts, histograms, scatterplots and a variety of other plots, just because the researcher produced them during an exploratory analysis of the data! It is sometimes useful to ask yourself 'What single display of the data conveys the information most relevant to the message I wish to convey?' If there is more than one thing to convey, then more charts may be needed, but at least this priority approach prevents you from spending too much time on less important details and, at the same time, encourages you to decide what really is important. Graphics, tables and textual description should be included in a report only if they add new and interesting information about the data.

Most displays of data make prominent one or more aspects of the data set. They generally reduce the prominence of other aspects of the data and usually also discard other information that was present in the raw data.

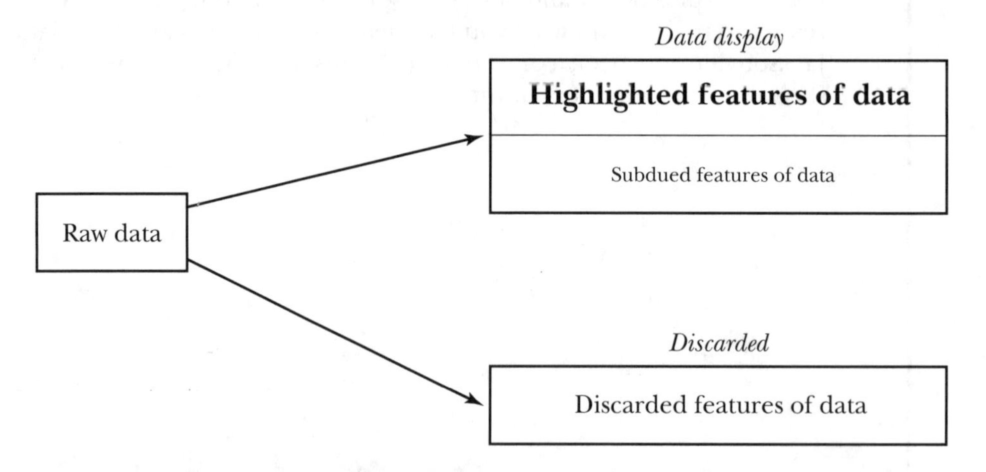

The features of the data that are of interest are called the *signal* in the data and other (less important) aspects of the data are called *noise*. If a display of the data is well designed, it highlights the signal in the data and omits or gives reduced prominence to the noise.

For example, when summarising two batches of numbers with box plots, information is both lost and obscured, but the main differences between the batches are displayed more prominently.

When box plots are used, there is an implicit assumption that the order of values in the data set is unimportant and that the distribution in each batch has a smooth shape, so that order and detailed shape information are noise. This is illustrated in the diagram at the top of the following page.

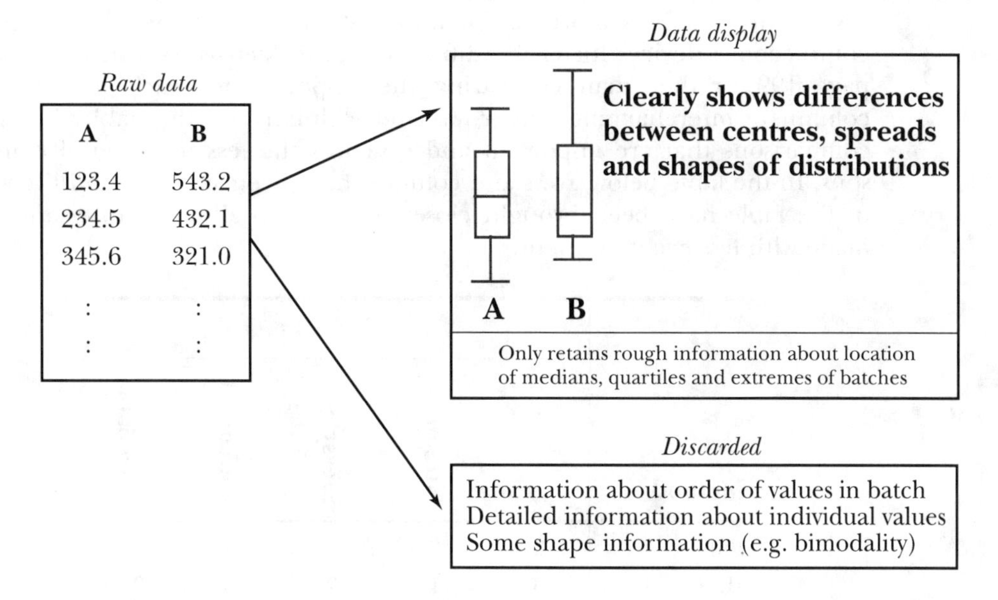

A well-designed data display makes the signal in the data very prominent and effectively removes, or at least minimises, the visual impact of data noise. Different displays which retain the same information may highlight different aspects of the data. Restructuring the information in a display can therefore be used to bring important features into prominence.

Consider the following table which describes attitudes by fecund women in Great Britain in 1970 towards various contraceptive methods.[5] The numbers are percentages of all respondents using particular contraceptive devices who experienced each of five problems with the methods.

	Method				
Problem	**Pill**	**IUD**	**Withdrawal**	**Condom**	**Diaphragm**
Interferes with love-making	2	3	53	44	21
Nuisance to use	3	11	13	39	52
Messy to use	1	7	20	27	50
Unnatural	17	14	42	32	19
Unreliable	4	28	69	31	30

In tables, the easiest comparisons for the eye to make are between adjacent values. Also, it is easier to compare values down columns rather than across

5. M. Murphy, 'The contraceptive pill and women's employment as factors in fertility change in Britain 1963–1980: A challenge to the conventional view', *Population Studies*, 47 (2), 1993, p. 226. London: Population Investigation Committee. Compiled from M. Bone, *The Family Planning Services: Changes and Effects*, London, HMSO (tables 4.6, 4.9 and R4.24). The rows and columns of Murphy's table have been interchanged to allow us to show the effect of such changes to a table.

rows. Since there is more useful information to be gained by comparing contraceptive devices for each attitude category (across rows in the table on page 199), rather than comparing the responses for each device (down columns), interchanging the rows and columns of the table highlights comparisons that are important and obscures the less meaningful comparisons. In the table below, rows and columns have been swapped and the values in the table have been brought closer together to allow comparisons to be made with less eye movement.

	Problem				
Method	**Interferes**	**Nuisance**	**Messy**	**Unnatural**	**Unreliable**
Pill	2	3	1	17	4
IUD	3	11	7	14	28
Withdrawal	53	13	20	42	69
Condom	44	39	27	32	31
Diaphragm	21	52	50	19	30

Scanning down columns shows quickly which contraceptive methods are most likely to cause each of the problems.

We can rearrange the table further to highlight the main features of the data. In tables such as this, where there is no natural ordering of the categories, it is best to order the categories by some overall measure of the size of the columns or the rows since 'similar' categories are brought together, allowing us to see more subtle differences more clearly. Ordering categories alphabetically should be avoided. Ordering the rows and columns of the above table by the row and column totals results in the following table.

	Problem				
Method	**Unreliable**	**Unnatural**	**Interferes**	**Nuisance**	**Messy**
Withdrawal	69	42	53	13	20
Condom	31	32	44	39	27
Diaphragm	30	19	21	52	50
IUD	28	14	3	11	7
Pill	4	17	2	3	1

Scanning down columns of the re-ordered table allows us to detect more readily which methods are 'out of order' for a problem category; this highlights problems that are particularly prominent for any particular method. Even for simple graphical displays like pie charts and bar charts, re-ordering categories in decreasing order of their proportions helps the eye to understand the data better.

A single data display does not need to contain the whole signal in the data set. Rich data structures often require several displays to adequately summarise the information in the data, with each display clearly highlighting a different aspect of the data. Effectively displaying complex data is a great challenge.

In section 6.3, the removal of data noise from displays will be discussed further. Not all noise in a display is data noise; noise in the form of 'artistic enhancements' is sometimes added to a display. Section 6.4 warns against this practice.

Finally, we should be aware of how our audience is likely to interpret our graphics. Some information is usually lost or obscured in a data display, and we must ensure that this lost information is really 'noise' or the display may mislead the reader. From the data display that we provide for the reader, he or she will 'extrapolate' to re-create an idea of what the underlying data must be like.

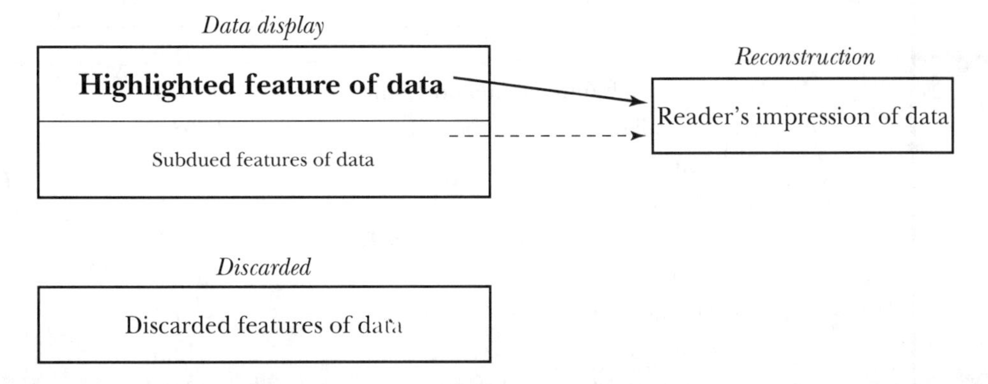

We should ensure that such reconstructions will conform to the actual data. For example, a reader presented with box plots of several batches of data will assume that the underlying data have a smooth unimodal distribution. If the data are bimodal or have outliers, the display would mislead.

There are many published instances of statistical graphics that support the statement (attributed to Disraeli) that there are 'lies, damned lies and statistics'. However, misleading graphics are not always the result of intentional deception. They can also arise from careless design of the data display. In section 6.5, there are several examples of published data displays from which the reader is likely to form an incorrect impression about the underlying data. It is sometimes difficult to assess whether such misleading graphics are a result of naivety or the desire to make a point that is not supported by the data.

EXERCISE 6.2

1. A time series is often smoothed with running means or running medians; this is a useful device for amplifying the signal and suppressing the noise. Will the smoothed series always capture the signal in the time series? Is the residual always noise?
2. A jittered dot plot is often used to portray a batch of numbers. If each data value is exactly represented by the position of its corresponding dot, is any noise suppressed?
3. Refer to the table of data in this section concerning contraceptive methods. The table was rearranged to make it more easily assimilated, by permuting the rows and columns.
   (a) Was any significant information lost in the process?
   (b) What is the general principle, used in the rearrangements just described, for improving a display?
   (c) Explain how this principle can be used to improve a bar chart of a categorical variable.
4. The table below describes the destination of New Zealand exports in 1989.[6]

Country/region	Exports ($million)
Africa	71.5
Asia (excluding Japan and China)	1738.7
Australia	2608.9
Canada	261.2
China	820.9
Eastern Europe (including USSR)	395.7
Japan	2660.3
Middle East	427.1
Oceania	413.8
South and Central America	404.9
United Kingdom	1036.2
United States	2008.1
Western Europe (excluding UK)	1812.4

Reorganise this table to be more informative. Draw a pie chart of the data and discuss.

6. *New Zealand Official 1990 Yearbook*, Department of Statistics, Wellington, 1990, pp. 601–602.

6.3 Removal of data noise

The main aim of statistical graphics is to present information about a set of data, so the most important principle is that the graphic should present this information about the data clearly. The amount of the display that is used to show the intended information (signal) should therefore be large in relation to the amount of the display that shows details that are unimportant in the data (data noise).

An important example is the presentation of too many significant digits in numerical displays. For example, the means of two batches might be reported as 10.4275 and 12.0338. The reader is unlikely to be able to get any useful information from the last three digits in each mean, so these digits are merely data noise. The differences between the batches are more clearly described and more easily digested if the means are reported as 10.4 and 12.0.

Removal of 'noise' digits from numerical values is especially helpful in tabular displays. Consider the following tabular presentation.

39.421	24.617	16.482	14.130
36.092	21.703	13.118	11.734
28.617	19.038	11.819	10.254
25.062	13.705	10.027	14.636

The revised version of the table, shown below, rounds the data values. This retains all the 'signal' information in the data, but removal of the noise digits makes it easier for the eye to make comparisons between adjacent values and therefore to detect the anomalous value in the bottom right of the table.

39	25	16	14
36	22	13	12
29	19	12	10
25	14	10	15

In a set of data, how do we determine which are the information-carrying and which are the noise digits? Leading digits always carry information about the 'centre' of the data and cannot be discarded because this would lead to loss of information about magnitude. However, if they do not vary among the data values, they carry no information about the differences between them. In the data set below, the first two digits of each value do not vary and, hence, carry no useful information about differences between the values.

12.142 12.738 12.419 12.385 12.016

Effective data displays should, as a rule of thumb, present two (or perhaps three) *varying* digits. The varying digits are those that differ between values that the eye is likely to compare in a data display. In some tabular displays, different columns (or rows) may contain values of different magnitudes, so the positions of the varying digits may be different within different columns (or rows). Different degrees of rounding may therefore be appropriate for different parts of the table.

When presenting summary statistics such as means, it is often appropriate to use an extra digit in a display, in addition to the number of digits used for the display of the raw data being summarised. This will be explained and further developed in chapter 8.

A similar kind and greater degree of rounding implicitly occurs in most graphical displays of data. It is usually impossible to read off values from displays such as scatterplots to any more accuracy than two varying digits; the lost information is not missed.

Most of the standard graphical displays that were described in chapters 2 to 5 also remove or obscure additional data noise. You should re-examine these chapters and consider what is lost in the data displays that were described there.

Provided it does not obscure the main signal, as much of the data as possible should be retained in the display. Consider the use of least squares lines to summarise the relationship between a stimulus and a response within two separate groups of individuals. The following display of the two least squares lines describes the fitted relationships.

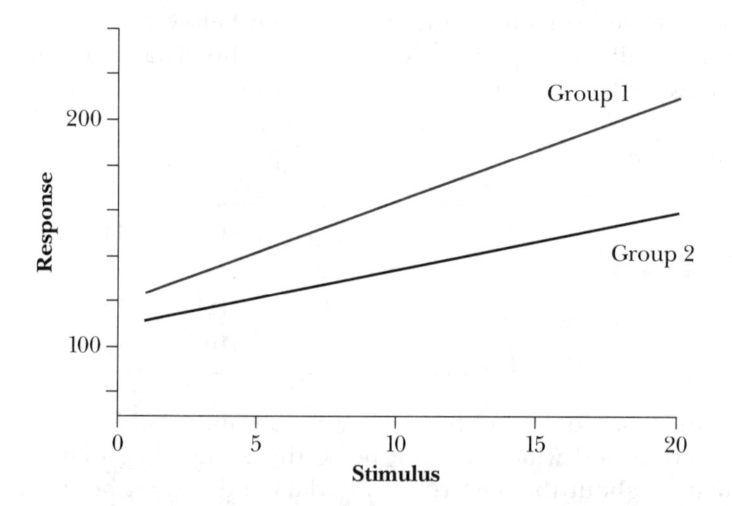

The fitted lines summarise some aspects of the relationship but do not indicate the strength of the relationship; nor do they tell us anything about the individual data values. Thus, information about the designed variation in the stimulus values and the observed variation in the response has been discarded. This information is not noise. For example, the display of the least squares lines does not allow us to distinguish between the following two data sets, both of which are summarised by the same least squares lines.

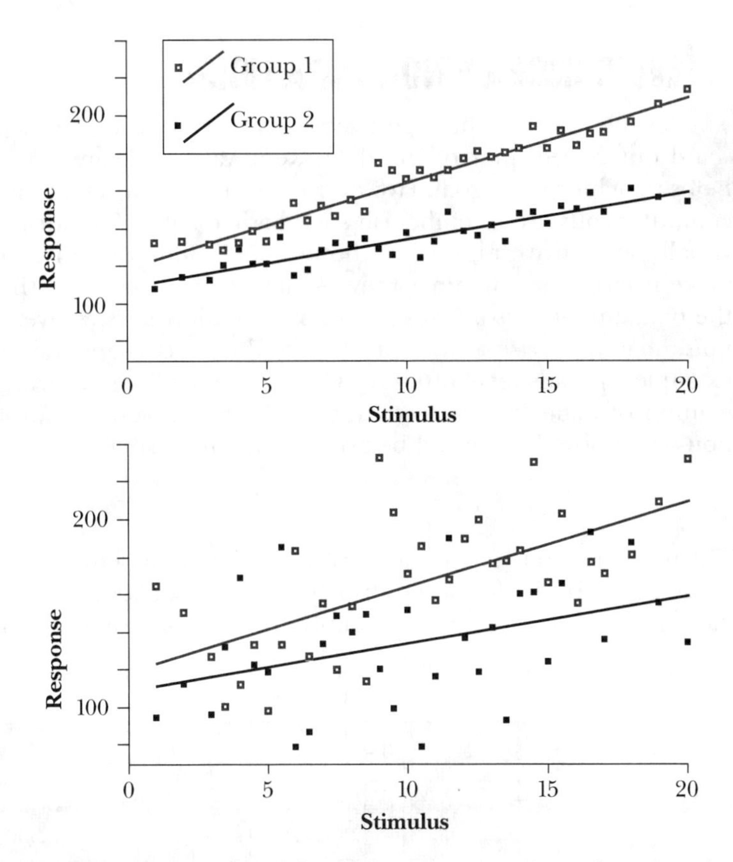

A further striking example of fitted lines not representing the whole truth about a relationship was shown in section 4.2. Anscombe's four data sets are all summarised by the same least squares line, but this line clearly does not explain the relationship between the two variables in three of the data sets.

Additional information about the data should always be retained in a display provided it is useful information and does not obscure the main signal that is being presented.

EXERCISE 6.3

1. Project 5 at the end of chapter 4 contained a table of data about vehicle production in the world in 1993.
   (a) Rounding the data accents the important differences between countries. Give a different reason for rounding these data.
   (b) Improve the tabular presentation of the data. Discuss any anomalous countries that are apparent in the table.
2. When the data points in a time series are plotted with a smoothed series, what information is provided by the data points that is not included in the smoothed values?

6.4 Avoidance of Non-Data Noise

The last two sections have stressed the need to make the 'signal' in a data set stand out in a display of the data. Removal (or de-emphasis) of 'data noise' helps to achieve this goal. However, there is a great temptation to use modern computer software to embellish such basic displays 'artistically' to create more visually attractive graphics. Although such additions to a basic display may make it more eye-catching, they usually reduce the strength of the 'signal' in the data and are also, from a data presentation perspective, 'noise'. Non-data noise may also arise as a by-product of the process generating the display; for example, spreadsheets often draw unnecessary lines separating each row and column of data. In this section, we will show several examples of redundant non-data noise that should be avoided in data displays.

Grids

The use of regular horizontal or vertical lines in data displays should be kept to a minimum. In the following display, it is almost necessary to send out a search party to find any data ink![7] Certainly, most of the non-data ink is unnecessary.

Relationship of Actual Rates of Registration to Predicted Rates (104 cities 1960).

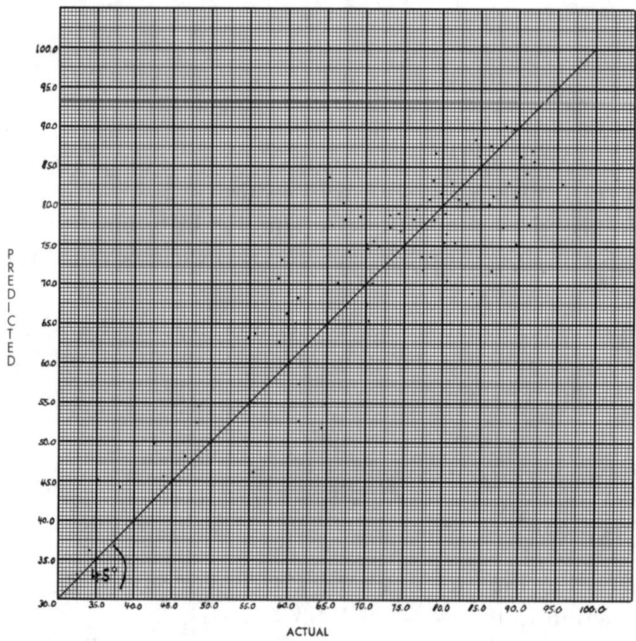

Less extreme examples often arise in published tabular data. The following display shows the fees for New Zealand university students in 1993, as published by the New Zealand Association of University Staff.[8]

7. E. R. Tufte, op. cit., p. 94.
8. *AUSNZ Bulletin*, no. 10, December 1992.

University	Category A		Category B		Category C	
	SR	NSR	SR	NSR	SR	NSR
Auckland	1589.50	1589.50	1589.50	1589.50	1589.50	1589.50
Waikato	1487.00	1487.00	1487.00	1487.00	1487.00	1487.00
Massey	1212.30	1747.30	1437.30	1867.30	1612.30	1997.30
Victoria	1602.00	1602.00	1602.00	1602.00	1602.00	1602.00
Canterbury	1398.00	1398.00	1398.00	1398.00	1398.00	1398.00
Lincoln	1232.00	1472.00	1424.00	1712.00	1712.00	2072.00
Otago	843.00	1503.00	1283.00	2383.00	1453.00	2723.00

SR = Study Right, NSR = Non-study Right. This categorisation was used by the New Zealand government in defining university funding entitlements. Students entitled their university to a higher level of government funding for their first three years of study (their *study right*), and some universities correspondingly reduced fees for these students.
Category A includes agriculture, arts, commerce, education and law;
Category B includes computing, music and science;
Category C includes applied sciences, health sciences, pharmacy, physical education and surveying.

The boxing of values in the table is unnecessary and only has the effect of making comparisons down columns more difficult. Where the rows or columns naturally fall into groups in some way, judicious use of 'white space' to separate groups is generally more effective than lines. After also rounding fees to the nearest $10 and re-ordering the universities, it becomes much easier to read information from the table and to make comparisons.

	Arts		Science		Applied Science	
	SR	NSR	SR	NSR	SR	NSR
Victoria	1600	1600	1600	1600	1600	1600
Auckland	1590	1590	1590	1590	1590	1590
Waikato	1490	1490	1490	1490	1490	1490
Canterbury	1400	1400	1400	1400	1400	1400
Lincoln	1230	1470	1420	1710	1710	2070
Massey	1210	1750	1440	1870	1610	2000
Otago	840	1500	1280	2380	1450	2720

See note to previous table for abbreviations.

In data displays, grids are useful only:

- as an aid to drawing the display by hand, in which case they must be removed before 'publication', or
- in the rare graphics where accurate values will need to be read off the display.

If a grid is included, it should be drawn with hairlines or, ideally, in grey.

The next graphic provides a wealth of information about the railway timetable for the Paris–Lyon line in the 1880s.[9] Some 10 years prior to the publication of Marey's graph (and apparently independently), Albert Prewett, then a clerk in a regional Traffic Manager's Office in Queensland, developed a similar graph specifically aimed at assisting the management and timetabling of trains on single track railways.[10] Modern railway timetables would do well to follow this clarity of presentation. Arrival and departure times can be easily read from Marey's timetable to an accuracy of 10 minutes or better, even from this shrunken version of the graphic. The grid lines are essential for reading detailed information from this display, but use a distractingly large amount of ink; thin grey lines improve the presentation.

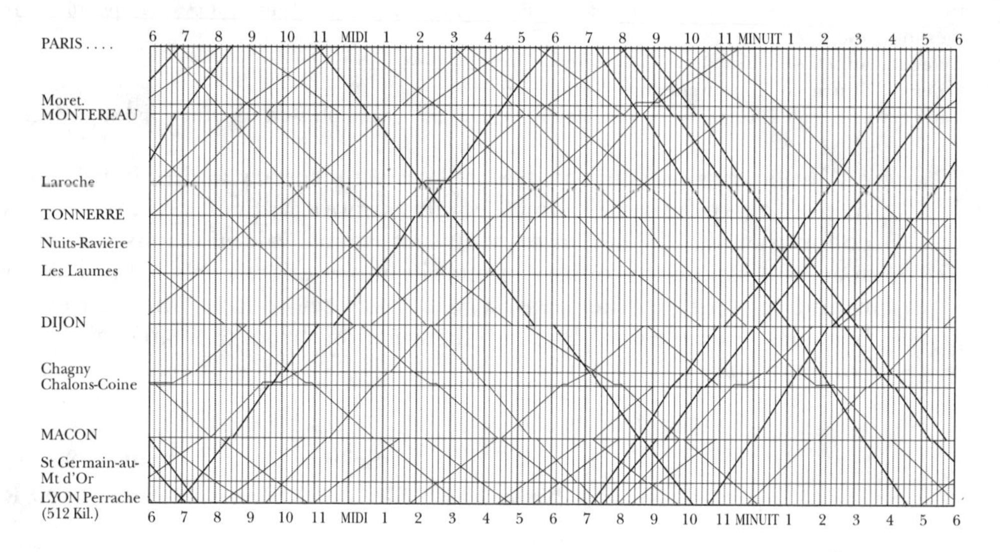

Axes

The axes on a graph should be labelled, but there is no need for large numbers of ticks or values on an axis. More than 10 ticks are rarely needed. The axes should usually only be drawn on the bottom and/or left of a display or may occasionally box the whole display.

9. E. J. Marey, *La Méthode Graphique*, Paris, 1885, p. 20.
10. J. Armstrong, 'Prewett's Train Working Diagram', *The Australian Railway Historical Society Bulletin*, no. 483, 1978, pp. 13–18.

Shading

Avoid filling regions of a display with patterns. If possible do not shade regions at all (for example, in bar charts). In many displays (for example, in pie charts), text and arrows are a better alternative to filling regions of the diagram with patterns and providing an explanatory key.

In the bar chart on the right, the shading of categories is not only unnecessary, but is also distracting.[11] If shading is needed to adequately distinguish regions, use greys rather than other patterns. Especially avoid diagonal stripes which can produce distracting moiré effects and can give a misleading impression of increases or decreases.

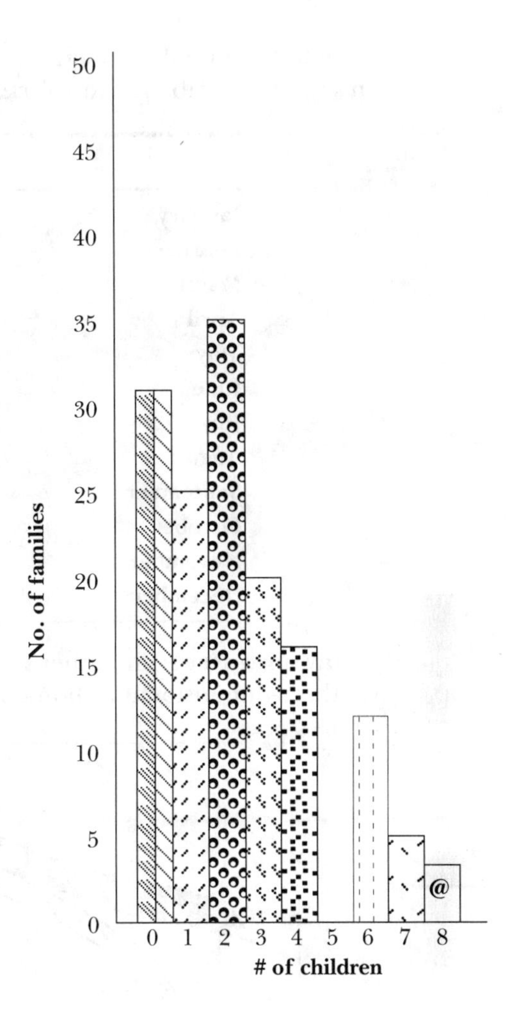

Three-dimensional enhancements

The term *chartjunk* is used to denote adornments to a basic graphic to make it appear 'more artistic'. Chartjunk is most often added when a graphic is used to display only a few values; a simple bar chart, pie chart, and so on, often seems 'bare' and without visual interest. It is therefore misguidedly 'enhanced' by artificially adding a third dimension to a two-dimensional display.

The temptation to add three-dimensional chartjunk is encouraged by the ease with which many computer programs can now produce it. The result is often more a display of computer technology than a display of the data. It can obscure the information that the basic display would show, and can never enhance it.

11. W. T. Federer, *Statistics and Society*, 2nd ed., Marcel Dekker, New York, 1991, p. 314.

The following table describes the average monthly temperatures in degrees Fahrenheit for Albany, New York, and Reno, Nevada.[12]

	Albany	Reno
January	21.1	32.2
February	23.4	37.4
March	33.6	40.6
April	46.6	46.4
May	57.5	54.6
June	66.7	62.4
July	71.4	69.5
August	69.2	66.9
September	61.2	60.2
October	50.5	50.3
November	39.7	39.7
December	26.5	32.5

Many spreadsheet and other software packages can readily produce an 'attractive' three-dimensional ribbon chart of the data such as that shown below.

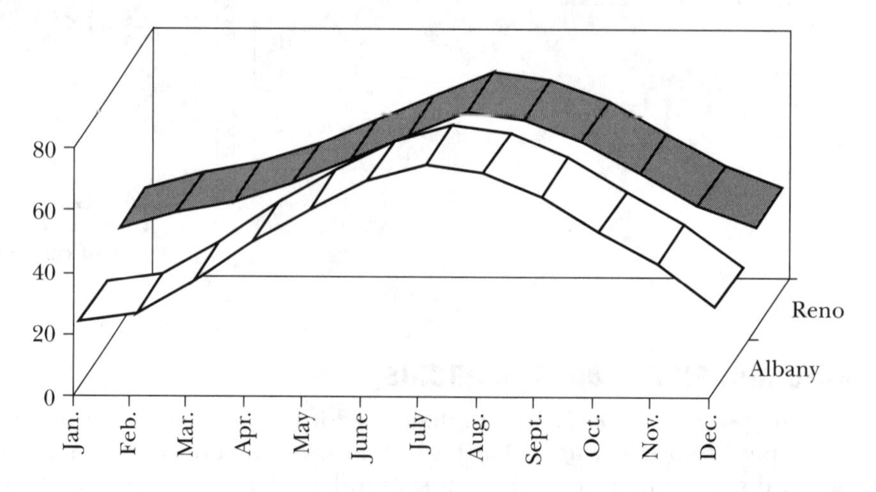

Although the ribbon chart looks more polished than the conventional time series plot that is shown at the top of the following page, it is far less informative about the data. The conventional chart allows accurate reading of temperatures and accurate comparisons of the two cities each month. The added depth distorts these values on the ribbon chart to the extent that it is impossible to see on that chart how the difference between the temperatures in the two cities changes over the year. In particular, the ribbon chart does not even show (in a form which is readily interpretable) which of the two cities has the higher average temperature in any month.

12. *Statistical Abstracts of the United States, 1994,* US Bureau of the Census, p. 217.

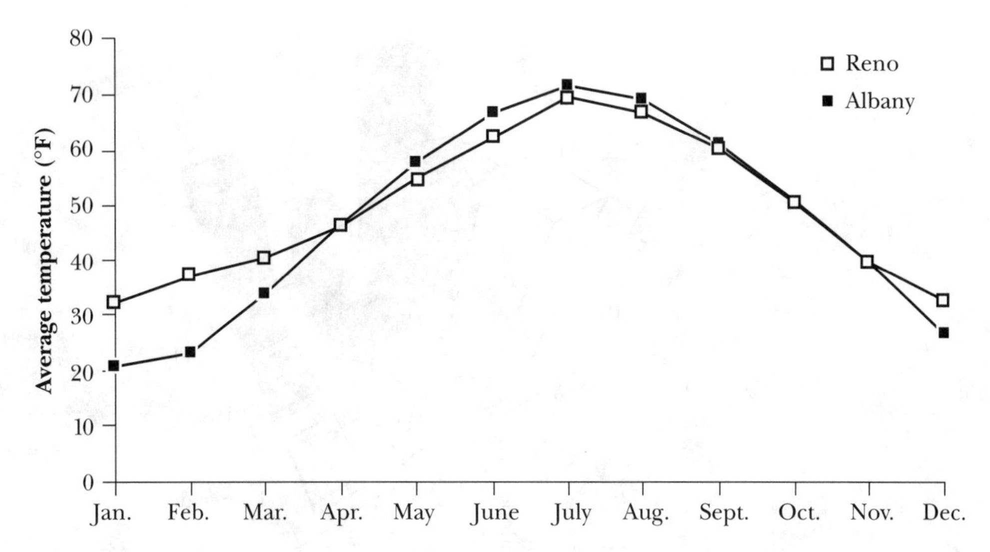

A further example of a three-dimensional embellishment of a two-dimensional display is given by the pie chart below. Such displays can again be misleading; although the volume of the cylinder has been correctly split into the correct proportions, the visual impact of the 'front' dark region is greater than those of the other two categories since the dark area on the paper is much greater than half the total area. Such subtle effects often occur with three-dimensional representations.

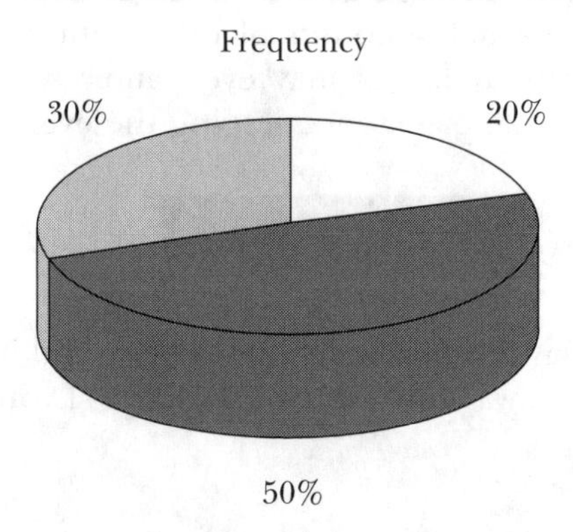

Note that, even after eliminating the three-dimensional effect, the circular pie has been distorted into an ellipse. This also distorts visual impressions since angles are no longer proportional to percentages.

The problems with interpreting three-dimensional displays become worse when the displays are drawn from a perspective viewpoint. Displays like that on page 212 are easily produced with many spreadsheets, but are less easily interpreted![13]

13. Although three-dimensional enhancements to two-dimensional data displays only add non-data noise, there are other situations where three-dimensional charts can effectively display information. For example, there are various three-dimensional displays that can allow us to visualise a surface.

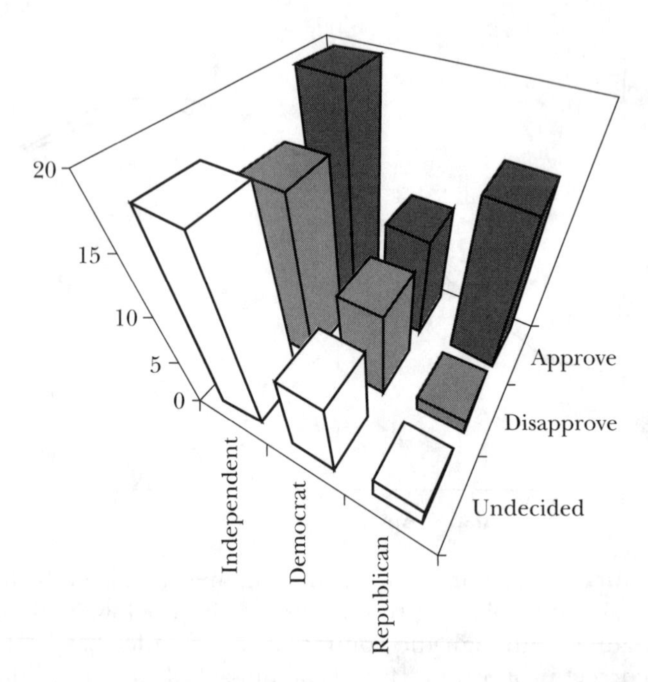

If there is little information to be presented in a data display, it is far better to present this information clearly in a small display than to embellish it with chartjunk to fill a page. In particular, very small pie charts and bar charts usually present their message as effectively as larger ones. Using small graphical displays to show small amounts of information rather than adding chartjunk to make the display larger may even allow several small graphics to be displayed on the same page, encouraging the eye to compare related displays.

EXERCISE 6.4

1. In what situations might grid lines be useful in a time series?
2. Why is the railway timetable of the Paris–Lyon line thought to be so clever?

6.5 FAITHFUL REPRESENTATION OF DATA

One great problem that arises in the display of data is distortion of the truth. Misrepresentation of statistical data is not only possible, it is common. It is sometimes deliberate. No wonder, then, that statisticians are often berated with the 'Disraeli' quotation or some other 'anti-statistics' sentiment.

It is important that a display does not misrepresent the data, whether or not the deception is deliberate. This misrepresentation of truth may arise from either the display of inappropriate data or the incorrect display of appropriate data. Although we are mostly concerned with the latter topic in this section, some examples of the former will be given first.

Display of inappropriate data

In the following example, close examination of the graph on the left reveals that the last time period (4 years) is shorter than all previous time periods (10 years).[14] Aggregate quantities must be aggregated over equal time periods, or be adjusted to reflect unequal time periods, to allow meaningful comparisons.

The diagram on the right below extends the data to 1980, so that the final entry describes the whole decade 1971–1980. This diagram paints a very different picture of the trend for the United States! The reason is not that the second half of the decade was any different from the first half, but that the number of awards in a 5-year period should not be compared to the number of awards in 10-year periods. If the data on the left are standardised, so that the vertical axis is labelled 'Number of awards *per year*', the relative height of all countries in the final 5-year period would be doubled. The display would then give a similar message to that obtained when the full data for the decade became available.

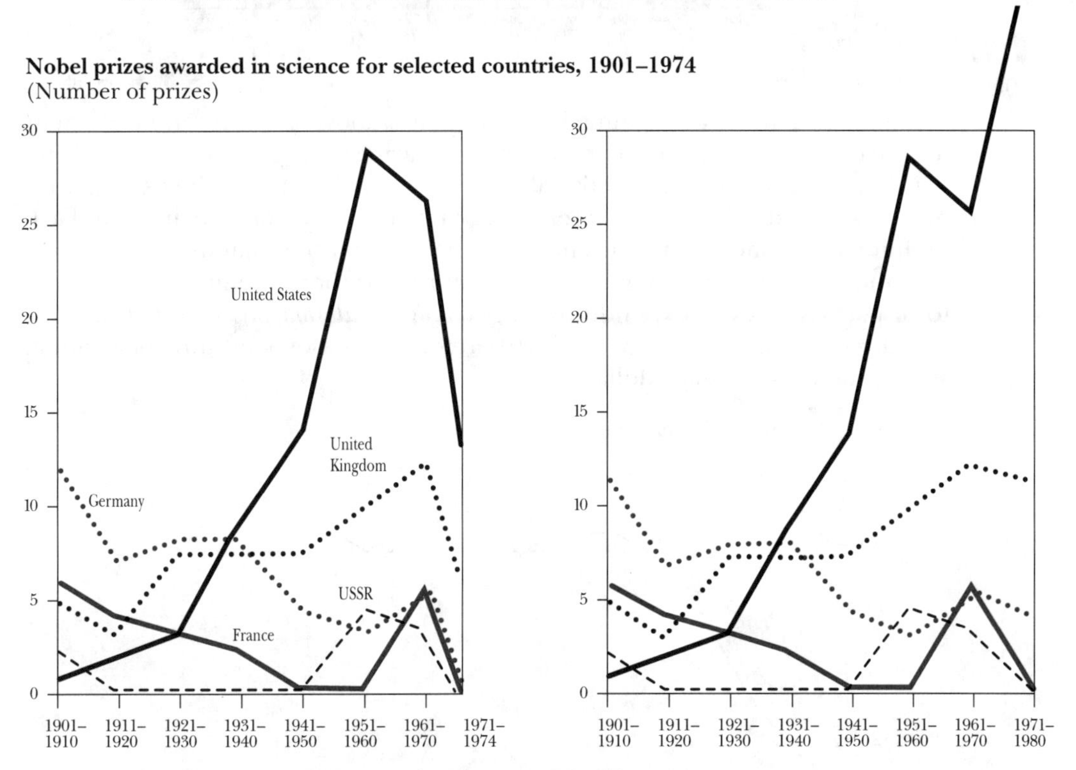

National Science Foundation, *Science Indicators*, 1974 (Washington, DC, 1976), p.15.

The graphical presentation on the following page contains several visual gimmicks (chartjunk) which are distracting and distorting.[15] A perspective viewpoint (shading to the right for the first half of the display and to the left for the second half) should be avoided. The shading is, moreover, all noise and therefore should be removed.

14. E. R. Tufte, op. cit., p. 60.
15. ibid., pp. 66–68.

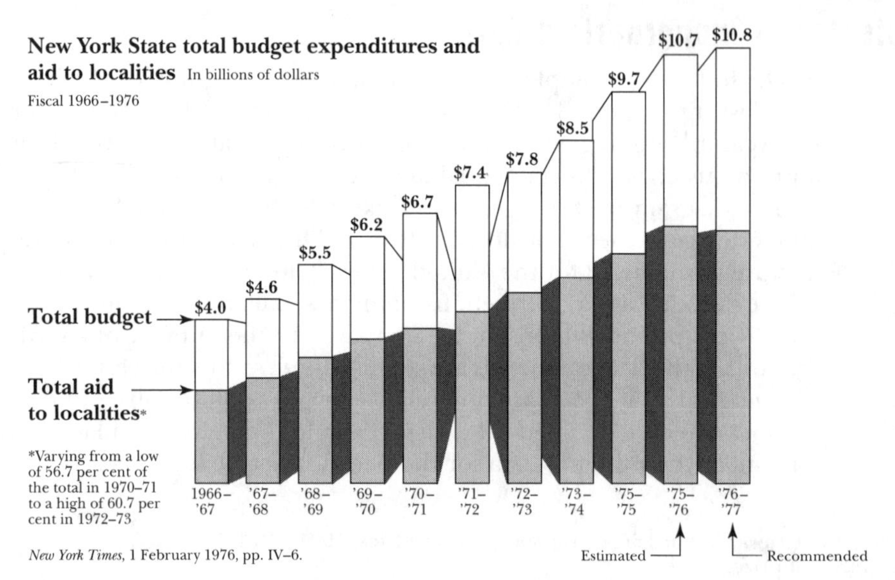

Eliminating the visual gimmicks in the graph above is, however, insufficient to reveal the 'true picture'. The display summarises an aspect of the underlying data that is not highly informative about what has happened to the expenditure over this period. For most time-series displays of money, deflated and standardised units of monetary measurement are better than nominal units.

Rather than summarising total expenditure, it is therefore more meaningful to display changes in expenditure *after taking population increases and inflation into account.* Tufte's revised display therefore shows expenditure per capita, standardised to constant dollars.

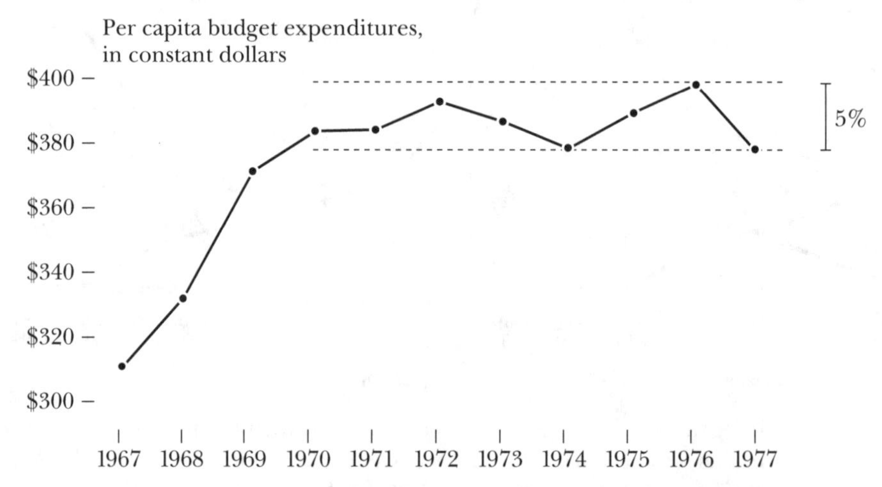

A related problem arises when, for political purposes, politicians alter their definition of 'the rate of inflation' or 'the unemployment rate' or some other measure to suit the message that they are trying to convey. It is important to be aware of this unfortunately common practice. It is never acceptable to adjust the data to match the message. We should always interpret the data honestly to define the message.

Displays that mislead

Sometimes, misrepresentation results from the way the display itself is constructed. The reader observes a display of data and uses it to form a more general picture of what the data are like; the reader will assume that any information that is not shown (or is hard to assimilate from the display) is either unimportant or at least unlikely to contradict the impression given by the display. The more highly summarised a display, the more opportunity for the summary to misrepresent the data.

For example, when presented with several box plots as summaries of different batches of numerical data, the reader will assume that each batch has no important internal structure other than that described by the box plot. In particular, it would be misleading to use a box plot to describe data whose distribution splits into two or more clusters (a bimodal distribution), since clustering is extremely important information that would be concealed. For a similar reason, the use of a straight regression line to describe a bivariate relationship would be misleading if the relationship was known to be curved.

Readers are likely to interpret quantities on a display as being proportional to the size of the corresponding data values that are being represented. In particular, representing a quantity with a three-dimensional object (and especially a perspective view) should be avoided since there is ambiguity over whether it is the area on the display or the volume of the object that is being used to represent the quantity. Consider the following display.[16]

This line, representing 18 miles per gallon in 1978, is 0.6 inches long.

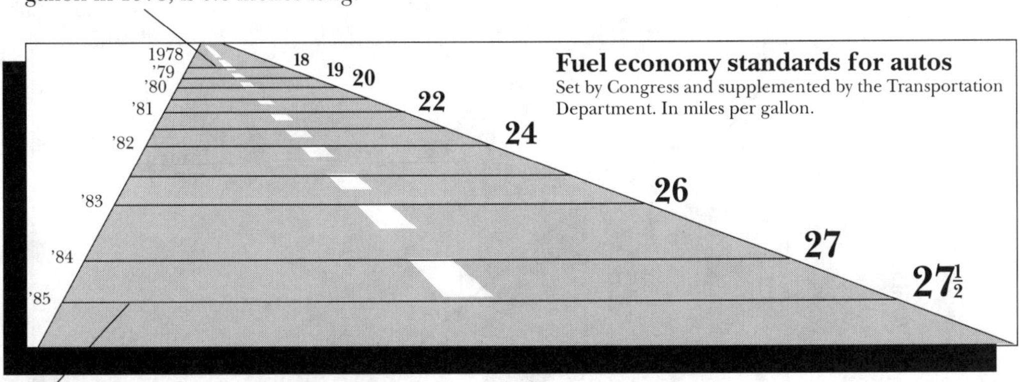

This line, representing 27.5 miles per gallon in 1985, is 5.3 inches long.

New York Times, 9 August 1978, p. D–2.

Some readers will interpret horizontal widths on the paper as being proportional to petrol consumption; others may assume that it is distances along the road that are being used. Still others will assume that equal distances along the road represent equal amounts of time. None of these interpretations is correct. The data are severely distorted.

16. ibid., p. 57.

The example on the right is even worse. In the graphic, dollars are represented by three-dimensional barrels.[17] The graphic is full of ambiguity. Should dollars be represented by the height of each barrel, its area on the paper or its volume? The display does *none* of these! Again, the perspective display further distorts the information.

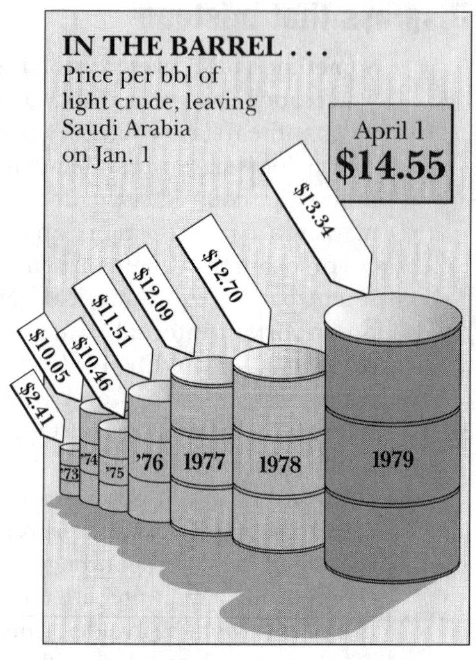

Time, 9 April 1979, p. 57.

Our final example in this section describes a related type of problem that may arise when 'quantity' information is displayed graphically. If the origin of the 'quantity' axis of the graphic is not zero, it is possible for the reader to misinterpret variation in the quantities through not carefully reading the scale. In the example below, which describes the level of the water supply serving Palmerston North during a dry spell in 1995, this is enhanced by the shading of the area under the line; a careless reader might interpret the drop in the water storage as being rather more extreme than was really the case![18]

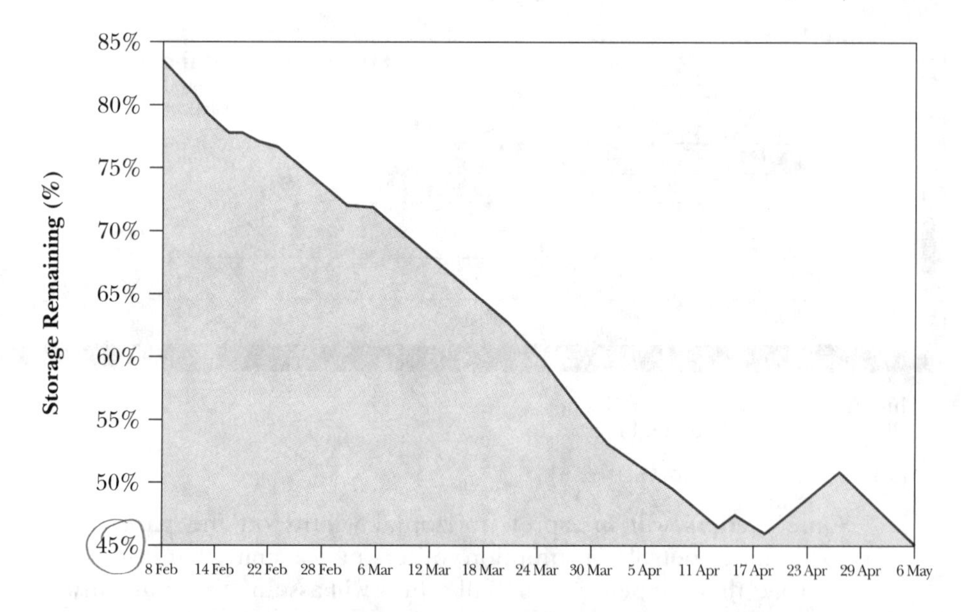

Palmerston North's main water supply, Turitea dam, remains at a low level.

17. ibid., p. 62.
18. *Guardian*, Palmerston North, 11 May 1994.

EXERCISE 6.5

1. A government claims to have increased funding of universities by 4% in the previous year. What extra information would you want to properly assess its claim of generosity?
2. A city council reports that rubbish collected has doubled over the past 20 years and wishes to use a diagram to describe this fact. Which of the diagrams below would be appropriate?

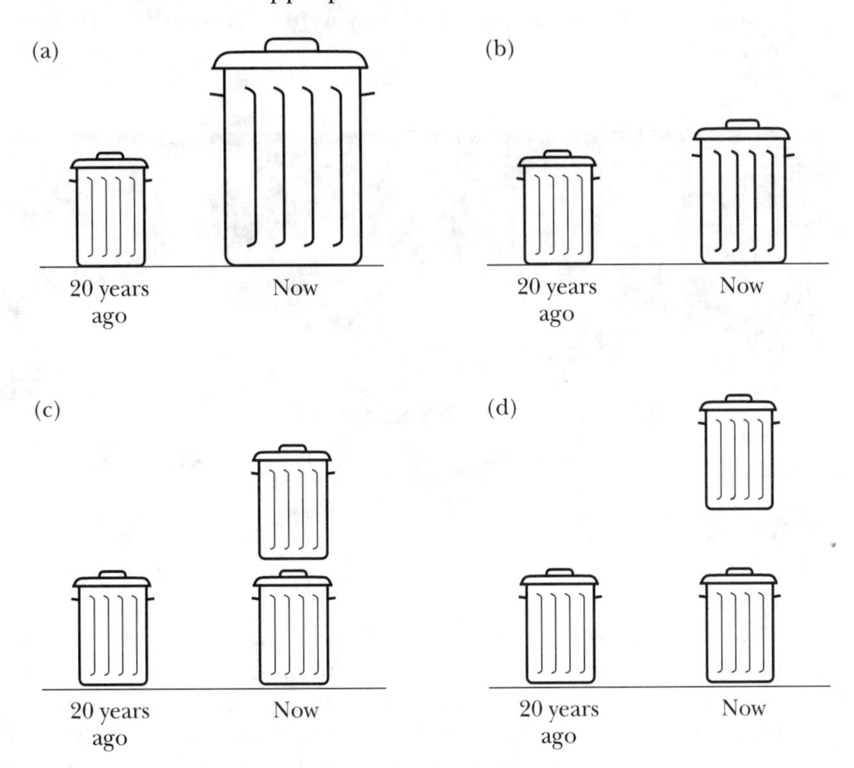

3. Criticise the display on the right which describes the decrease in some measurement related to smoking between 1970 and 1990.

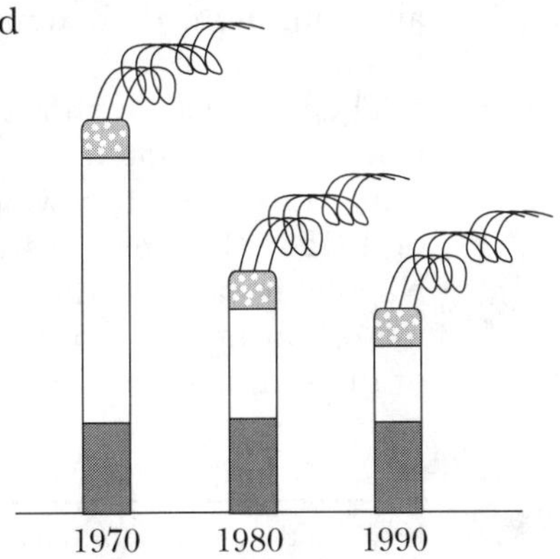

6.6 GRAPHICAL EXCELLENCE

The previous sections in this chapter have described several principles by which data displays can be judged. There is a wide range of display types that conform to these general principles and can clearly show data features. The more complex the data structure, the more scope for innovative displays to display the data effectively. The best examples of statistical graphics can effectively portray large amounts of data and allow the reader to understand complex interactions between different aspects of the data. A good example of this is the New York city weather record for 1980 which is shown below.[19]

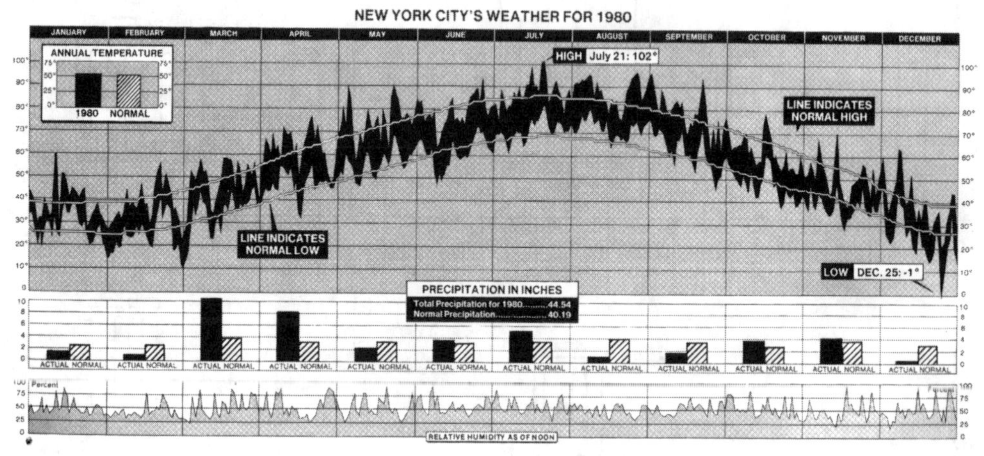

New York Times, January 11, 1981, p. 32.

This clearly shows five time series (actual and normal daily high and low temperature and noon daily relative humidity) and 13 bar charts, plus useful annotations. It also allows the reader to easily compare features in the various time series. For example, the daily temperature range (maximum–minimum) was typically lower in the winter than in the summer (shaded vertical distance between the minimum and maximum time series). Also, the period from late August to early September was much warmer than normal. (Both maximum and minimum daily temperatures tended to be higher than normal during this period.)

The next display shows that graphical excellence is not a recently developed concept and that there is plenty of scope for originality. In 1861, Charles Minard used the following graphic to portray Napoleon's disastrous campaign of 1812–1813 when his army invaded Russia. After laying siege to Moscow, Napoleon was forced to retreat by the harshness of the Russian winter. On the map showing the movements of the army, Minard has superimposed a display of the number remaining alive. These are indicated by the changing width of

19. From the *New York Times*, 11 January 1981, p. 32. In E. R. Tufte, op. cit., p. 30.

the band tracing Napoleon's route (grey in advance, black in retreat). Temperature during the retreat is also shown in a linked display. This graph is both elegant and effective and Tufte claims that it 'may well be the best statistical graphic ever drawn'.[20] Many statisticians would argue the claim, but all would agree with the sentiment.

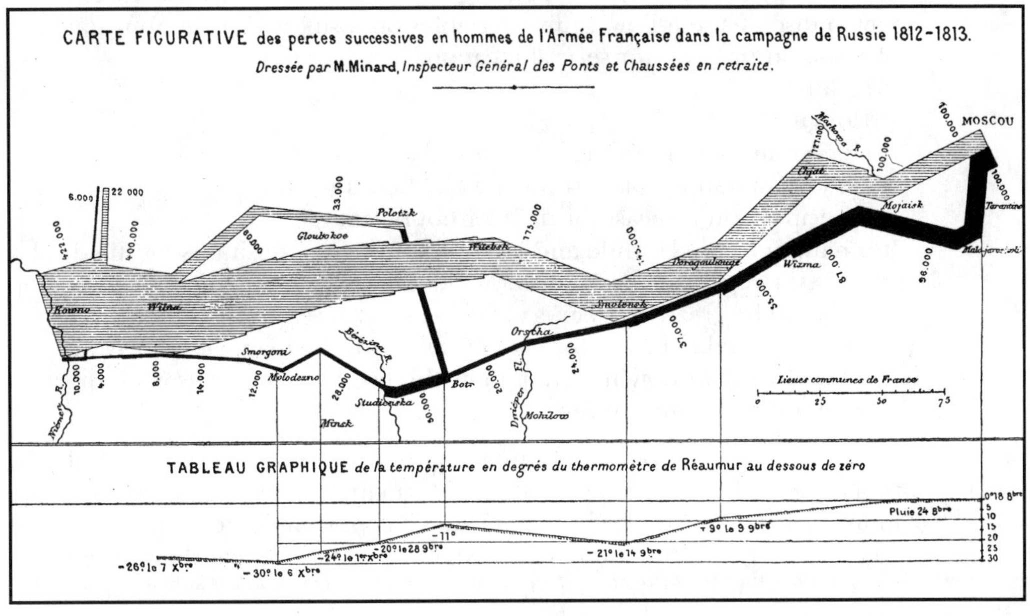

Xbre = December   9bre = November   8bre = October

EXERCISE 6.6

1. Smooth the actress age data from section 3.1 with 5-point running medians. Now apply 5-point running *minimums* and 5-point running *maximums* to the data and plot all three series on the same plot. Does addition of the running minimums and maximums add useful information to the display?
2. What is the main message of the chart of Napoleon's Russian campaign?

20. ibid., p. 41.

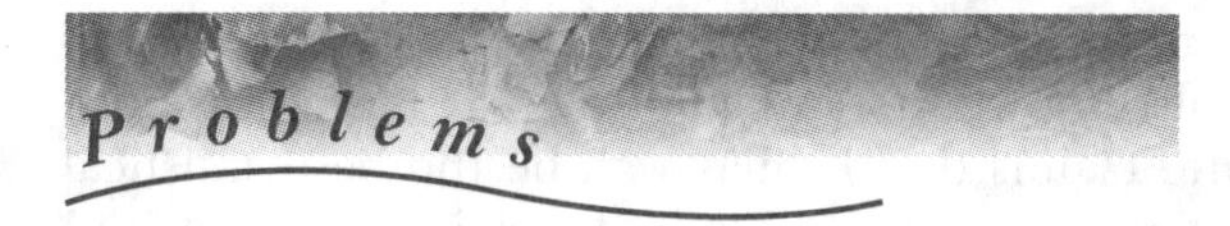

1. The great variety of techniques for graphical display of data provides opportunity for creativity in devising effective displays. Devise an effective way to display the following five variables on a single display. Your data display should present, for several locations in your country:
   (a) latitude
   (b) longitude
   (c) high temperature forecast for next 24 hours (°C)
   (d) low temperature forecast for next 24 hours (°C)
   (e) precipitation forecast for next 24 hours (mm).
   It is natural to use latitude and longitude to define positions on the display (that is, to represent each location by some symbol or picture at its map position). The creativity begins with how you represent the other three variables in the symbol drawn for each location on the map. Describe your proposal and explain how it would allow the eye to quickly absorb important features of the weather forecast.

2. Some authors have suggested modifying the standard stem-and-leaf plot by replacing each leaf digit with other information about the corresponding individual. For example, in a stem-and-leaf plot of some characteristic of countries, a two-letter country code might replace the leaf digits; the countries Australia, Britain, Canada, France, Germany, New Zealand and USA might be shown as follows:

12	FR
11	CA, US
10	GE, BR
9	NZ
8	AU

   Northern hemisphere countries might be distinguished from southern hemisphere countries by different colours. The following problem involves a different kind of modified stem-and-leaf plot.

   The Paris-Lyon train timetable presented in section 6.4 is an effective display for a small number of trains, but would become cluttered and much harder to read if there were more trains. When trains come and go every few minutes, a different solution is required. One display of arrival and departure times at a particular station was devised for the Tokyo–Kyoto train using modified stem-and-leaf plots. Devise a display that shows the arrival and departure times of trains at a particular station. Travellers should be able to read off the times of trains to the nearest minute from the display.

   The display used on the Tokyo–Kyoto line contained no information about how long any train took to get to a particular destination although, on this line, all trains ran at roughly the same speed at any time of day, so the timed route schedule of each train was not necessary. There was no assumption that the intervals between trains was regular, since train frequency adapted to the intensity of commuter traffic.

If some trains are express trains, how might this be indicated on your display?

3. A 5 kilometre cross-country race is organised for school children of various ages. The winning times and the average times to complete the race are calculated separately for each sex and age group (10, 11 and 12 years). You are not required to collect actual data for this problem, but should make up plausible values for these 12 means.

   Devise a well-designed table displaying your data that facilitates comparison of boys and girls and comparison of different ages. Contrast this table with a badly designed table of the same data.

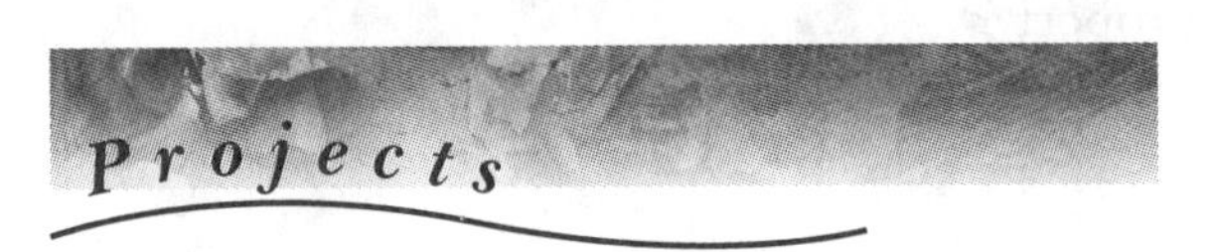

Projects

1. **Data displays in magazines**
   Select a magazine or newspaper that interests you and that may have data-based charts either in its articles or in its advertising. Choose three charts from the magazine and criticise them based on the principles you have learned from this course so far. Include a photocopy of your selected charts with your report of this project. (Magazines like *Time*, *The Economist* or *Discover* have many charts. Some financial newspapers are also good sources.)

2. **Aircraft emissions**
   In section 2.1, the weights and CO_2 emissions of various aircraft were presented. CO and NO_x emissions from the same aircraft were presented in section 4.1.

   Investigate the relationships between these variables. Consider ways to represent more than two variables in a single display. Then write a concise report with no more than two graphical displays, describing your findings to an environmentalist who is interested in whether large aeroplanes are worse polluters than small aeroplanes and which aeroplanes are relatively clean or dirty.

3. **Mutual funds**
   Financial papers often show the relative performance (equivalent annual rate of increase in value) of mutual funds (unit trusts) over a five or ten year period. The aim of this project is to compute the performance corrected for inflation of several funds and to summarise the 'real' performance of the funds. You will need to find a source of fund performance and of inflation rates for this project.

   Select one fund for the first step of the project. Obtain data on inflation for the years in which performance is measured. Compute the total inflation factor for the period in question — for example, if the annual inflation rates for three years are 1.3%, 2.4% and 1.5%, compute $1.013 \times 1.024 \times 1.015 = 1.053$. Also, convert the reported annual performance over the period (three years in the example) to a comparable factor: 7.1% per annum becomes $1.071 \times 1.071 \times 1.071 = 1.228$. Now compute:

$$\left(\frac{1.228}{1.053}\right)^{\frac{1}{3}} = 1.053$$

where the fractional power is determined by the number of years in the reported performance data. The real rate of increase is then 5.3%. Do this calculation on one fund. Give your answer the 'reasonableness' check before proceeding to the final step.

The final step is to choose 10 funds for which performance data are available for your chosen time interval (3, 5 or 10 years, for example). Then compute the real performance of the 10 funds and present your result in a tabular form. How do the real returns compare with the enjoyment of owning real property?

CHAPTER 7

Data collection — the importance of design

Observation is a passive science. Experimentation is an active science.

Claude Bernard (1813–1876)

7.1 AVAILABLE DATA

The preceding chapters have focused on examining data that have already been collected. A wide variety of techniques have been presented which allow us to effectively explore a data set, to extract useful information from it and to effectively present that information in tabular and graphical form to others. These can be useful techniques, but none of them serves a useful purpose if the data being examined do not contain information relevant to the aims of the investigation. Furthermore, there must be sufficient quantity of relevant information to make the investigation worthwhile.

The most convenient data to collect are data that someone else has already collected. From library resources or the World Wide Web, data at various levels of credibility may be available. Unscrupulous researchers may choose to make up data they cannot find, and we must always be on the lookout for fabricated data. However, available data are usually recorded honestly, and we should not ignore this source of information even though we must keep in mind the possibility of both deliberate and unintentional biases.

Many investigations have their origins in anecdotal evidence. Information often attracts the attention of someone because it is unusual or atypical. But, in a world of variability, the unusual *will* happen from time to time. Conclusions should not be hastily drawn from a few individual cases. Too often, a series of unusual events (such as murders, bankruptcies or plane hijackings), which is given high publicity by the news media, is interpreted by the public as evidence that such events are becoming more frequent; politicians often react to such anecdotal evidence by introducing legislation to reverse the perceived increase in frequency.

Anecdotal evidence should not necessarily be ignored. Major scientific advances can begin with such anecdotal data. Recently in Australia, a new strain of HIV, the virus which causes AIDS, was discovered because the long time survival of a HIV infected blood donor was linked with similar long-term survival of a group of recipients of blood from that donor. This anecdotal link led to a scientific investigation which showed that these patients were infected with a mutant strain of virus which lacks the capacity to replicate in the human body in the way of the common strains of the virus. This has provided great hopes for the development of an AIDS vaccine.

Available data may come in many other forms. In the age of information, a researcher undertaking a study may have access to vast resources of data which were not specifically collected for the purpose of the study. Immediate access to available data is extremely attractive compared to the time, cost and effort needed to plan and carry out data collection of your own. Available data can provide a very valuable resource.[1]

---

1. Available data can also be valuable to suppliers. A substantial income stream is provided to official statistics agencies, pollsters and other companies involved in data collection through the sale of data for uses not directly related to the original purpose of collection.

Often, no available data meet the needs of the study at hand. Use of available 'related' information instead of more relevant information that is not immediately available can lead to false conclusions. For example, there is a vast quantity of available data on crime, but crime rates in different countries (or over different times) may not be comparable from these data. The precise definitions of crimes such as assault or rape may differ from country to country, as do the proportions of such crimes that are reported to police. Direct use of published data may therefore misrepresent the differences between countries and may lead to erroneous conclusions.

If seeking to interpret history, there is little choice but to rely on the data which are available, although good historians will use their research skills to find a variety of data sources and to interpret the available data in the light of constraints imposed by the nature of the available data.

Available data should be carefully scrutinised for relevance and adequacy before being used in a scientific study. If the limitations in these data are severe, or if no relevant data are available, data must be collected specifically for the study.

EXERCISE 7.1

What is the role of anecdotal data in scientific investigations?

suggest hypotheses

7.2 DATA COLLECTION

If we are interested in a question which cannot be answered using available data, we must collect new data to answer the question. Careful design of the data collection procedure is extremely important. A haphazard or poorly designed data collection procedure is likely to result in unrepresentative data and may therefore lead to conclusions that are grossly wrong. Well-designed data collection procedures result in data that contain information that will throw light on the questions of interest in the study. For the remainder of this chapter, we will focus on design.

Because of their reliability and efficiency, designed statistical studies have had a great influence on the advancement of knowledge in many fields of endeavour and have changed the practice of many disciplines. A design specifies the procedure by which we collect data on more than one individual or unit.

There are two important classes of statistically designed studies: *observational studies* and *experiments.* In an observational study, information is collected as unobtrusively as possible from a process or group of individuals; the result is a description of characteristics of this process or group of individuals.

In contrast, the investigator has a more active role in an experimental study. In an experiment (sometimes called a scientific study), experimental conditions are adjusted by the investigator to determine their effect on the process or individuals being examined. Because experiments involve intervention, it is easier to infer from their results that changes to some variables *cause* changes to others, and to predict the effect of future changes. The goal is often to draw conclusions about how characteristics of a process can be changed to obtain the best 'output' from the process.

For example, consider a study to investigate the effect of vitamin C on health. An observational study may collect information about vitamin C intake and number of days off work in the previous year from a random sample of individuals in a city. It is, however, difficult to infer a *causal* relationship between vitamin C intake and days off work from such a study, even if those who consume little vitamin C are observed to have more days off work than those whose intake is higher. An observed relationship could perhaps be caused by some racial or social groups having diets containing less vitamin C and also being more likely to take days off work; other genetic or environmental factors could affect both characteristics, again causing an observed relationship between vitamin C consumption and health that is not a causal one. It is difficult to conclude that, if an individual consumes more vitamin C, that individual's health is likely to improve.

In contrast, an experimental study about vitamin C would use a group of similar individuals. Half of the individuals would be given a daily tablet of vitamin C, and the other half would be given a daily dose of a tablet which looked and tasted similar but contained no active ingredient (a *placebo*); none of the individuals would know whether the pills they consumed contained vitamin C. The number of days off work over a year in each group would be recorded. Any systematic differences between the health of the groups can now be attributed to vitamin C.

Experimental studies usually allow clearer conclusions to be reached, but are not always possible for ethical or practical reasons. In particular, ethical constraints limit the types of experiment that can readily be conducted on humans. Investigations relating to the dangers of cigarettes, lead, asbestos or other chemicals cannot use data collected from experiments on human subjects; either experimental data must be collected from animals or observational data must be obtained from some group that has (accidentally) been exposed to the substance. Even in studies that do not involve humans, experimentation may not be possible. For example, it would be very nice to collect your own data on the effect of earthquakes on flooding in earthquake prone areas, but the experiment to settle the causal link is simply not feasible; an observational study involving time series of earthquake activity and flooding would be the best one could do.

Observational studies are (usually) passive data gathering instruments in that they gather data on an existing population in such a way that gathering the information is intended not to influence or change the characteristics or measurements on individuals included in the sample.

Experiments are active. In an experiment, the output of some process is observed under varying conditions. Deliberate changes in conditions are employed by the experimentation in order to observe a response to that change. The goal is to assess the effects of various treatments or interventions so as to draw conclusions about how best to operate the process. The design of experiments is further discussed in section 7.4.

EXERCISE 7.2

1. What is the important difference between an observational study and an experiment?
2. Why is the distinction in question 1 important for scientists?

7.3 SAMPLE SURVEYS

Censuses and samples

The most common type of observational study is a *survey*. A survey is a process that involves collection of information about a specified population (of people or things). The population may be a specified group of people; it may also be a group of households, a group of businesses, the cars at a wrecking yard, the trees in a plantation or the apples in a supermarket. The individual members of the population are referred to as *units*.

The largest and best known surveys are those in which we all participate. These are censuses. In a *census*, an attempt is made to survey every individual in the population. In many countries there are legal requirements that government statistical agencies collect census information from the population at regular intervals, for example every 10th year.

Data collection through censuses is expensive and infrequent. Censuses also require large and complex administrative support structures and processing the information that is collected is expensive and slow (although, by using modern technology, it is much cheaper and quicker than in the past). Finally, because they are infrequently conducted, the most recent census information may no longer be current.

Conducting your own census has similar problems, even if the population is considerably smaller. Further, in some industrial applications (for example, testing explosives or testing wine bottle corks for tainting), recording information from a unit may be destructive; a full census is not an attractive option in such a circumstance!

A *sample* can be a very attractive alternative to a census. A sample is part of the population, as shown in the diagram on page 228. The purpose of sampling is to gather information from part of the population so that conclusions can be drawn about the whole population. Using information from a sample to make valid conclusions about a population requires that the sample is as representative as possible of the population. Some methods of designing representative sampling schemes are more efficient than others. Good sample design will be discussed later in this section.

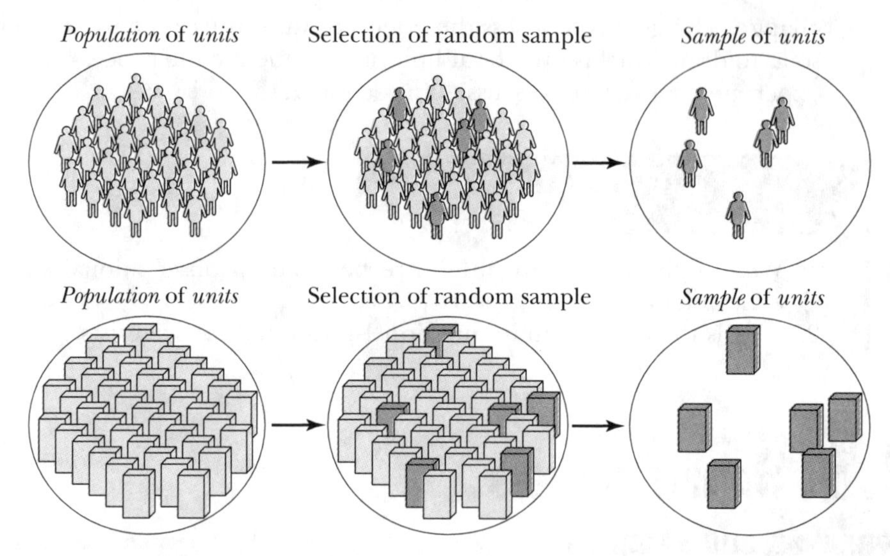

Sampling has the advantages of simplicity, cost reduction and timeliness. Perhaps surprisingly, a small, well designed and constructed sample can give more accurate information than a larger sample because it can be much better managed and the data collected on individuals may, accordingly, be more reliable. For example, in the 1936 American presidential election there were two candidates, Roosevelt and Landon. The *Literary Digest* conducted a poll, aiming to predict the result of the election; its procedure was to mail questionnaires to 10 million Americans (using names from telephone books and club membership). From the 2.4 million replies, they made the following prediction:

	Per cent for Landon	**Per cent for Roosevelt**
Literary Digest's prediction	57	43

Despite the very large sample size, the *Literary Digest* poll later proved to be very inaccurate because the responses were not representative of the whole population. Only individuals in households with telephones or in clubs were sampled, and only 24% of these replied. These groups would have different voting characteristics than the whole population.

Another pollster, George Gallup, conducted a much smaller survey. He selected only 50 000 people, but put more effort into making his sample representative. (Each interviewer was given a quota of people to select in each of various categories: males, females, old, young and so on.) Despite the much smaller sample size, his poll proved to be a far more accurate prediction of the election results.

	Per cent for Landon	**Per cent for Roosevelt**
Literary Digest's prediction	57	43
George Gallup's prediction	44	56
Actual result	**38**	**62**

Surveys must be well planned; they do not just happen. Sample design is generally far more important than selecting a large proportion of the population for the sample (a large *sampling fraction*). Before we look at the good ways of taking a sample from a population, we will discuss some more fundamental aspects of the survey process.

Design and development

There are many questions which need to be addressed while developing a survey.

- What is the purpose of the survey?
- What are the issues on which the survey is to focus?
- How will the information obtained be used?
- What action will be taken as a result of the survey?

If there is no clear idea of the objectives of the study, it is unlikely that the data collected will contain suitable information. Even in well thought out surveys, when analysis of the survey results begins, it is often discovered that insufficient information has been collected to answer some questions. ('If only I had asked ...'.)

Many novices embark on data collection without clear objectives and only the hope that some useful information will emerge. (These are sometimes referred to as 'fishing expeditions'.) They often involve a large expenditure of time, money and effort without producing much useful data.

We will discuss further details of survey design by posing some more questions. Once all of these questions have been addressed, it is then useful to reflect on whether the survey should be run at all. If differing survey results will not lead to different actions, perhaps there is little or no purpose in conducting the survey.

- What is the population you wish to survey?

The population of interest needs to be clearly specified. The conclusions that are drawn from the survey can be applied only to the population from which the sample is selected. They cannot be readily generalised to a wider population.

Should the population for a retail survey, for example, include businesses which are also wholesalers, those which open for business or close down during the survey period? How should a business with multiple outlets be treated? What is the area of interest; is it the whole of a city or just the central business district, however defined?

We therefore need a definition of what constitutes a unit in the population of interest. Many sampling designs further require a list of all such units in the population. It may be found that the only available lists of population units are incomplete, out of date or inaccurate for some other reason.

- Are there other sources of the information being sought?

If existing information is available, a survey may not be necessary. More likely, the available information will be incomplete, but it may be possible to make use of this information in conjunction with freshly obtained survey data. Combining data from two or more sources requires careful prior planning to ensure that the data are compatible. That is, it must be possible to sensibly relate the data from the different sources.

- What resources are available?
- Are there time or other constraints?
- What precision is required?

Surveys are generally used to throw light on numerical characteristics of the population from which the sample is drawn. For example, we may be interested in the proportion of students at a university who regularly listen to a particular radio station, or the mean weight of cornflakes in packets manufactured on a particular day. In each case, the corresponding characteristic of the sample provides an *estimate* of the population characteristic.

Clearly, the higher the sampling fraction, the greater the precision of a survey if all other aspects of the survey remain the same. However, surveys are expensive to run. A decision must be made fairly early in the design of the survey about the balance between the resources employed (time and money) and the precision of the estimates that are likely to be obtained from the survey.

The resources required to run a survey are, in the main, money, people and time. There are many important resource issues which should be addressed. These include staff recruitment and training, organisation of data collection, data entry and analysis (including equipment needs), consumable and overhead costs and whether some resource issues and costs must be explicitly recognised as being linked to the survey or can be absorbed into the general business costs of the survey organisation. The true cost of conducting a properly designed complex survey can be about $30 per surveyed unit (in 1996), sometimes more.

- What questions should be asked to address the important issues?
- How should those questions be framed?

Care must be taken to word questions carefully. It is only too easy to present someone with a list of alternatives and forget one. For example, in a transport survey, the only responses allowed to a question about the main mode of transport used to travel to work were ‘car’, ‘bus’, ‘train’, ‘foot’, ‘motorbike’ and ‘bicycle’. Some respondents travelled to work by taxi and added this category to their questionnaire, whereas others who travelled by taxi presumably picked ‘car’ or ‘bus’ as being the nearest category. Subsequent analysis of the taxi users was therefore impossible! It is equally bad to use questions that, by their phrasing, either give offence or are 'leading questions' that encourage one particular answer.

It is much harder than you might imagine to design a questionnaire, and this should be a major part of the design process. If the questionnaire is too long, a high proportion of those sampled will not bother to complete it. Too many questionnaires are filled with questions for which the answers are not used. It is also essential that questions are not omitted if answers to those questions may be needed to address the main issues involved.

A *pilot survey* (a small scale trial survey) is often conducted to assess questionnaire design and also to try out other aspects of the survey process.

- What sampling design should be used?

By 'sampling design', we mean the method by which the sample is to be selected from the population. The success of the sampling strategy, with all the benefits it provides, depends critically upon the way the sample is selected. We will describe various sample designs and their characteristics in the next section.

Sampling and non-sampling errors

To understand sampling design, we need to have a clear idea of how the sample will be used to supply information about the population. The most common uses of sample data are to use a sample mean as an *estimate* of the corresponding population mean or to use a sample proportion to *estimate* the corresponding population proportion. If the sample data are the yields of milk from a sample of cows, then the mean yield in the sample provides an estimate of the average yield for the whole population of cows. In a public opinion poll, the proportion intending to vote for a particular party in the sample is used as an estimate of the proportion intending to vote for this party in the whole population.

Since the whole population is not sampled, there will usually be an *error* in any estimate based on a sample. The reason sampling design is so crucial is that it determines the likely size of the error incurred by using the estimate. If we are going to use the mean milk yield from a sample of cows to estimate the mean milk yield from the whole herd, we must be able to predict something about the size of the error.

Whatever the sampling design, the data in one sample will vary from the data in another. The unavoidable error arising because one sample has been chosen rather than another is called *sampling error.* For example, the simplest sampling design is called *simple random sampling,* or sometimes just *random sampling.* If n items are to be selected from the population and all possible selections of n items have the same chance of being selected, the sampling method is random sampling. If this sampling method is properly implemented, the difference between the sample mean and the population mean is the sampling error.[2]

2. Chapter 8 describes in detail the characteristics of the sampling error in simple random sampling.

There are various ways in which a sample designed to be a simple random sample from a population of interest can fail to be one. These failures are called *non-sampling errors*, to stress the fact that they are not due to the random sampling procedure but to a failure to actually do the intended random sampling. Non-sampling error can be more serious than sampling error and its likely magnitude is harder to assess. Non-sampling errors tend to *bias* a sample — they cause the sample characteristic to be systematically higher (or lower) than the population characteristic being estimated. For example, a lecturer teaching a large class asks a random sample of 20 students attending a particular lecture whether they find the course easy. There is sampling error because only 20 students have been selected; selecting a different random sample of 20 students would result in a different proportion finding the course easy. However, there is also non-sampling error because all students in the course do not attend the lecture. The students who are not there may be absent because they find the course easy; if so, the proportion in the sample who find the course easy will usually underestimate the corresponding proportion in the whole course.

Sampling from part of population

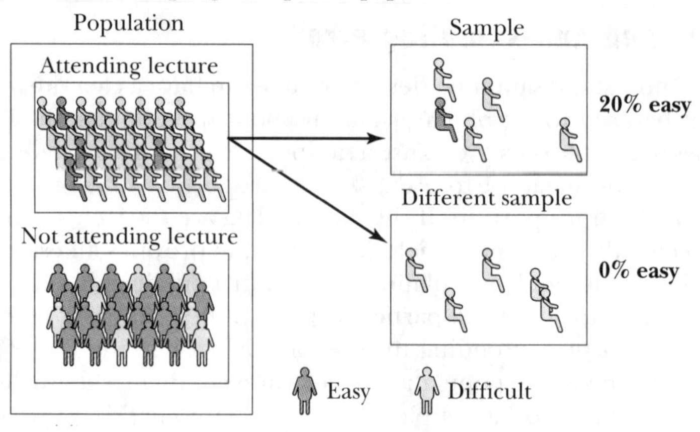

Sampling from whole population

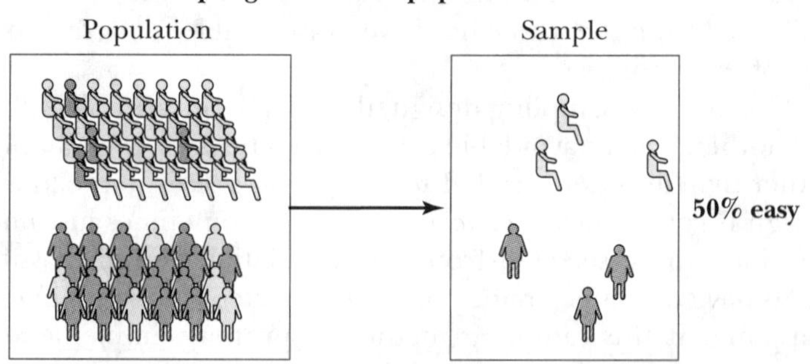

Good sampling design can help to reduce bias. In considering the sampling design, we need to keep in mind the possibility of the following non-sampling errors that can bias our sample.

Coverage error

Coverage error is caused by some units in the population being excluded from the list of units from which the sample is drawn (the *sampling frame*). An example is the exclusion of households with neither telephones nor club memberships in the *Literary Digest* poll.

Non-response error

Non-response error occurs when some units selected for the sample could not be contacted or chose not to participate in the survey. Non-response rates vary widely, depending on the nature of the survey, the length of the questionnaire, the mode of data collection and the amount of follow-up.

Non-response does not always lead to bias, but usually does. The extent of the problem is determined both by the proportion of units selected for the sample that respond and by the degree to which the non-respondents differ from the respondents. Although some non-response bias is inevitable when dealing with human subjects, consideration of this issue in designing the approach to the respondent can reduce it.

Instrument error

The word 'instrument' in sampling survey jargon means the device by which we collect data — usually it is a questionnaire filled out by the respondent. Different wording of a question can lead to different answers being given by a respondent. When a question is badly worded, the resulting error is called *instrument error*. The wording of the question may be such as to elicit some particular response (a *leading* question) or it may simply be carelessly worded so that it is misinterpreted by some respondents. Questionnaires must be carefully designed.

Some instruments would not involve a questionnaire: alternatives are face-to-face interviews, telephone interviews or even observations in which the respondent is unaware of being observed. Each method of data collection may lead to an instrument error specific to it.

Interviewer error

Interviewer error occurs when some characteristic of the interviewer, such as age or sex, affects the way in which respondents answer questions. For example, questions about racial discrimination might be differently answered depending on the racial group of the interviewer.

Sampling designs

In many surveys, the non-sampling error greatly outweighs the sampling error. It is therefore especially important to use a good sample design that minimises the non-sampling error. We will next describe some commonly used sample designs and their properties.

Convenience sampling

The simplest type of sampling scheme is a *convenience sample*, in which data are collected from whichever population units are easily accessible. A common form of convenience sample is voluntary response (or self-selection) sampling, often in the form of a write-in or phone-in response. This is certainly a cheap method of sampling but is almost sure to be strongly biased because voluntary respondents tend to have very different opinions or characteristics from those who do not choose to respond.

Probability sampling

Probability sampling designs are preferred, since such designs give each member of the population a known (sometimes constant) chance or probability (see chapter 8) of being selected. The use of chance in selecting a sample is an essential principle of good sample design. Any probability sample requires a list of the population units (a *sampling frame*) as well as a mechanism for random selection.

In simple random sampling, each unit in the population has the same chance of being selected. Other, more sophisticated probability sampling designs include stratified random sampling (in which a simple random sample is selected from each of several groups, or *strata*, in the population but the proportion of the units sampled is different in different groups). Like block designs for experiments (see section 7.4), stratified sampling designs allow efficiency gains through taking advantage of the known or anticipated variability between groups.

Systematic sampling

When there is a readily available list of the population units, it is sometimes both convenient and cheap to sample regularly spaced units. For example, every 100th item produced by a machine may be inspected for defects or every 10th household in a street may be given a questionnaire. This is referred to as *systematic sampling*. It is often essentially equivalent to simple random sampling, but sometimes can induce biases. This happens, for example, if every 10th unit happens to be a house on a corner block or if the interval between samples from a process coincides with some natural cycle.

Quota sampling

Conducting a survey using a probability sampling scheme can be difficult and expensive. Some less rigorous methods such as *quota sampling* are often used. Quota sampling does not require the construction of a sampling frame. While no list of the population is used, known population demographics are used to ensure that, in terms of key demographic variables, the sample is representative of the population. Each interviewer is given a *quota* of individuals with different characteristics to sample. In terms of other variables, including those addressed in the survey questions, the sample may not be representative of the population and, hence, may be biased.

Sample size

A major decision in conducting a sample survey is choosing the number of units from the population that will be selected in the sample — the *sample size*. There is often a requirement that the estimate of a population characteristic (mean or proportion) will be likely to be within a certain amount of the true, unknown population value. As the precision of an estimate depends on the sample size — the larger the sample size, the smaller the sampling error — the decision about sample size will depend on these precision requirements.

- The required sample size depends on the required precision of the estimates.

Details of how to decide on an appropriate sample size will be described later in this book. As precision depends not only on sample size, but also on the variability of the characteristic of interest within the population, a pilot survey is often needed before deciding on the sample size.

The original sample size is often reduced by non-response or invalid responses, and the resulting smaller sample provides less precise estimates. If we know something about the likely size of the non-response, we can make some provision for it. For example, if we decide that 400 respondents will provide adequate precision, and we know that the response rate will be only 50%, then we will have to select an original sample of size 800.

The strategy of selecting more sample units than we need as actual respondents solves the problem of increased sampling error due to non-response, but it does not solve the problem of non-response error. It is wise to try to keep the non-response rate to a mimimum. Follow-up mailings or telephone call-backs will usually produce more responses and reduce non-response error. However, in most situations where humans are sampled, some non-response is inevitable. The most resistant non-respondents may differ markedly from the majority of the population, so the non-response bias may still be considerable.

Finally, cost is a major consideration. It is often found to be too expensive to obtain the accuracy that a researcher really wants.

EXERCISE 7.3

1. Why was the *Literary Digest* poll so biased in predicting the outcome of the 1936 American presidential election? What kind of non-sampling error did it make?
2. What response rate (percentage of intended respondents who actually respond) is large enough that response bias is not a serious problem?
3. A limited amount of money is available to conduct a survey within a town. Taking into account the different types of sampling error, when would a telephone survey be more accurate than a survey conducted by face-to-face interviews?
4. A women's magazine asks readers to complete a questionnaire about feminism and send their responses to the editor. If the magazine intends to report on current women's attitudes to feminism, what types of non-sampling error are likely to be most important?
5. What are the advantages and disadvantages of a long questionnaire?
6. When is it not possible to select a random sample from a population of interest? Give an example.

7.4 Designed experiments

Experiments are studies in which something is done to a system to see how it responds. Experiments are aimed at discovering cause-and-effect relationships, or comparing the relative effect on a response of known causes. Good experimental design allows us to extract useful information from the results of such experiments, whereas it may be impossible to gain any information from a badly designed experiment.

A designed experiment involves a sequence of *runs* (or *trials*) in which deliberate and carefully planned variation is built into the process so that change in the output of the process can be observed and its source identified. These runs may be conducted on a process in some temporal order (such as measurements from the output of a water purification plant at regular intervals) or applied to physically distinct *experimental units* (such as trees, animals or small plots of land). Through the use of such experiments it is therefore also possible to choose the optimal combination of input variables or factors.

Designed experiments are used in such diverse applications as:

- assessing new crop varieties for maximum yield
- determining the best combination of input settings for a manufacturing process so as to minimise the variability of the output
- assessing the efficacy of drugs for medical treatments
- evaluating the effects of tyre construction on fuel consumption
- testing the efficacy of various learning strategies in the classroom
- taste-testing foods and drinks.

Objectives

An experiment may be conducted to address a variety of questions; these all relate to how the factors whose variation is controlled in the experiment (temperature, crop variety, dose of drug and so on) affect one or more measured outputs from the experiment (yield of wheat, blood pressure, exam marks and so on). The controlled variables are often called *input* or *explanatory* variables and the output variables are called *responses*. The objectives of an experiment may include answering questions such as those that follow.

- Which input variables most influence the response?

Many studies do not start with a clear idea of which of several factors will have most or any effect on the response measurement of interest. A psychologist may want to assess the relative effects of diet, sleep, age, sex and prior training and various other variables on the ability of a rat to learn some task. If a new strain of bacteria is discovered, we may want to discover which of a wide range of antibiotics will be most effective against it. The experimenter wishes to understand the process by discovering which variables control it and the relative importance of each variable.

- Which setting of the controllable variables best allows the optimal (or desired) response?

We often know which inputs to a process affect the output, but do not know which combination of input values will optimise the output. Temperature, pressure, reaction time and the concentrations of various catalysts will affect the purity of the output in many chemical reactions. How should the process be operated to obtain output of optimum purity? Similarly, we might ask which brand of motor oil is most effective in minimising engine wear.

In many situations, some settings of the controlled variables are more expensive than others, so the notion of optimality may also involve cost.

- Which setting of controllable variables minimises the variability of the response?

Often, we are interested in maximising the value of the output variable. Sometimes, however, we want to reduce the variability of the process. In manufacturing, much of quality control is aimed at reducing the variability of the output to obtain a more consistent product.

- Which setting of controllable variables minimises the effects of uncontrollable variables?

In many processes, the responses are affected by various factors that cannot be controlled by the experimenter. Crop yields are affected by weather; the results of experiments on rats may be affected by uncontrollable genetic features; the output of an industrial process may be affected by the quality of raw materials. We want the output to be affected as little as possible by such uncontrollable variables, thus making the process *robust*.

Designed experiments have a broad range of applications. They represent a crucial part of the scientific method and are thus a fundamental means of finding out how processes and systems work.

Some processes may be understood by making use of a mathematical model suggested by scientific theories. Comparing the model with the data helps to validate the model or refute it. An experiment in which the control variables proposed by the scientifc theory are varied deliberately, and the response observed, will provide the data with which to compare the model.

More often, there is no simple means of formulating a mathematical model. In such circumstances an empirical model must be developed and tested; that is, an experiment must be conducted so that a model describing the results can be found. Development of such ad hoc models is usually an iterative process that involves an initial guess about the form of a suitable model, collection of data to calibrate the model, assessment of the adequacy of the model, a series of refinements to the model and further testing of how well the model describes the process.

Models help to simplify the complex real world. A model is useful because of this simplification. When a model fails to predict exactly what response a system will produce for a given stimulus, this does not mean the model should be thrown out — if the simplification it provides helps us to understand and control the system, it is a useful model. Just like the architect's model of an urban development, it is judged to be a good model or a bad model by whether it helps people to visualise what the final development will be like and not whether the streets in the model are wide enough for real cars!

Model fitting and assessment are done through running designed experiments.

A well designed experiment can result in:

- improved process performance (as measured by output)
- reduced variability and closer conformance to desired levels
- reduced development costs and time
- processes that are simpler to run and more reliable
- better understanding of the process, leading to better control of it.

Statistical design

The purpose behind planning and designing experiments is to answer scientific questions as efficiently and reliably as possible. Experimental material (such as plots of agricultural land or rats) is usually variable and there is usually additional uncontrollable variability in the measurement process (such as measurements of crop yields or rat lifetimes). It is critically important that the experimental data allow the scientific effects of interest (such as the effect of a fertiliser or drug) to be separated from the underlying variability. This is *only* possible through statistical design.

There are three principal concepts underlying statistical design: replication, randomisation and blocking.

Replication

Replication simply means repeating all or part of the basic experimental design. There is generally some natural variability in the process under investigation that cannot be controlled. Repeating the experiment under the same conditions (as closely as can be managed) would result in different response measurements. This natural variability affects the accuracy of any conclusions drawn from the experiment. Replication allows variability to be estimated and therefore allows the likely precision of any conclusions to be described. Replication also allows averaging of the response, which produces more precise estimates.

> Replication provides increased precision from averaging and also allows the sampling error to be assessed.

Randomisation

Experiments are usually conducted on a group of distinct experimental units (plants, animals, plots of land, bottles of wine); the individual experimental units in the group usually have different characteristics that may affect the response being measured, but which cannot be controlled or avoided. It is desirable to select experimental units to be as similar as possible (for example, specially inbred strains of white mice), but some variation is inevitable.

If the seeds of two species of plant are to be planted in 20 greenhouse plots, 10 for each species, then the actual planting location in the greenhouse of the 10 seeds of each species has to be decided upon. This process is described as the *allocation of species to plots.* The generic term for the factor whose affect is to be observed and compared is a *treatment,* so we call this process *allocation of treatments to experimental units.* This allocation process is called random if the assignment of particular treatments to experimental units is determined by a random process such as the toss of a coin.

Random allocation of treatments to experimental units is an essential part of a designed experiment. In the greenhouse example, the fertility of the planting soil, the amount of light, the exposure to air currents or the closeness to irrigation outlets are all things that might make the greenhouse planting locations differ in their effect on plant growth. If the 10 greenhouse plots allocated to species *A* are all adjacent, then any observed differences between the species may be caused by systematic differences between the plots rather than by differences between the species. Randomly deciding which 10 plots will be used for species *A* minimises the chance of systematic differences between the plots being mistaken for differences between the species (called *confounding*). Furthermore, this method of treatment allocation allows us to predict the effect of the uncontrolled variation between the experimental units, as will be described in chapters 8 and 9.

> Randomisation helps separate treatment effects from other uncontrollable factors affecting the response.

Blocking

Sometimes, aspects of the natural variability of the experimental units can be determined before the experiment is conducted. For example, fertility may vary over a field; if the field is split into rectangular plots of land (the experimental units), adjacent plots would have similar fertility. If experiments are conducted at different times on an industrial process, there may be differences in process characteristics at 9 a.m. compared with 12.00 noon.

Rather than randomising allocation of experimental units in the experiment, it is preferable, if possible, to group the experimental units into *blocks,* so that the units are more homogeneous within blocks than overall. As the variability *within blocks* is reduced, comparisons within blocks can be made more efficiently. For example, it is more efficient to compare the weight losses of two people using different diets if they have similar characteristics (initial weight, age, sex, type of employment) than to compare the corresponding weight losses of people whose characteristics differ. A well designed experiment which incorporates blocking can increase the statistical precision of a design and can lead to significant efficiency gains.

An important example of blocking occurs when the experimental units are grouped into pairs of similar units. The pairing may be done on the basis of measured characteristics of the experimental units (for example, weights of pigs) or may have played a part in the choice of the experimental units themselves (for example, the use of identical twins in psychological and medical studies). If two different types of experimental condition are being compared (for example, two brands of tyre), then the two experimental conditions are (randomly) allocated to the two units in each pair. The efficiency gains from using such *paired experiments* can be large.

> Blocking reduces the response variability attributable to differences between experimental units.

Designing an experiment

There is a well-defined sequence of planning steps in designing a good experiment.

> 1. Define and state the questions that the experiment will be used to answer.

A clear idea of what the experiment will be conducted to achieve is essential before any other aspects of the experiment are designed. In defining the goals of the experiment, it is important that people with intimate knowledge of the process or subject area are included in the team which is charged with designing and running the experiment.

Many experiments are intended to solve a problem. After recognising that there is a problem, it needs to be carefully defined. Stating the problem clearly and obtaining general agreement that this statement really does describe the problem is imperative. Quite frequently, a clear statement of the problem can lead to process improvement without any experimentation, simply through creating a greater understanding of the process.

> 2. Choose the experimental units.

As discussed earlier, it is desirable for the experimental units to be as similar as possible, so every attempt should be made to make the experimental units homogeneous. We should therefore characterise the process in terms of uncontrollable variables and endeavour to find ways of minimising this variability for the experiment.

Often, however, the experimenter has little influence on the choice of experimental units and must contend with whatever variability exists. If possible, the experimental units should be grouped into blocks which can be used later in the design process to obtain more precise answers to the questions of interest.

3. Choose response variable (or variables).

Sometimes there is a single obvious response measurement from an experimental unit (for example, crop yield per square metre, concentration of impurities, exam mark), but often there are several variables which can be considered response measurements.

4. Choose the factors or variables and levels for controllable inputs.

Thought needs to be given to which variables need to be controlled (the input variables) and what settings should be used for these variables in different experimental runs. In an agricultural experiment, do we want to assess only the difference in yields for three crop varieties or do we simultaneously want to determine the effects of different levels of application of fertiliser? (And if so, what levels should be used?)

5. Choose the experimental design.

A decision must be made next about how many experimental units will be used in the experiment (the sample size) and how the various experimental conditions (values for the input variables) will be allocated to the experimental units.

Replication, randomisation and blocking need to be taken into consideration at this stage. Good design here will allow questions about the process to be answered efficiently, whereas bad allocation of variable settings to experimental units may even prevent some questions from being answered at all.

A detailed experimental protocol should be produced which spells out precisely which experimental conditions are used for which unit and (where appropriate) the run order.

6. Carry out the experiment, analyse the results and draw conclusions.
Make recommendations.

A single experiment is often insufficient. Pilot experiments may be necessary to help define the problem and to set up sensible conditions for the main experiment. There may also need to be a continuing cycle of experimentation. Experimentation is most commonly interactive. The conclusions from each experiment are input to the planning and design of the next experiment.

While very complex designs can be appropriate and some of these require very sophisticated analysis techniques, a simple design involving analysis techniques which do not go far beyond the scope of those presented in this book, coupled with appropriate data presentation and summary, can be very effective.

Simple experimental designs

Completely randomised design

The simplest type of design is one in which two or more settings for the input variables (*treatments*) are compared in a simple comparative experiment. The treatments usually represent different levels of a single factor; these levels may be either numerical or categorical. Treatments are allocated randomly to experimental units in what is referred to as a *completely randomised design.*

Usually, each treatment is replicated the same number of times. In some experiments, all treatments other than one (the *control* treatment) involve some change in conditions for the experimental units; the design sometimes allows greater replication of the *control* treatment. For example, if three different feed supplements for chickens are being tested, the control group (which gets none of the supplements) may contain more experimental units than the other groups.

The data resulting from a completely randomised design may be displayed using the methods in sections 2.7 to 2.9 (if the treatment is categorical and the response is numerical), chapter 4 (if both the treatment and response are numerical) or sections 5.3 and 5.4 (if the response is categorical).

Randomised block design

Sometimes there will be substantial variation in the experimental units. To minimise the effect of this, the units are grouped together in homogeneous blocks. In some applications, such as agricultural experiments, a block consists of a group of physically adjacent units, such as neighbouring plots of land. In most randomised block designs, each block contains the same number of experimental units and the treatments are replicated equally often within each block. Typically, each treatment occurs just once in each block. Allocation of the treatments to experimental units within each block is done randomly.

Factorial designs

Factorial designs are used when there are two or more different input variables. In order to assess the simultaneous effects of these variables, it is not sensible to carry out a sequence of 'one-at-a-time' experiments since the effects of the variables may interact with each other. (The effect of temperature on the output of a fermentation process might be different if the sugar content of the grapes is low instead of high.) Even if the input variables do not interact in this way, there are more efficient ways to estimate the effects of each variable than one-at-a-time experiments.

In a factorial design, every possible combination of the chosen levels of all of the factors is used. Thus, in an experiment with three factors, each at three levels, the number of possible combinations is $3 \times 3 \times 3 = 27$.

Often, it is also not practical to carry out experiments at every desired combination of levels of the variables. A *fractional factorial design* provides an efficient solution to such problems through careful selection of a (balanced) subset of all possible combinations of the factor levels. This avoids the cost of huge experiments if there are many factors (even if the number of levels for each is as small as two). Detailed discussion of these, and other complex designs, and their analysis are beyond the scope of this book.

Practicalities

There are a few practicalities that complicate experimentation with human subjects. For ethical reasons, experiments involving potential danger to the subjects are not possible. Even if there is no known danger, the subjects should be aware of what is involved in the experiment and must give informed consent.

There are many instances where an experiment is intended to measure the effect of a particular treatment, such as the improvement of a medical condition caused by administration of a particular drug. A naive experimenter may record the value of some variable (for example, concentration of some chemical in the blood) before medication is commenced and also after the medication has been used for a week. However, any improvement in the condition may have resulted simply from the passage of time and it may not be related to the drug. In order to assess the effect of the treatment, some (randomly selected) subjects who have *not* received the treatment should also be included in the study — differences between the improvements of the two groups can then be attributed to the treatment. Subjects who receive no treatment are called *controls.*

Unfortunately, the act of administering a treatment to a human subject may itself affect the response, irrespective of the treatment effect. For example, if a drug is being assessed for its ability to reduce headaches, the knowledge that medication has been administered may make the subject feel better, even if the drug has no active ingredient. To avoid the psychological effect of the treatment on the subject being confounded with the effect of the drug, an indistinguishable 'treatment' with no effect may be given to the control group of subjects; this is called a *placebo.* For example, two batches of pills of similar size and taste may be prepared, with only one batch containing the active ingredient being assessed. Any difference between the control group and the treatment group can therefore be attributed to the treatment.

A further complication may arise when the act of measuring the response from each subject may be affected by knowledge of the treatment applied. If the experimenter knows which treatment has been applied to each experimental unit, there may be a subconscious tendency to systematically over- or under-assess one treatment. To avoid this potential problem, the experimenter may be unaware of which experimental units received which treatment until after the experiment. If the treatment being used is unknown while the experiment is being conducted, the experiment is called *blind*.

The experiment is called *double-blind* if neither the experimenter nor the subject knows which treatment has been applied. For example, a third party may randomly decide which of two drugs will be given to each subject and package the appropriate pills for each subject in unlabelled containers. The experimenter would administer the treatments and record results without knowing which subjects were receiving which treatments. Again, the aim is to ensure that other factors do not confound comparisons of the treatments.

EXERCISE 7.4

1. A simple cake recipe is to be varied to determine if it can be improved. Small cakes will be baked in a fan-forced oven large enough to hold 6 of these cakes at a time. A fairly even temperature is kept in different parts of the oven, but the temperature may vary during the experiment.
   (a) Suggest a 'response' measurement that might be used to assess each cake.
   (b) How might you define blocks in this experiment?
   (c) Suggest two 'treatments' that may be of interest. Suggest different levels for these treatments.
   (d) In an experiment investigating a single ingredient, how would you make use of blocking?
   (e) In the experiment in part (d), how would you randomise, and what would this achieve?
2. How would you make the cake recipe experiment in question 1 a double-blind experiment?
3. What is the purpose of randomisation of treatments to experimental units?
4. In the cake baking example, if 2 levels of sugar, 2 levels of water and 2 levels of baking powder were to be used in a factorial experiment, how many runs would the experimenter need in a single replication?
5. An experimenter who is interested in comparing 3 varieties of wheat sows 1 field with each, then measures the yield of wheat from each field.
   (a) What are the problems with the design of this experiment?
   (b) An improved design involves splitting each of the 3 fields into 6 subplots. Explain how you would use replication, randomisation and blocking to conduct the experiment.
6. R. A. Fisher is often referred to as the father of the discipline of statistics. In commenting on a certain study design, he is reported to have said 'That's not an experiment, that is an experience!' How would you paraphrase this quotation in the language of this chapter?

Problems

1. In a study of traffic accidents, it is determined that 40% of accidents are caused by male drivers between the ages of 16 and 21. This would tend to support the widespread belief that young male drivers are more reckless than other drivers.
   (a) What other explanations, other than recklessness of males in this age group, might logically explain the study outcome?
   (b) How would you further investigate your explanation in part (a), using data that might already be available?
   (c) Discuss the feasibility of a designed experiment to determine if young male drivers really are involved in more than their share of accidents.

2. A petrol company wishes to test the relative effectiveness of two proposed incentive strategies on the quantity of petrol sold. One strategy is to give away a free glass with each purchase of 20 litres of petrol, and the other is to give away a lottery ticket with each such purchase. The company has 125 petrol stations but wants to try out the incentive scheme on only six of them, extending the more successful strategy to the additional 119 stations in one week's time. The supply of lottery tickets and glasses is ready to go and the whole experiment is expected to be completed in one week.
   (a) Propose details of the experiment under the following headings:
      (i) statement of the research question
      (ii) selection of experimental units
      (iii) allocation of treatments to experimental units
      (iv) measurement of the response. (*Hint*: Is your suggestion of a response measurement comparable between treatments?)
   (b) How might the average weekly volume of petrol sold over the past year at each petrol station be used as a blocking variable?

3. In order to study the effect of eating a high carbohydrate meal just before an athlete runs a 10 kilometre race, the following research designs are proposed:
   (a) Select 20 participants in the race who are willing to participate in a study which will require them to eat a small meal one hour before the race. Then assign these 20 at random into two groups of 10 participants, where one group gets the high carbohydrate meal and the other gets a low carbohydrate meal.
   (b) Ask 10 participants in the race to eat a certain high carbohydrate meal one hour before the race and compare their race results with all (100, say) of the other participants.
   (c) Ask 20 participants to fill out a short questionnaire about what they have eaten in the two-hour period prior to the race. Divide the group of 20 into two groups of 10 with one group having the highest carbohydrate consumption prior to the race.

   Discuss the advantages and disadvantages of these proposed designs.

4. In an experiment on pole-vaulting techniques, three types of pole (*A*, *B*, *C*) are used. Nine jumpers volunteer to participate in the experiment. For each of the nine jumpers, one of *A*, *B* or *C* is selected at random. The jumpers try several jumps with that pole and determine the maximum height that they can clear. Each jumper has three attempts at each of a succession of heights until a height is reached that cannot be cleared in the three attempts. The distributions of the recorded heights for each type of pole, *A*, *B* and *C*, are compared to determine which type is the most successful.

   Criticise the design of this experiment under the following headings:
   (a) control of extraneous factors
   (b) control of sample size for each treatment
   (c) the measurement of the response variable
   (d) the order of the jumps
   (e) anything else you think is important.

5. Explain how you would design a research study that would compare the effectiveness of two weed killers. Both weed killers *A* and *B* are known to be effective, but it is not known which is more effective. Assume that the weed killers must be tested in an area in which several different weeds are common and which includes many different conditions of soil, moisture, exposure to sun and so on. (For example, you may have to use your neighbour's lawns, even though they vary in many relevant characteristics.) No record exists of the particular characteristics of the potential experimental plots. Your design should include some suggestions for the analysis of the data. (*Hint*: A simple design should be adequate here.)

6. A commercial nursery has an automatic system to deliver water to the plants through small hoses. The distribution of the water fed to the plants is hard to control exactly, and some plants inevitably receive more water than others. An observant plant nursery attendant notes that the plants in the greenhouse seem to droop when they are given more water than usual. This observation is confirmed by careful measurements, taken over a one week period, of the amount of water applied (to a series of containers of fixed size) and a subjective measure of the 'droopiness' of the plants. The results seem to show a clear correlation between the excess water and the droopiness. Is there any doubt that the excess water caused the droopiness? Discuss briefly with reference to the design of the study.

7. Design an experiment to compare the effect of alcoholic intake on the night vision capability of adults. That is, write a short proposal outlining your objective and method of data collection. Keep in mind the following constraints:
   (a) There may be a dependence of the response on age and sex.
   (b) Subjects not used to drinking alcohol may not wish to participate.
   (c) Experiments with human subjects must satisfy a review of ethical standards.
   (d) Experimental supplies may be expensive.
   (e) Food intake may lessen the intensity of the response.
   (f) Visual perception cannot easily be measured objectively, since subjects must report what they see. Eyeglass use may be a factor.

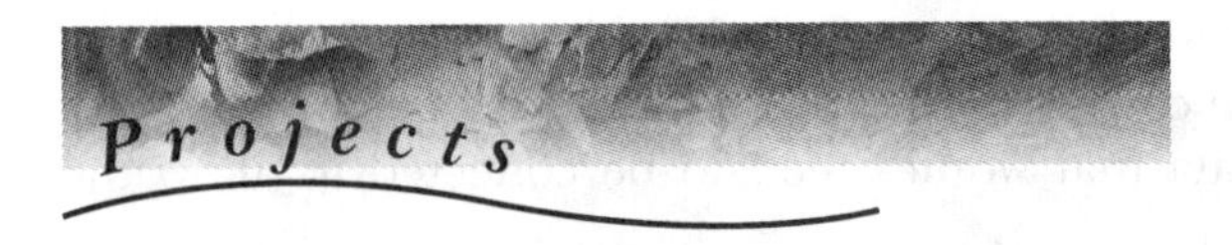

Projects

1. **Paper aeroplanes — time aloft**
   Design and carry out an experiment to determine the design factors which enable a paper aeroplane to stay aloft as long as possible. Use the same size and weight of paper to construct all aeroplanes, but vary some aspects of the design as described below. Factors to assess in your experiment should be:
   (a) horizontal area of the wing
   (b) size of trim tabs (vertical tabs at the back of the wing — see diagram below)

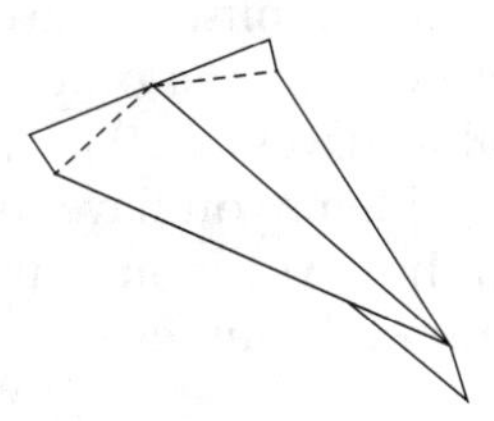

   (c) position of centre of gravity.
   Use just two values (*levels*) for each variable. The levels should be fairly close to ones you think will work best, so that designs that do not 'fly' will be avoided. You may need a definition of what qualifies as 'flying' if you have any doubt.
   You should use a standard launching technique and do your experiment in a large room (or outside if you have a very dry, windless day). A stop-watch would be helpful to count the seconds until touchdown, since a typical time will be in the range 1.0 to 3.0 seconds. Report your results in the following format:
   - statement of objective
   - definition of the levels of each variable used
   - what you have done to minimise unexplained variability between runs of your experiment
   - average time aloft for each of the eight designs, and the standard deviation of times aloft for each design
   - a tabular or graphical display that makes it easy to assess which levels of the factors make the aeroplanes stay aloft longer
   - conclusions.

2. **Research news**
   Look through some newspapers for articles about the results of a data-based research study. Pick one that interests you and consider how the data were collected to enable the researchers to make the conclusions reported. It is likely that the data collection methods will not be described in the newspaper article, but you can still discuss what sort of data collection was likely done. Newspaper articles usually do make some comment on the conclusions of a research study, if they report it at all. The question you should attempt to answer is: 'How could the data be collected that would support the conclusion that was drawn?'

Your report should include the following components:

(a) a brief summary of the conclusion reported in the newspaper report

(b) a suggestion of the data that would need to be collected to allow the conclusion to be drawn

(c) a comment on how likely it is that the required data were actually collected for the study reported.

3. **Student relationships**

In this project, you should design a survey to assess the proportions of students in your university or college who are married, engaged, have 'steady' partners or are unattached. You should consider how you would select your sample, how to approach them, the wording of your questionnaire and how you will present your results. You are encouraged to attempt this project in collaboration with one or two other students.

(a) Conduct a pilot survey of 20 individuals.

(b) Discuss any problems you have identified with your survey methodology and consider how you could try to reduce them. (However, do not conduct a large-scale survey.)

CHAPTER 8

From populations to samples — sampling phenomena

> *Certainty generally is illusion, and repose is not the destiny of man.*
>
> Oliver Wendell Holmes (1841–1935)

8.1 INTRODUCTION

Extracting information from data is both an art and a science, as you will realise from the earlier chapters. A complication in interpreting data is that they exhibit random variation — variation that has no immediate explanation — and this variation or 'noise' often obscures the information or 'signal' in the data. One way of coping with this randomness is to imagine an underlying *population* of measurements that could be obtained by extending data collection to a much larger number of measurements. Sometimes this population has a real existence (such as when students are sampled from a large class), but in other cases the population is imaginary (as in measurements of alcohol concentration in homebrewed beer). In both cases the population has a fixed, unchanging nature. In contrast, our data are a *sampling* of values from this population and each sample will be different. However, each sample exhibits and reveals the variation within the population and so can be used to infer something about the population.

Why is the discipline of statistics so interested in describing an unobserved population (which may not even exist), instead of the very real data in our hands? Science is interested in describing phenomena that are reproducible, so we need to assess the extent to which our sample can be relied upon for reproducible information.

We begin our discussion of sampling by considering situations in which the population is real and finite, and we demonstrate that selections from such a population also produce samples which exhibit variation; opinion polls are perhaps the most widely known use of samples of this kind. However, you should realise from the very start that our model of populations and samples equally well describes any measurements which exhibit unexplained variation; that is, variation observed when a repeated measurement does not necessarily give the same value and there is no particular explanation for why this is so. We will describe random variation as the result of random sampling from a population, even when the population may not exist as a real object.

EXERCISE 8.1

1. Which of the following populations are real and which are hypothetical?
   (a) population of potential voters in an election
   (b) population of potential measurements of carbon from a soil sample
   (c) population of areas of leaves from a particular tree
   (d) population of wheel diameters produced by a forging process
2. For each of the following contexts, why is it useful to conceive of a population that is not real?
   (a) The average time aloft is to be determined for a particular design of paper aeroplane, based on several flight trials.
   (b) Samples of water are to be taken from a river, both downstream and upstream from a certain town, over a one week period, at random times and locations. The aim is to assess the amount of acidity in the river.
   (c) The yield of tomato plants from a particular batch of seed is to be assessed, based on the yield of 25 plants planted in a greenhouse laboratory.

8.2 Populations and sampling

Populations

To a layperson, populations of interest might include a population of voters in a certain electoral district, a population of stray cats within a certain municipal boundary or a population of farmers in a certain rural area. In statistical work, however, the populations of interest need not be people or animals and would include a population of oak trees in a certain national park, a population of bottles of acid in a chemistry lab or even a population of soil in a farmer's field. In fact, the most common population considered in statistics is a population of numbers (for example, the population of household incomes for all households in a certain country). As you can see, the everyday notion of a population has been extended considerably.

In statistical studies, the population is usually the entity of interest, but it is often impractical to measure the population completely. For example, prior to an election, polls are conducted almost daily to gauge voting intentions and responses to the policies of the political parties. If each poll had to include the entire voting public, it would be too costly and time-consuming to be practical. Fortunately, such *census* polling, in which the entire population is polled, is unnecessary. A properly selected sample from the electorate will provide almost as reliable information as a census, yet may sample only 1000 individuals instead of millions. A census is therefore usually not only impractical but also unnecessary.

Random sampling

Sampling strategies are a very efficient way of obtaining information about a population. However, in conducting sample surveys, the method of selecting the sample from the population is of crucial importance. We want the sample to be 'representative' of the population in its important characteristics. As described in chapter 7, poorly selected samples can lead to very misleading results.

Suppose it is your job to examine the costs of textbooks used in current courses at your university — perhaps the student council is grumbling about the increasing cost of tertiary study! You decide to sample this population of textbook costs and want your sample to be as representative of the population as possible. You want a sampling method that produces a list of 25 courses in a way that is 'fair' — any particular selection of 25 courses should be as likely to be sampled as any other particular selection. A sample collected according to this prescription is called a *simple random sample* of 25 courses from the population of all courses; the resulting 25 textbook costs from these courses will similarly be a simple random sample from the population of all textbook costs. A great deal of statistical theory is based on samples collected with this sampling scheme.

There are several ways to select a random sample from a population. One way that works is to create a list of the members of the population and number them from from 1 to N, where N is the number of members in the population. A computer can repeatedly generate 'random' numbers in the range 1 to N, until 25 numbers have been selected without repeats. These 25 numbers identify the 25 population members that would be sampled.

Applying simple random sampling in this way to the list of first year courses would generate 25 courses whose textbook costs would be collected; statistical processing would be conducted on the resulting 25 textbook costs. The sample of costs (in dollars) might look like the following:

47	86	111	0	95	51	86	88	17	20	75	91	81
24	19	0	31	54	19	0	0	70	62	73	62.	

These data provide a basis for commenting on the costs of textbooks in first year courses. Whether or not the basis is a reliable basis will be discussed in section 8.3.

In the sampling scheme above, we specified that the 25 courses in the simple random sample should be selected 'without repeats'. This type of sample is called a simple random sample *without replacement* (*SWOR*). In some circumstances, sampling without replacement is not practical, and we allow the same population member to be selected two or more times. For example, when sampling wildlife populations such as fish that are released after capture, it may be costly to ensure that the same individual is not subsequently recaptured. When each individual in the population is equally likely to be selected at each stage of the sampling process, the sampling scheme is called a simple random sample *with replacement* (*SWR*).

SWOR is generally preferred when practical, but for various reasons we will concentrate on SWR in the remainder of this book. Firstly, if the population is large enough, it does not matter whether we use SWR or SWOR; we get the same kinds of samples in either case, because the chance of a repeat is negligible. Secondly, statistical theory is much simpler for SWR. You should note, however, that if the population is small, it can make a difference. For example, a population of numbers {1,2,3, ... 10} can produce samples of size 5 like {1,1,1,1,1} or {1,1,2,2,2} even though these are not very likely.

To allow us to avoid the compexity of SWOR, we will deal only with sampling situations where the size of the sample is small compared to the population size. To obtain a sample like {1,1,1,1,1} from a population like {1,2,3, ... 100} is *very* unlikely, and it is of little consequence if we use a SWR model to describe what is actually a SWOR process. In opinion polls, we would not ask the same person their opinion twice, even in the unlikely event that they were selected twice; however, it will be convenient to think of the sampling process as SWR, for technical simplicity.

Descriptions of populations

Before examining the properties of simple random samples, we will consider how to describe populations of numbers efficiently, without presenting the entire list of N numbers. This is not only a convenience; it enables us to focus on characteristics of a population called *parameters*, which help to make precise our notion of population information.

Chapter 1 provided many ways to summarise a set of data, and all of these summaries can be used to describe a population of N values. Two features of numerical populations of special interest are the location and spread of the numbers. As discussed earlier, the most common summaries of these aspects of a distribution are the mean and the standard deviation. When descriptive of populations, the mean and standard deviation are called *population parameters.* The corresponding sample summary numbers are called *sample statistics.*

To reduce the potential confusion from having population summaries and sample summaries with the same name (and measuring similar characteristics), we conventionally use Greek letters to describe the population parameters. The population mean is traditionally denoted by the Greek letter μ (pronounced 'mu'), while the population standard deviation is denoted by σ (pronounced 'sigma'). We conventionally denote the population size by N and the size of a sample from that population by n.

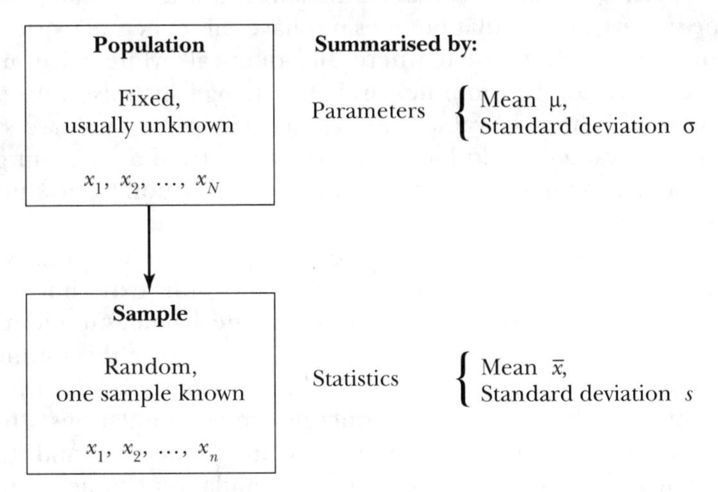

The formula used to compute μ is the same as the formula to compute $\bar{x}$, except that for μ we average all the population values instead of just the sampled values. Similarly, the formula for σ is similar to that for s; the only difference being the use of $n - 1$ in the denominator of s instead of the N in the denominator of σ (when the population size, N, is large, the distinction is negligible).

	Mean	**Standard deviation**
Sample statistic	$\bar{x} = \dfrac{\sum_{i=1}^{n} x_i}{n}$	$s = \sqrt{\dfrac{\sum_{i=1}^{n} (x_i - \bar{x})^2}{n-1}}$
Population parameter	$\mu = \dfrac{\sum_{i=1}^{N} x_i}{N}$	$\sigma = \sqrt{\dfrac{\sum_{i=1}^{N} (x_i - \mu)^2}{N}}$

In spite of the close similarity between their formulae, it is important to distinguish between the sample statistics and the population parameters. In most practical situations, $\bar{x}$ and s are known values (computed from available sample data) while the values of μ and σ are unknown. Although the population parameters are of most interest, our only information about their values comes through the sample — in particular, the sample statistics.[1]

Infinite populations

In some situations, such as opinion polls or our example of textbook costs, there is a well-specified finite population from which a random sample is selected. (We do not know the population values but we can, at least in theory, write down a list of the individuals comprising the population.) There are, however, many other situations in statistics where a 'sample' is collected, but the underlying population does not have this concrete existence. For example, consider an experiment where 10 'identical' white mice are subjected to a loud noise for two minutes and the change in pulse rate of each mouse is recorded. These 10 values (the changes in pulse rate) are sample data since different values would have been recorded from a different group of 10 mice in the same experiment, but what is the population from which the sample has been selected?

In statistics, we define the underlying population for this type of situation to be the hypothetical results from repeating the experiment indefinitely with more and more 'identical' white mice. The 10 measurements in our data can be considered to be a simple random sample from this infinite population.

There is no practical difference between our use of a hypothetical infinite population and the earlier concrete finite populations. In both cases, the population is unknown, both in shape and in mean, μ, and standard deviation, σ. The only window we have into the population characteristics is through the single set of sample data that has been collected.

Although μ and σ are important characteristics of a population, they do not completely describe its frequency distribution. Finite populations can be displayed graphically using the summaries from chapter 2, but how can we graphically summarise the frequency distribution of an infinite population? The graphical displays that we use for an infinite population are closely related to the displays that are appropriate for samples from that population.

If the population is discrete (e.g. counts), a bar chart of the population provides a graphical description of the distribution's shape. For example, consider a data set consisting of the numbers of eggs in nests belonging to birds of a particular species. The collected data could be considered to be a

---

1. Although the mean and standard deviation of a population are its most important characteristics, note that they do not fully describe the distribution of values in the population. For example, the two sets of seven numbers:

0, 3, 3, 3, 3, 3, 6
and:
1, 1, 2, 3, 4, 5, 5

each have mean 3 and standard deviation 1.7321, but the 'patterns' of their distributions are very different.

random sample from the hypothetical population of numbers of eggs from that species in general, and this population could be displayed with a bar chart, as shown below.

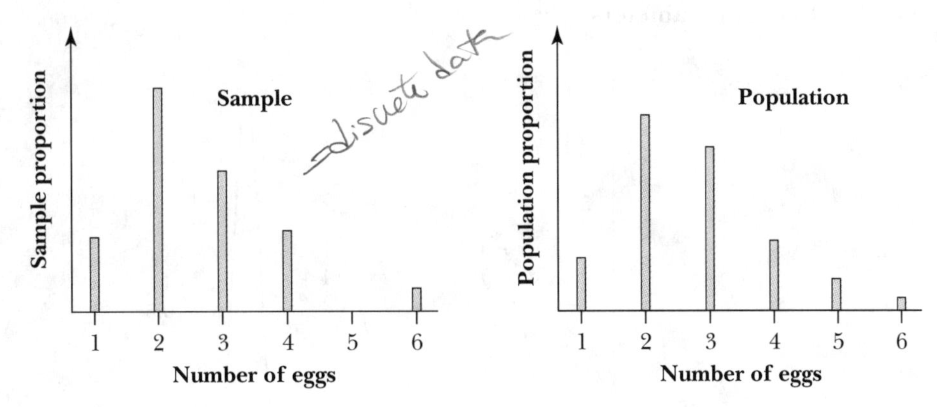

If the sample (and population) values are continuous, rather than discrete, histograms would be appropriate graphical summaries of the sample and population. However, as noted in chapter 2, the more values in a data set, the narrower class widths can be made while still retaining smoothness in the display. Taking this to the extreme of an infinite population, the population histogram can usually be shown as a smooth curve. For example, the change in pulse rate of white mice might be represented by the following sample and population histograms.

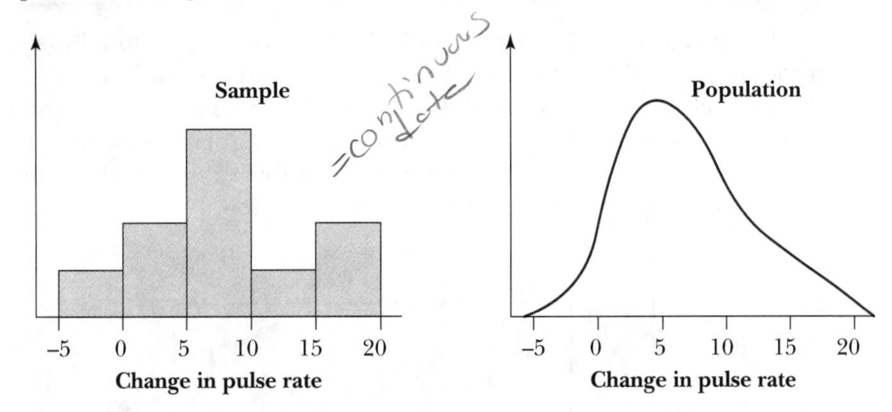

As we noted earlier, the shape of a population distribution is usually unknown in practical applications. Fortunately, a few shapes are quite common in practice, and one of the most famous families of distributions is called the family of *normal* distributions.[2] Normal distributions are appropriate for continuous measurements; they are all symmetrical

2. Do not confuse the term 'normal' in this technical sense with the usual English meaning of the word. There are many populations that are quite ordinary and 'normal' in the usual sense that are not 'normal' in the technical sense used for statistics.

(around their mean) and all have the same bell shape. A normal distribution of numbers is completely determined by its mean, μ, and standard deviation, σ. The diagram below shows four normal distributions with different parameters.

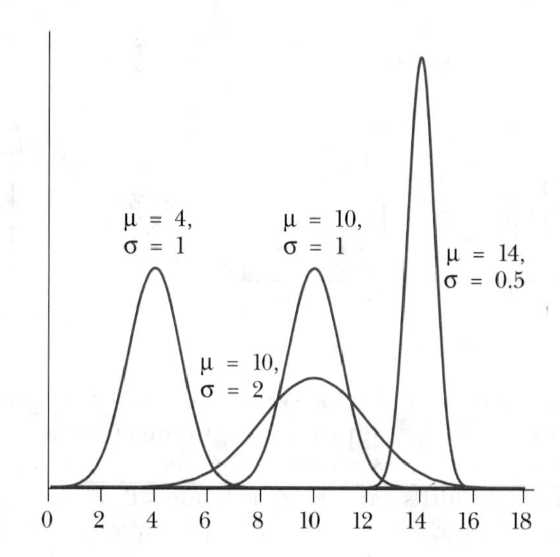

Note that changing the parameter μ shifts the distribution, whereas changing σ adjusts its spread. In all cases, however, the distribution is symmetric and bell-shaped.

Although a normal distribution is the most common model that is used for a population, there are other shapes of distribution that are also used. Another class of symmetric distributions is the family of *rectangular* distributions. For example, if a population has a rectangular distribution defined on the interval between 1 and 6, then every possible value between 1 and 6 occurs with equal frequency. Its histogram is shown below.

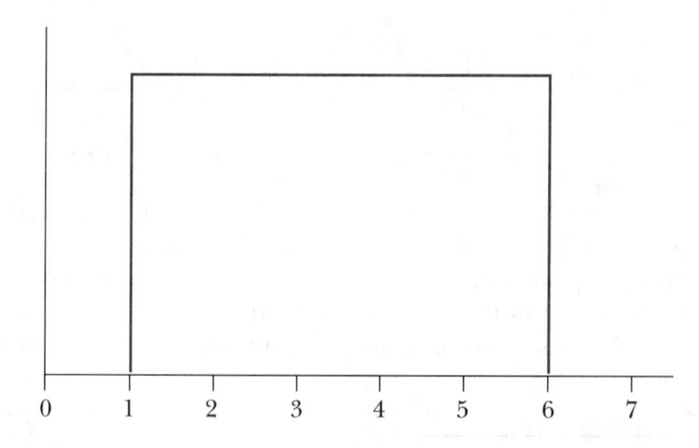

A third family of distributions that is sometimes used for infinite populations of continuous measurements is the family of *exponential* distributions, some of whose shapes are shown in the following diagram.

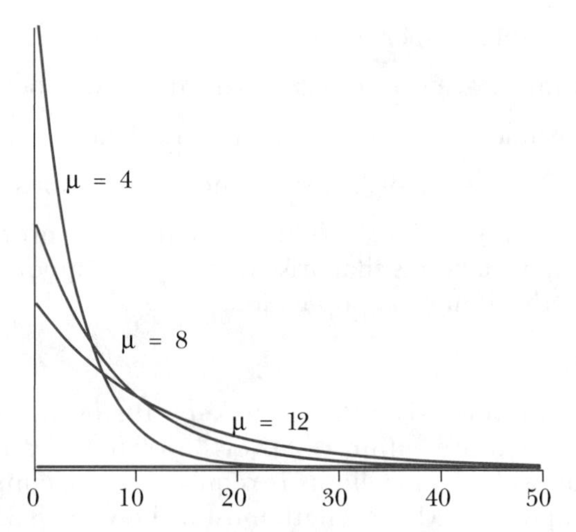

There are clearly many shapes of distribution that may describe an infinite population of values. You should note that while the population mean, μ, and standard deviation, σ, describe two important characteristics of a population, the shape of the distribution is something completely separate from this. Each of the three infinite populations shown below has mean $\mu = 4$ and standard deviation $\sigma = 4$, but the shapes of their distributions are quite different.

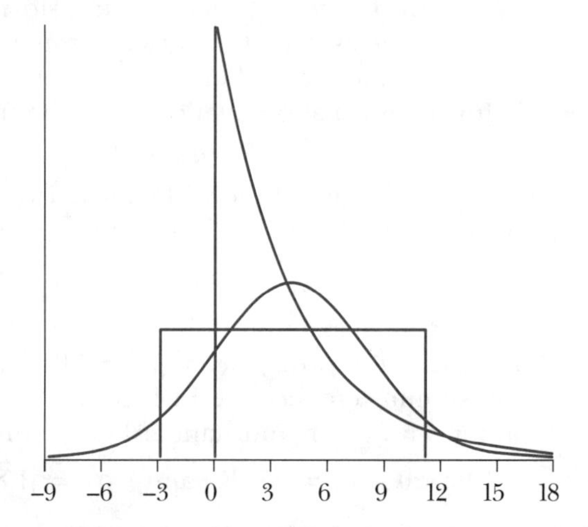

Indeed, the finite population consisting of only the two values 0 and 8 also has the same mean and standard deviation.

Given this warning, it is perhaps surprising that the normal distribution describes the shape of many empirical data sets. This is a very useful empirical finding, because we can describe a normal population completely by specifying that it is normal, and quoting its mean and standard deviation.

The following properties hold for all normal distributions:

- 68% of values are within one standard deviation of the mean
- 95% of values are within two standard deviations of the mean
- 99.7% of values are within three standard deviations of the mean.

The comparable properties of the tails of the distribution are that:

- 16% of values are over one standard deviation above the mean
- 2.5% of values are over two standard deviations above the mean
- 0.15% of values are over three standard deviations above the mean.

In section 2.9, the '70-95-100 rule' noted that these properties also approximately hold for many data sets that arise in practice, supporting the wide applicability of an underlying normal population.

Normal tables

In many situations, it is sufficient to know the proportion of values in a normal population that are within σ, 2σ or 3σ of the mean, μ, but occasionally the proportion within other limits is required. For example, a manufacturer produces components whose length must be between 8.21 cm and 8.32 cm for use in a particular application. If component lengths have a normal distribution with mean $\mu = 8.26$ cm and standard deviation $\sigma = 0.024$ cm, what proportion of them will be rejected as unsuitable for this application?

Some statistical software can directly answer questions of this form. However, to obtain an answer without a computer, we need to understand an important property of normal populations. In section 2.5, we noted that a linear transformation of values in a batch results in a transformed batch whose distribution's shape is the same as that of the original values, but whose location and scale are different. The same result holds for infinite populations. In particular, if the values, X, form a normal population, then the transformed values:

$$Y = aX + b$$

also have a normal distribution. Indeed the mean, μ_Y, and standard deviation, σ_Y, of Y are related to those of X with the simple formulae:

$$\mu_Y = a\mu_X + b \quad \text{and} \quad \sigma_Y = |a| \times \sigma_X.$$

For example, if the daily maximum Celsius temperature in a city in March, X, has a normal distribution with mean $\mu_X = 20°C$ and standard deviation $\sigma_X = 3°C$, this maximum temperature in Fahrenheit, $Y = 1.8 \times X + 32$, has a normal distribution with mean and standard deviation:

$$\mu_Y = 1.8 \times 20 + 32 = 68°\text{F} \quad \text{and} \quad \sigma_Y = |1.8| \times 3 = 5.4°\text{F}.$$

We can use this result to transform *any* normal population into a common *standardised* form, by subtracting that distribution's mean, and then dividing by its standard deviation:

$$Z = \frac{X - \mu}{\sigma}.$$

Whatever the distribution of X, the distribution of Z is normal with mean and standard deviation:

$$\mu_Z = \frac{\mu - \mu}{\sigma} = 0 \quad \text{and} \quad \sigma_Z = \frac{1}{\sigma} \times \sigma = 1.$$

This distribution is called the *standard normal distribution.* The transformed measurement, Z, is often called a *z-score* and simply re-codes the values as numbers of standard deviations from the mean.

At the start of this section, we described an example in which components are produced whose lengths have a normal distribution with mean $\mu = 8.26$ cm and standard deviation $\sigma = 0.024$ cm. We are interested in the proportion of components whose lengths are greater than 8.32 cm, or less than 8.21 cm. The value 8.32 cm is:

$$z = \frac{8.32 - 8.26}{0.024} = 2.50$$

standard deviations above the mean. Similarly, the value 8.21 cm is:

$$z = \frac{8.21 - 8.26}{0.024} = -2.08$$

standard deviations above the mean or, in other words, 2.08 standard deviations below the mean.

Any question about the original measurements can be rewritten as a question about z-scores. For example, our question about the proportion of components outside the range 8.21 cm to 8.36 cm is equivalent to asking for the proportion of z-scores outside the range −2.08 to 2.50. This simplifies problems since we now only need to deal with a single distribution — the *standard* normal distribution — and tables are available to help.

We can therefore obtain the proportion of components with lengths less than 8.21 cm by determining the proportion of z-scores, in a standard normal population, that are under −2.08. The tables in the Appendix allow us to read off this proportion.

.09	***.08***	**.07**	.	.	.	**.01**	**.00**	*z*
			.	.	.		.0002	**−3.5**
.0002	*.0003*	.0003	.	.	.	.0003	.0003	**−3.4**
.	.	.	.	.	.	.	.	.
.	.	.	.	.	.	.	.	.
.	.	.	.	.	.	.	.	.
.0143	*.0146*	.0150	.	.	.	.0174	.0179	**−2.1**
.0183	*.0188*	*.0192*	.	.	.	*.0222*	*.0228*	***−2.0***
.0233	*.0239*	.0244	.	.	.	.0281	.0287	**−1.9**

Therefore:

$$\Pr(X \leq 8.21) = \Pr(Z \leq -2.08) = 0.0188.$$

Similarly, the proportion of components less than 8.36 is the same as the proportion of z-scores less than 2.50, which can be read off the second page of tables in the Appendix, as is illustrated at the top of the following page.

z	.00	.01	.	.	.
0.0	.5000	.5040	.	.	.
0.1	.5398	.5438	.	.	.
.	.	.	.	.	.
.	.	.	.	.	.
.	.	.	.	.	.
2.4	.9918	.9920	.	.	.
2.5	.9938	.9940	.	.	.
2.6	.9953	.9955	.	.	.

Therefore:

$$\Pr(X \leq 8.36) = \Pr(Z \leq 2.50) = 0.9938.$$

The proportion of values greater than 8.36 is therefore 1 – 0.9938 = 0.0062 and the proportion of rejected components is 0.0062 + 0.0188 = 0.0250.

In general, to obtain the proportion of values in a normal population within any range, *a* to *b*:

1. transform the range into z-scores, $z_a = \frac{a-\mu}{\sigma}$ and $z_b = \frac{b-\mu}{\sigma}$
2. look up the z-scores, z_a and z_b, in standard normal tables
3. find the difference between the tabulated values. (The proportion of values in the range is the difference between these values.)

Sampling from normal populations

Now we know how to describe certain infinite populations, namely normal populations, by specifying two numbers — the population mean, μ, and the population standard deviation, σ. We will use this shorthand to study the kinds of samples that random sampling produces.

Suppose that a class of students studying probability is annually taken to a casino as part of their course. Each student is given \$100 and told to gamble no more than \$2 at a time. After one hour, most students are expected to have left from the \$100, on the basis of earlier field trips to the casino, a balance of between \$50 and \$130. If we interpret the word 'most' to mean 95%, a reasonable model for the money a student has left after one hour's gambling would be a normal population with mean \$90 and standard deviation \$20.[3] Now let us see what happens when a class of 25 students gambles for an hour, assuming that the claimed population is a correct description of the money a student has left. We can generate the student balances by drawing 25 values from this normal population (by computer), with a random sampling scheme.

3. We noted earlier that approximately 95% of values in a normal population lie within 2σ of the population mean, μ.

The results are shown, in order of finishing student balance, in the following table.

Rank	Student	Balance ($)	Rank	Student	Balance ($)
1	Clifton	136.17	14	Kitching	82.30
2	Caldwell	131.47	15	Wong	79.99
3	Neale	117.67	16	Wells	78.78
4	Taylor	117.64	17	Cuff	78.37
5	Plier	112.93	18	Hodge	78.15
6	Glasgow	111.49	19	Meads	75.01
7	Boniface	104.99	20	Allen	74.82
8	Page	103.46	21	Sones	73.71
9	Hardacre	102.93	22	Louie	70.70
10	Jenkins	101.25	23	Riordan	65.59
11	Fink	99.64	24	Duffy	60.71
12	McQuarrie	84.20	25	Trow	46.96
13	Scott	83.10			

We can see that James Clifton had most money left, with Nicki Caldwell a fairly close second place. Ian Trow lost badly during the hour.

Because the students play games of chance, the outcomes are quite 'random'.[4] Another hour's gambling, again starting with an initial balance of $100, has been simulated by drawing another 25 values from our normal population, leading to the results shown in the table below.

Rank	Student	Balance ($)	Rank	Student	Balance ($)
1	Hodge	124.17	14	Caldwell	85.87
2	Sones	119.93	15	Duffy	84.88
3	Taylor	117.32	16	McQuarrie	83.37
4	Clifton	116.54	17	Riordan	83.18
5	Wong	111.43	18	Hardacre	80.59
6	Jenkins	106.59	19	Meads	78.18
7	Wells	102.25	20	Allen	73.94
8	Page	101.34	21	Plier	71.61
9	Glasgow	96.07	22	Kitching	71.32
10	Neale	93.27	23	Scott	66.95
11	Louie	89.13	24	Cuff	53.05
12	Fink	88.26	25	Boniface	52.32
13	Trow	86.14			

There is a very different story to tell this time, with Michelle Hodge way ahead of the rest, and Mike Cuff and Tanya Boniface trailing.

4. If there were a 'skill' element in the games, some students might systematically perform better than others, so the outcomes could no longer be simulated by a simple random sample.

The performances of the *individuals* in the class are clearly different in the two hours' gambling, but there are common features in the two data sets. The table below summarises the results of the two data sets.

	Hour 1	Hour 2
Mean	90.88	89.51
Standard deviation	22.53	19.59
Minimum	46.96	52.32
Maximum	136.17	124.17

The order of the students was certainly very different from one hour to the next, but the summary statistics for the 25 balances were quite similar for the two hours. In particular, the mean and standard deviation are very similar and are close to the mean and standard deviation of the normal population from which we generated the data.[5]

Dot plots of the two samples are shown below. It is again clear that while some characteristics of the two samples are different, others are fairly stable on repeated sampling.

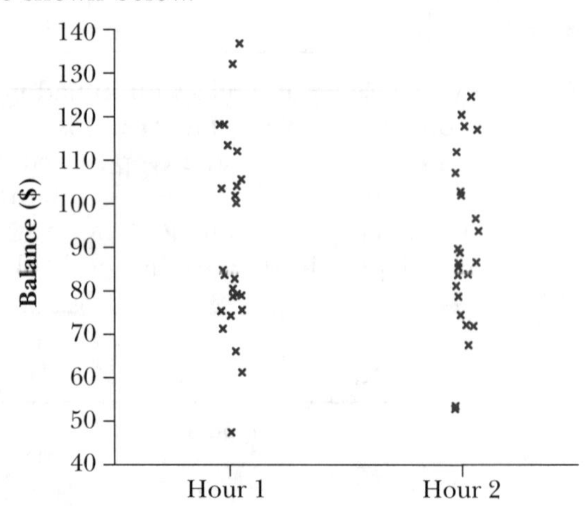

In section 8.3, we will examine more closely this phenomenon of sample variability and try to express the amount of variability in an objective and quantitative way. This will prepare us for taking account of the variability in samples, when we want to make use of them for describing populations, in chapter 9.

EXERCISE 8.2

1. From this textbook (or another one, if you wish), select a random sample of 20 pages by using a computer to randomly generate three-digit numbers until 20 different numbers have been obtained within the range of page numbers of the book.[6] Count how many of the 20 pages have a diagram or graphical display of data.
   (a) On the basis of your 20 page sample, guess the number of pages in the entire text that contain a chart. (If you are curious, you can check your guess by leafing through the whole book!)

5. We will see in chapter 9 that the sample mean and standard deviation are therefore good *estimates* of the population mean and standard deviation.
6. The three-digit numbers can alternatively be generated randomly with a deck of cards. Remove the face cards, shuffle, then deal a card three times, treating an ace as '1' and a 10 as '0'.

(b) Can you explain why we did not allow the same page to be selected twice in the sample? Is this sampling without replacement or sampling with replacement?

(c) If each page number had been generated by shuffling 10 cards labelled '0' to '9', then picking the top three cards in the deck, would this have produced a random sample?

(d) Could you use the same approach as in part (a) to estimate the number of italicised words in the text?

2. A bottle filling machine fills two-litre bottles with a carbonated drink. The machine is designed to slightly overfill the bottles, but there is unavoidable variation. The volume of drink has a mean of 2010 millilitres and standard deviation of 5 millilitres; moreover, it has a normal distribution.
   (a) Is there a use for a hypothetical normal population in this context?
   (b) What are the parameters of this normal population?
   (c) If one filled bottle was selected at random from a large batch of filled bottles, and the contents of this bottle carefully measured, in what range of values would you expect to find this measurement?
3. Identify the class of infinite population that the following samples are a sample from — one is normal, one is rectangular and one is exponential.

```
               :
               :  ..  : .
     .      .: :  ::..: :.:  .      .
(a) +---------+---------+---------+---------+---------+-------

           : .:
     :. :..: :::  :. .    .  ..         ..            .
(b) +---------+---------+---------+---------+---------+-------

             .             :
    .::  :..::: :     .  ..::...
(c) +---------+---------+---------+---------+---------+-------
  0.00      0.70      1.40      2.10      2.80      3.50
```

4. In the gambling example described in this section, the two simulations were summarised with mean, standard deviation, minimum and maximum. Based on this very small-scale simulation, does the sample mean or sample maximum exhibit more sampling variation?

8.3 VARIABILITY OF SAMPLES

The example of students' gambling balances that we used to illustrate the sampling process in section 8.2 was unusual; the goal in that case was the sample itself, so we could see which students fared best in the simulated gambling session. More commonly, the sample itself is only an intermediate step on the way to gaining information about the population from which the sample data were drawn. The variability of the samples provides a challenge to this plan. In this section we look at this variability more closely.

When applied to sample data, all the graphical and numerical summaries that were introduced in chapters 2 to 5 are associated with uncertainty. Such summaries are variable, in the sense that they would be different if they were applied to a freshly collected data set. This variability was evident in the dot plots of the gambling data in the previous section. We naturally, however, hope that the 'main' characteristics of the sample summaries will remain the same, and that these 'main' characteristics can be used to provide information about the underlying phenomenon; but what characteristics of a random sample provide reliable information about population features?

We will examine the stability of sample summaries by considering the following problem. For effective planning of resources, it is important to ascertain school rolls at the beginning of a school year. Rather than surveying all 363 high schools in New Zealand, it is decided to randomly sample from these schools and to obtain the rolls of the resulting sample of schools. Although we cannot demonstrate what would happen at the start of a future year, we can use the known rolls from mid-year (July 1994) to demonstrate the kind of variability that is likely to be observed.

The table below shows the rolls of all New Zealand high schools in July 1994.[7]

New Zealand high school rolls (July 1994)

632	802	1141	110	430	928	208	725	616	99	560	1026
466	963	1240	799	356	598	444	1165	484	513	473	251
193	1629	1367	307	649	168	1020	202	1053	161	851	280
593	727	1181	980	652	127	92	847	940	243	334	855
884	1162	1071	609	420	256	690	410	536	1092	1193	648
933	583	826	1534	392	162	511	949	507	1236	121	976
849	173	1229	999	800	631	116	1505	1065	1068	748	274
219	684	162	1200	408	607	862	793	1243	2033	326	625
597	532	1688	481	534	390	613	573	732	1325	314	355
391	1579	364	1295	536	610	1127	535	1102	519	1190	444
355	1387	1239	408	727	650	888	550	805	254	661	708
332	481	806	280	745	1114	228	1138	229	132	376	345
441	1139	1006	828	591	663	783	673	361	1188	431	579
224	886	770	944	830	763	133	915	249	897	635	622
528	531	1109	655	527	643	357	538	886	202	490	493
347	939	477	595	1247	351	124	901	959	439	226	297
81	1358	344	468	249	372	1271	515	621	445	669	785
151	1374	287	242	845	387	1125	420	1087	487	486	590
712	866	798	467	882	179	686	466	876	1613	561	256
755	825	206	355	735	650	426	283	869	246	260	750
992	1611	1228	299	1248	875	302	182	306	748	204	1147
245	447	441	445	1211	225	245	615	138	513	374	311
1039	508	589	492	239	687	227	731	255	543	272	438
1148	1435	127	375	907	477	735	134	1001	186	488	260
1904	557	354	671	145	974	272	389	466	974	312	757
131	378	1453	437	580	461	490	235	643	1069	397	733
144	1000	697	467	1138	1125	595	601	150	202	436	322
1896	1589	348	794	1042	1021	114	251	128	542	721	207
720	542	661	456	234	226	578	210	141	409	945	433
714	744	1376	1149	844	376	778	697	206	254	210	191
774	1044	1508									

7. *New Zealand Schools 1994*, New Zealand Ministry of Education, Wellington, 1995.

The diagram below shows jittered dot plots from several random samples of size $n = 10$ from this population of school rolls.

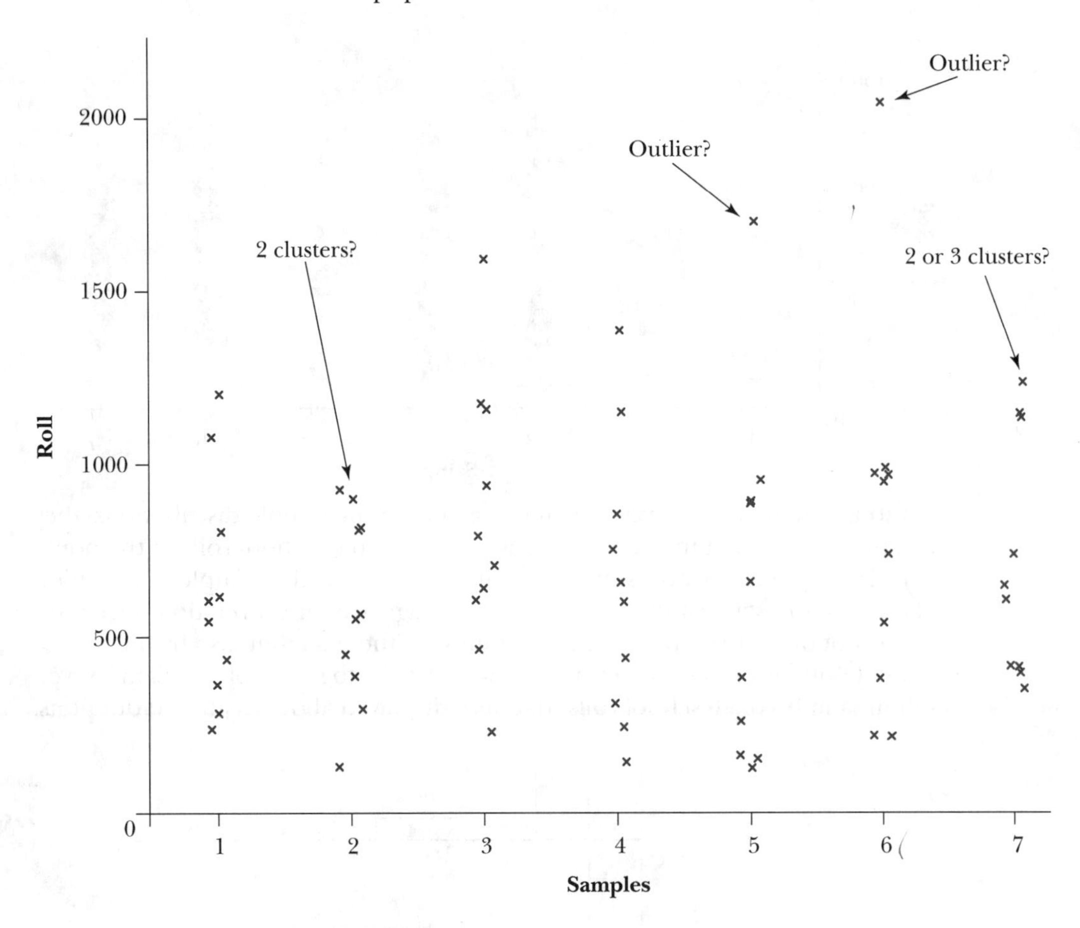

When the sample size is small, there is considerable variability between samples; features such as outliers or clusters in the sample may not reflect similar features of the population. Features such as outliers, clusters or skewness must be *strongly* evident in small samples before they should be interpreted as reflecting similar population features. For example, the outlier in the sixth sample is highly separated from the remaining data; if this were our only sample of 10 schools, the outlying high school would be investigated to determine whether it had further unique features to make it stand out from other schools.[8]

With a greater sample size, random samples are less likely to give a misleading impression of population characteristics. For example, the diagram on page 266 shows several samples of $n = 40$ high schools.

8. The outlier is Burnside High School, the school with the highest roll in New Zealand in July 1994. In this particular sample, the sample maximum happens to be the same as the population maximum.

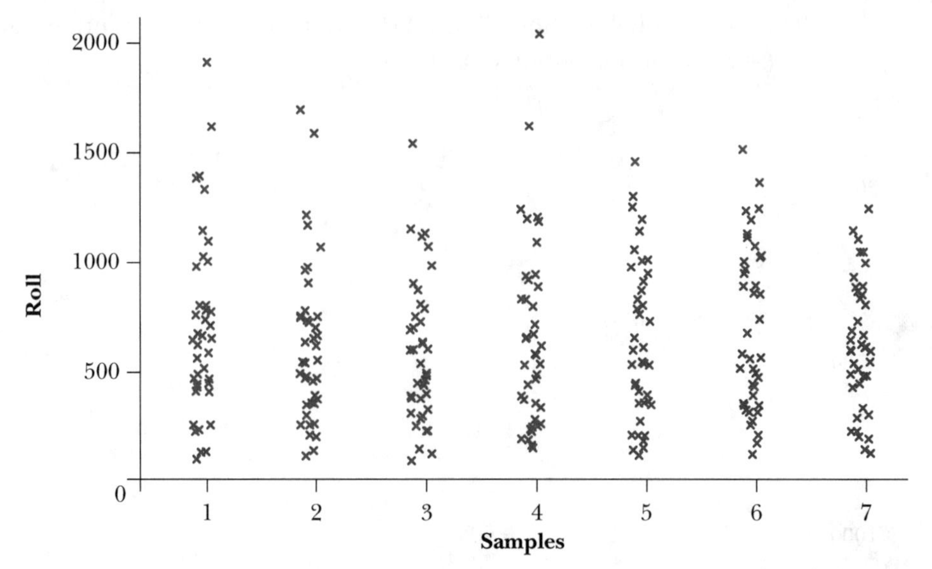

Although there is still variability in the shapes of the sample distributions, they give a more consistent impression of the distribution of school rolls in the population. Histograms provide summaries of the shape of the sample distribution that are no more stable and may indeed encourage over-interpretation since their shape is not only affected by the randomness of the data, but also by the choice of class boundaries. The diagram below shows histograms of the same seven random samples of 40 school rolls that were displayed above as jittered dot plots.

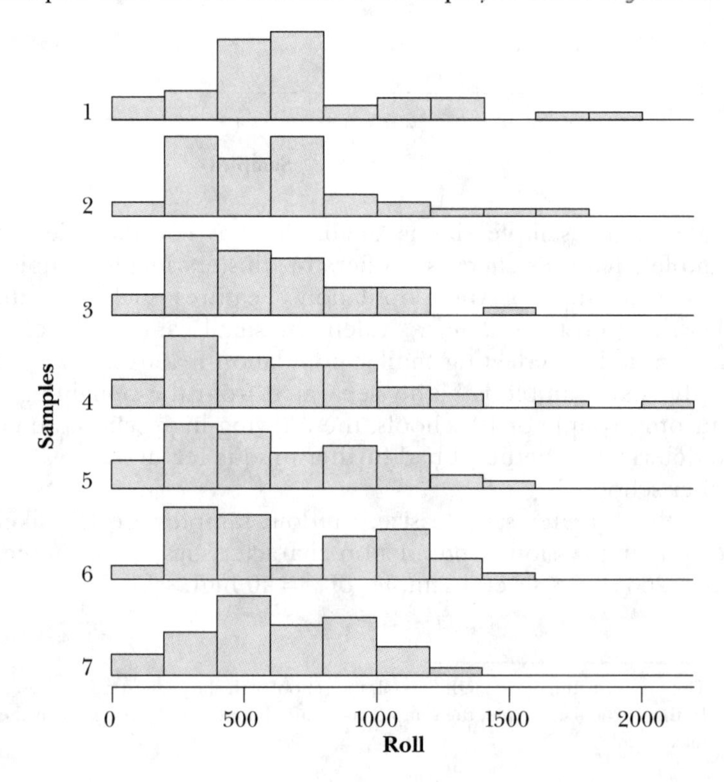

You may be wondering whether *any* reliable information about a population can be obtained from a random sample. Detailed information about the shape of a distribution can be obtained only from large samples, and *you should be careful not to over-interpret shape information from small samples.*

The more highly we summarise the sample data, the more stable this summary usually becomes and therefore the more reliable it is as an indication of population features. For example, box plots of the seven random samples of 40 school rolls are shown below.

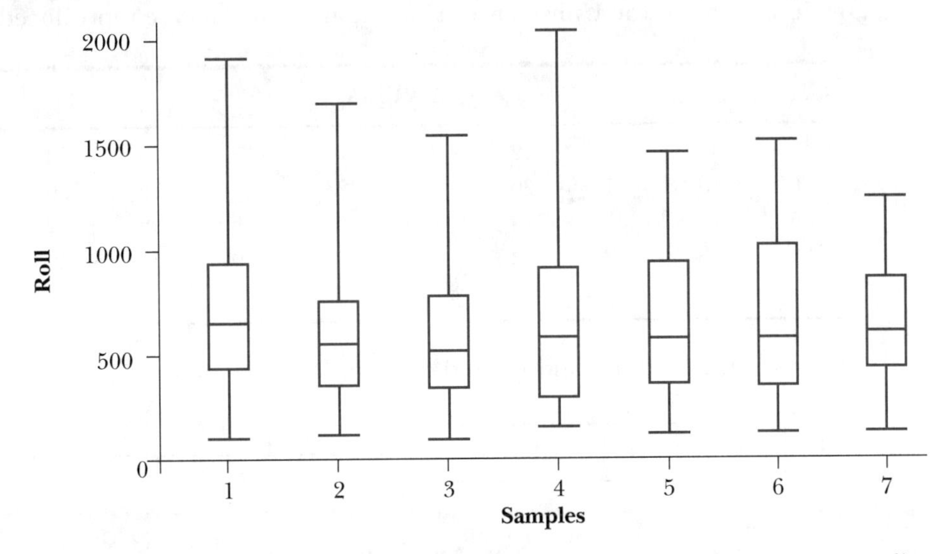

Although the maximum rolls in the samples are quite variable, the median, quartiles and minimum are reasonably stable. Again, the larger the sample size, the more stable the summary tends to be.[9]

The mean and standard deviation provide the most highly summarised information about location and spread in a set of data, so we would expect them to be relatively stable from sample to sample. The table below shows the means and standard deviations from our seven random samples of school rolls of size $n = 10$ and $n = 40$.

	Means		Standard deviations	
Sample	$n = 10$	$n = 40$	$n = 10$	$n = 40$
1	606.9	699.0	324.1	414.6
2	572.4	620.2	273.0	362.1
3	820.4	583.8	395.6	324.7
4	644.5	654.9	398.1	423.3
5	607.5	638.8	503.8	361.9
6	793.1	690.0	535.6	373.1
7	701.8	621.7	342.8	295.6

9. The only exceptions that you are likely to meet to this rule are the extremes of populations with long tails; the maximum and minimum of a random sample may remain highly variable even when the sample size is large.

The sample means and standard deviations are quite variable when the sample size is $n = 10$, but become more reliable indicators of the mean and standard deviation of the rolls of *all* New Zealand schools when the sample size is larger. In the next section, we will investigate the properties of sample means further.

In practice, we have only a single random sample available, so how do we assess its features? For example, in section 2.1, we described a data set containing the 'maximum voluntary isometric strength' (MVIS) of a group of 41 male students at the University of Hong Kong. The data are reproduced below.

MVIS (kg)									
33	40	44	12	20	36	15	13	14	18
16	18	21	26	31	23	15	25	20	23
35	54	29	10	12	22	16	26	22	26
33	18	12	12	19	20	20	41	19	19
47									

Here are the box plot and jittered dot plot for these data:

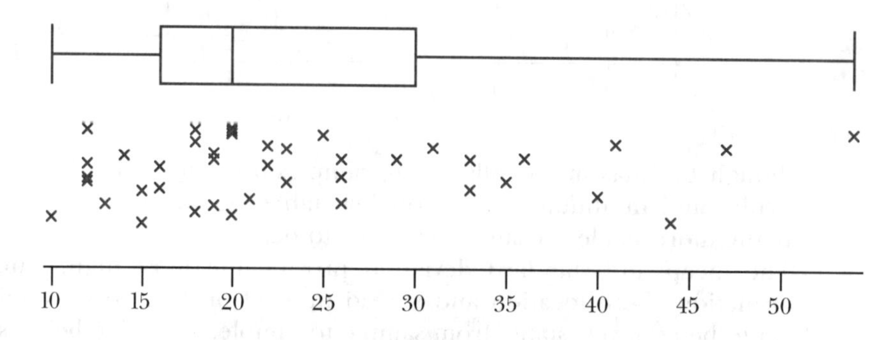

How can we assess whether the apparent skewness in this sample is an indication of skewness in the underlying population of MVIS in male students *in general* at the University of Hong Kong? We cannot conduct the experiment repeatedly to assess the variability of these displays — only one data set is available!

Prior knowledge of the nature of MVIS may suggest a skew distribution, but a more objective analysis is possible. One way to assess the data is to examine the variability of similar displays in random samples of the same size from a known symmetric population with *similar* characteristics. In particular, we might examine the variability of random samples from a normal distribution with the same mean and standard deviation as the sample, $\mu = 23.78$ and $\sigma = 10.530$.

The diagram on page 269 shows a jittered dot plot of the data, together with 10 simulated random samples selected from this normal population with parameters $\mu = 23.78$ and $\sigma = 10.530$. None of the simulated samples exhibit as much skewness as the real data, so it seems unlikely that the population underlying the data is symmetric.

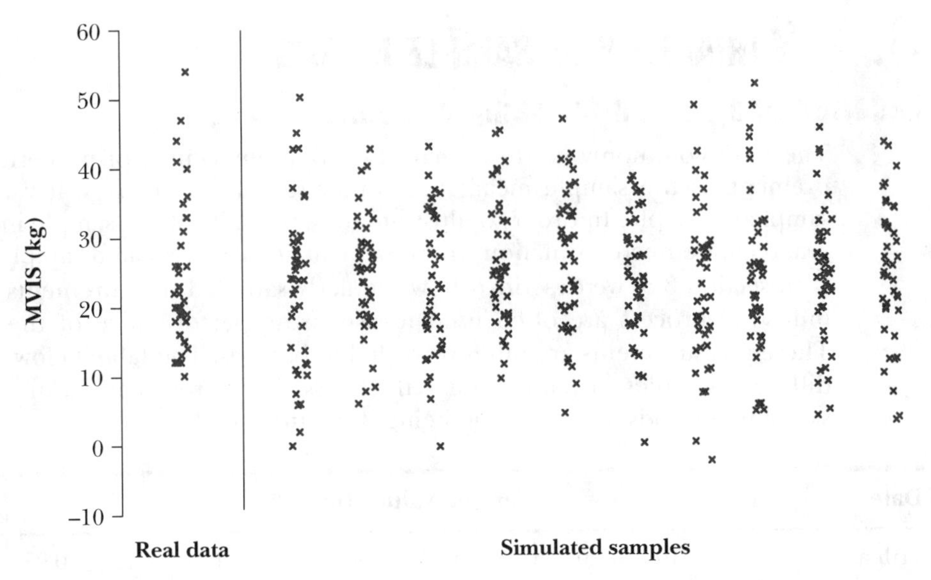

Similar simulations can be used to assess the size of outliers, the appearance of clusters and any other features of a sample summary. However, we will see in the next section that simulations are unnecessary to assess and describe the variability of the sample mean.

EXERCISE 8.3

1. Refer to the New Zealand high school rolls data in this section.
   (a) From examination of the dot plots of the samples of size 10, which varies most, and which the least, from among the maximum, minimum and mean?
   (b) Are the means for samples of size 10 systematically larger or smaller than the means for samples of size 40?
   (c) Are the means for samples of size 10 less stable than the means for samples of size 40?
   (d) Are the standard deviations for the samples of size 10 systematically larger or smaller than the standard deviations for the samples of size 40?
   (e) Are the standard deviations for the samples of size 10 less stable than the standard deviations for the samples of size 40?
2. A normal distribution with the same mean and standard deviation as the isometric strength data was used as a population to generate 10 simulated samples the same size as the data ($n = 41$). Dot plots of the data and the simulated samples are displayed at the top of this page. Use them to describe the difference in the character of the simulated data distributions from the actual data distribution.
3. Refer back to the graph at the end of section 8.2, showing the balances of a group of students after two different gambling sessions of one hour. The distribution for the first hour has a distinct gap in the middle, suggesting that there may be some underlying distinction between the two groups of gamblers. How should this gap be interpreted?

8.4 Variability of Sample Means

Location and spread for sample means

The most commonly used summary of a random sample of numerical measurements is the sample mean. Like other sample summaries, it varies from sample to sample. In order to illustrate the variability of the sample mean with real data, we need a situation where we actually have repeated samples.

In section 3.5, we described how regularly sampled measurements from an industrial process are often used to assess the performance of the process. These measurements are often recorded in batches. The table below shows 20 different samples of paint primer thickness that were obtained in successive sampling periods, the sampling being done twice daily.

Date	Sample values (mils)*										Mean
11th a.m.	1.30	1.10	1.20	1.25	1.05	0.95	1.10	1.16	1.37	0.98	1.146
11th p.m.	1.01	1.10	1.15	0.97	1.25	1.12	1.10	0.90	1.04	1.08	1.072
12th a.m.	1.22	1.05	0.93	1.08	1.15	1.27	0.95	1.11	1.12	1.10	1.098
12th p.m.	1.08	1.12	1.11	1.28	1.00	0.95	1.15	1.14	1.28	1.31	1.142
13th a.m.	0.98	1.30	1.31	1.12	1.08	1.10	1.15	1.35	1.12	1.26	1.177
13th p.m.	1.12	1.30	1.01	1.20	1.11	0.93	1.02	1.25	1.05	1.10	1.109
14th a.m.	0.92	1.10	1.13	1.02	0.93	1.17	1.24	0.98	1.34	1.12	1.095
14th p.m.	1.04	1.14	1.18	1.12	1.00	1.02	1.05	1.34	1.12	1.05	1.106
15th a.m.	1.08	0.92	1.14	1.20	1.02	1.04	0.94	1.05	1.12	1.06	1.057
15th p.m.	1.20	1.13	1.19	1.16	1.03	1.25	1.20	1.24	1.10	1.03	1.153
18th a.m.	1.25	0.91	0.96	1.04	0.93	1.08	1.29	1.42	1.10	1.00	1.098
18th p.m.	1.24	1.34	1.40	1.26	1.13	1.15	1.08	1.02	1.05	1.18	1.185
19th a.m.	1.13	1.16	1.12	1.22	1.12	1.07	1.04	1.28	1.12	1.10	1.136
19th p.m.	1.08	1.31	1.12	1.18	1.15	1.17	0.98	1.05	1.00	1.26	1.130
21st a.m.	1.08	1.26	1.13	0.94	1.30	1.15	1.07	1.02	1.22	1.18	1.135
21st p.m.	1.14	1.02	1.14	0.94	1.30	1.08	0.94	1.12	1.15	1.36	1.119
22nd a.m.	1.06	1.12	0.98	1.12	1.20	1.02	1.19	1.03	1.02	1.09	1.083
22nd p.m.	1.14	1.22	1.18	1.27	1.17	1.26	1.15	1.07	1.02	1.36	1.184
23rd a.m.	1.07	1.05	0.97	1.05	1.16	1.02	1.02	1.14	1.07	1.00	1.055
23rd p.m.	1.13	0.90	1.12	1.04	1.40	1.12	1.15	1.01	1.30	1.14	1.131

* 1 mil = 0.0254 mm (A mil is an old imperial measurement, and is one-thousandth of an inch.)

We will assume that the process is 'in control', so that each sample of 10 measurements is a random sample from the same unknown underlying hypothetical population of paint thicknesses.

When data such as these are analysed to assess process stability, the individual measurements are not usually examined. Rather, analysis is usually based on an examination of the means of the successive samples (and often, separately, their standard deviations or ranges).

The diagram below displays the individual measurements comprising each sample and also the sample means.

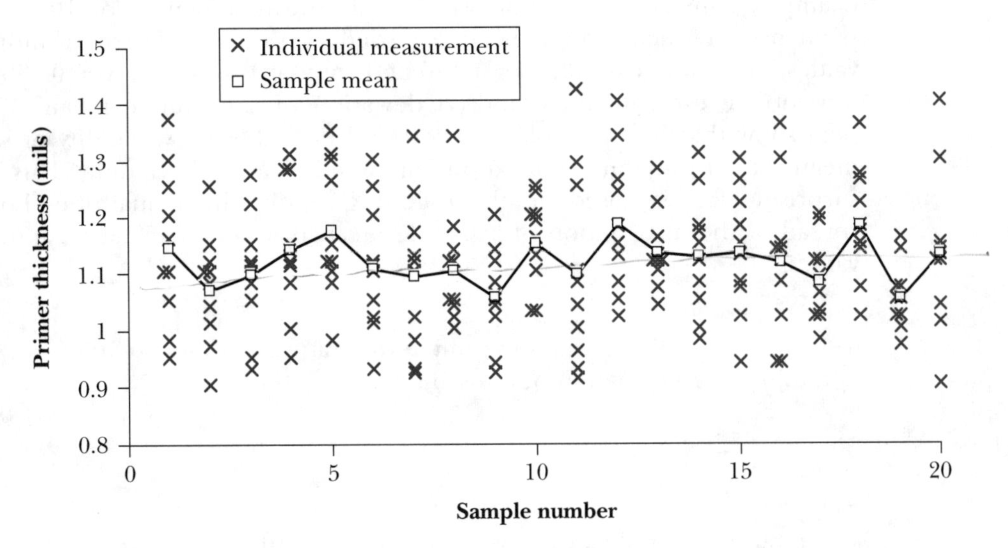

It should be clear from this diagram that the sample means are also random quantities — they vary from sample to sample — and they therefore also have a mean and standard deviation (and indeed a distribution whose histogram could be sketched).

What is the distribution of the sample mean? It is not known exactly, but we can see from the diagram above that the sample means seem to be centred near the population mean, but they are much less variable than the individual measurements. The diagram below shows this graphically with box plots of the 200 individual measurements and of the 20 sample means.

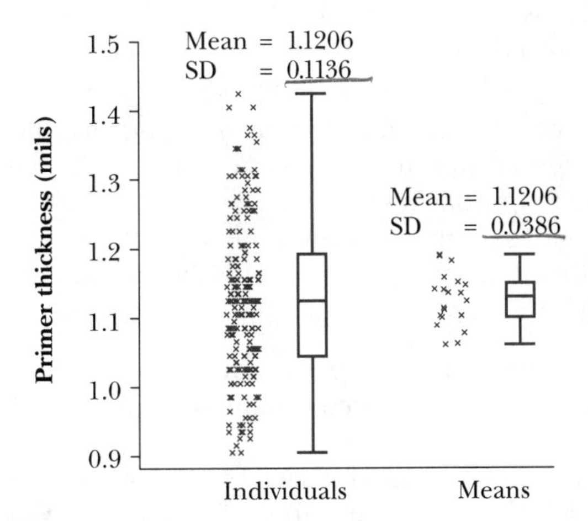

The 20 sample means of these 20 random samples of size $n = 10$ have mean 1.1206 and standard deviation 0.0386, and these provide estimates of the location and spread of the distribution of the mean of a sample of 10 paint primer thicknesses.

In practice, however, we usually only have a single random sample, so we cannot estimate the characteristics of a sample mean in this way. For example, consider a situation where only the first sample of 10 values of paint primer thickness has been recorded (11th a.m.). These 10 individual values have sample mean $\bar{x} = 1.146$ and standard deviation $s = 0.136\,32$, so how do we estimate the standard deviation of the sample mean? We will next show that it is possible to assess and describe the variability of sample means based only on information contained in a single sample. This rather surprising fact is based on the following result which quantifies how the spread of the distribution of a sample mean is lower than that of individual values.

If $X_1, X_2, \ldots, X_n$ is a random sample from any distribution with mean μ and standard deviation σ, the sample mean:

$$\bar{X} = \frac{\sum_{i=1}^{n} X_i}{n}$$

is a random variable with mean and standard deviation:

$$\mu_{\bar{X}} = \mu$$

$$\sigma_{\bar{X}} = \frac{\sigma}{\sqrt{n}}.$$

For example, consider random samples from a population with mean $\mu = 10$ and standard deviation $\sigma = 2$. Means from samples of size $n = 16$ will have a distribution with mean $\mu_{\bar{X}} = 10$ and standard deviation $\sigma_{\bar{X}} = \frac{2}{\sqrt{16}} = 0.5$. If the sample size is increased to $n = 100$, the distribution of the sample mean will remain centred at $\mu_{\bar{X}} = 10$, but its standard deviation will be reduced to $\sigma_{\bar{X}} = \frac{2}{\sqrt{100}} = 0.2$. As the sample size increases (as $n \to \infty$), the standard deviation of $\bar{X}$ therefore decreases, and its distribution becomes more concentrated around μ, the mean of the underlying process. In other words,

as $n \to \infty$, $\bar{x}$ approaches μ.

Applying this result to our first sample of 10 paint primer thicknesses, we would estimate that σ has the value $\sigma = 0.136\,32$ and, therefore, that $\sigma_{\bar{x}}$ approximately equals $\frac{0.136\,32}{\sqrt{10}} = 0.043\,11$. This value is reasonably close to the value 0.038 59 that was found from the standard deviation of the sample means observed in the 20 samples, but has been obtained by only considering one single sample of 10 values.

The important thing to notice is that our estimate of the standard deviation, 0.043 11, was determined from the *internal* structure of a *single* sample of size 10; it is very close to the standard deviation of the sample means, 0.038 59, that was observed by comparing *several* sample means. This is no accident and reflects the real utility of the general result in the box above.

Large sample normality

Although we have general formulae for the mean and standard deviation of $\bar{X}$, there is no general result giving the shape of its distribution. However, there is an extremely important general result that gives the *approximate* shape of the distribution of $\bar{X}$ when the sample size, *n*, is large.

Given a random sample, $X_1, X_2, \ldots, X_n$, from a distribution with mean μ and standard deviation σ then, irrespective of the distribution of the X_i, the distribution of $\bar{X}$ approaches the shape of a normal distribution with mean and standard deviation:

$$\mu_{\bar{X}} = \mu \text{ and } \sigma_{\bar{X}} = \frac{\sigma}{\sqrt{n}}$$

as the sample size increases.

This property of sample means is called the *central limit theorem*. It is the main reason for the importance of the normal distribution in statistics.

For example, the diagram below shows the combined distribution of all individual paint primer thicknesses; it appears to have a slightly skew distribution. The central limit theorem suggests (correctly) that the mean of a sample of size $n = 10$ will have a distribution that is closer to normal.

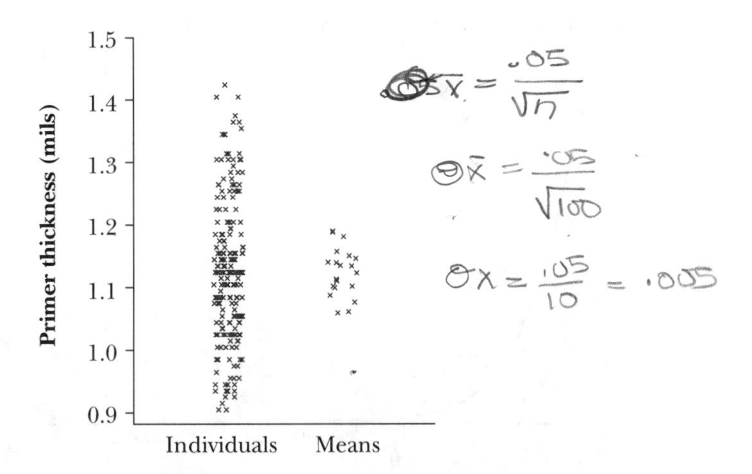

Sample sizes do not need to be particularly high before the distribution of sample means becomes approximately normal, even when the samples are taken from very skew or even bimodal distributions. For example, the diagrams on page 274 show the exact distributions of sample means of various sizes from a population of values that is highly skew.

When the sample size is $n = 1$, the distribution is simply the distribution of a single measurement and, in fact, is the population distribution. By the time n has increased to 8, the distribution of the sample means is very close to the symmetric normal bell-shape.

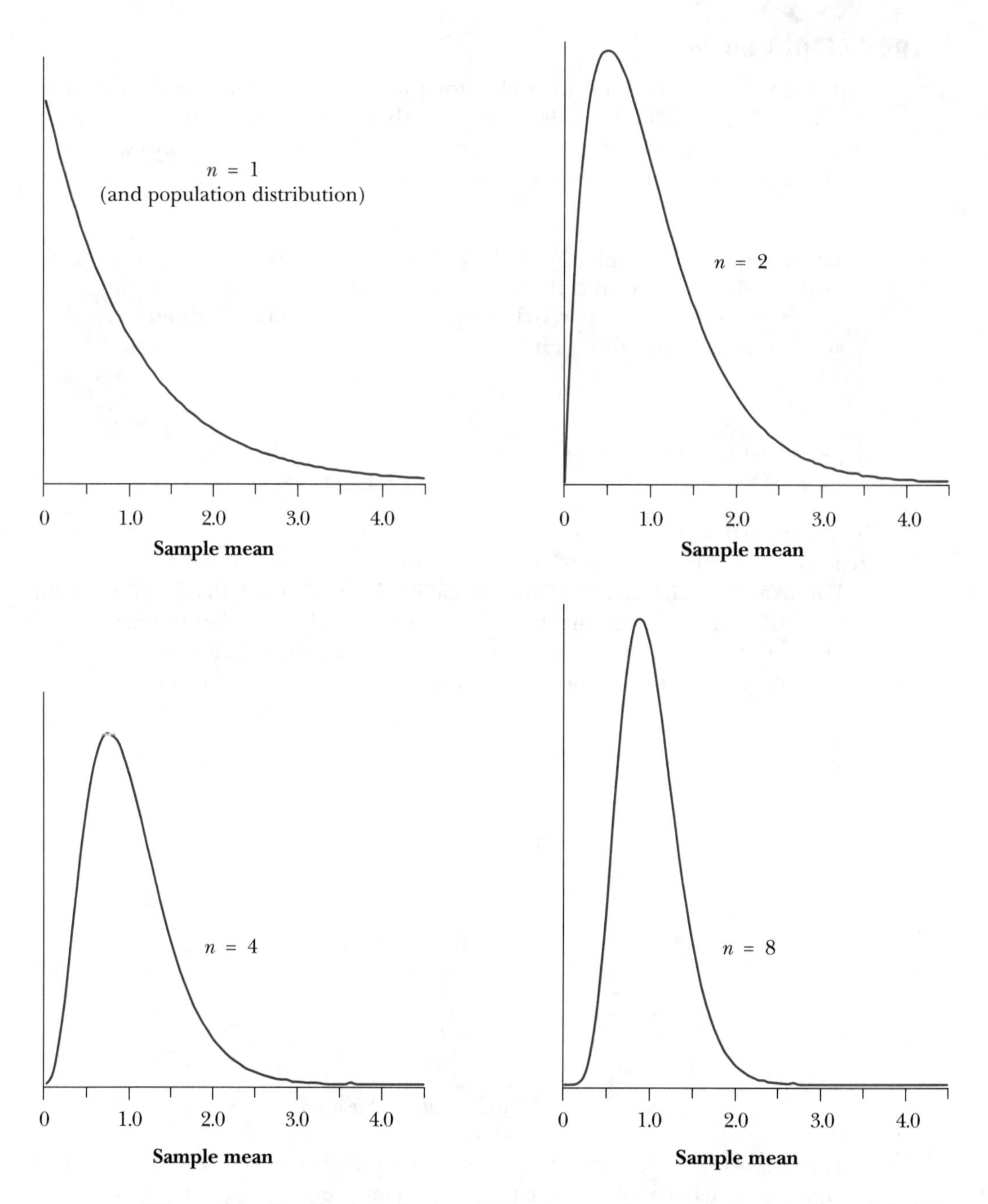

When the distribution from which the random sample is selected is less skewed than this population distribution, the distribution of sample means is often close to normal even when n is as small as 3 or 4.

As the paint primer data are not highly skew, their sample means will have a distribution that is very close to normal. We saw earlier that the parameters of this normal distribution are approximately $\mu_{\bar{X}} = 1.120\,55$ and $\sigma_{\bar{X}} = 0.035\,59$.

EXERCISE 8.4

1. Examine the box plots showing the distribution of the 200 individual paint primer thicknesses and the distribution of the sample means for samples of size 10.
   (a) Compute the ratio of the standard deviation of the individual measurements and the standard deviation of the sample means. How does this compare with $\sqrt{10}$? Explain.
   (b) By measuring the lengths of the two boxes with a ruler, compute the ratio of the lengths. How does this compare with $\sqrt{10}$?
   (c) The paint primer data (individual values) have some apparent outliers near 1.4 mils, revealing a mildly skew distribution, yet the distribution of the means appears not to be so skew. Why?
2. Examine the four distributions of the sample mean for random samples of size $n = 1$, 2, 4 and 8 from a skew distribution. One way to think of these curves is to imagine that stacked dot plots fill in the region under the curve and that the curve traces the tops of the stacks. With this understanding, the proportion of dots to the left of a certain value, say 2.0, is approximately the same as the proportion of the area under the curve that is to the left of that value. For the skew distribution of a single measurement (which is the mean of a sample of size $n = 1$), the proportion is about 0.86.
   (a) What is it for $n = 2$, 4 and 8?
   (b) Guess the mean of each of the distributions shown. Are they getting smaller as n increases?
   (c) Apart from looking more and more like a normal distribution as n increases, what else is happening to the shape of the distribution? Why?

8.5 VARIABILITY OF SAMPLE PROPORTIONS

In chapter 5 we discussed the distinction between quantitative data, like age or temperature, and qualitative data, like sex or race. While the most usual way of summarising quantitative data numerically is with means, in the case of qualitative data the usual way is with proportions.

For example, if we record the sex of customers as they enter a sports shop:

M, M, F, M, F . . .

then a natural summary would be the proportion of males and females. (Proportions and percentages are equivalent in this context.) If we noted the sport of primary interest of each customer:

tennis, jogging, football, jogging, squash, football, football . . .

then a numerical summary of the following form would be appropriate.

Sport	Percentage of customers
Jogging	29%
Football	43%
Squash	14%
Tennis	14%

If the order in which the sample is collected is of no relevance, these proportions, with the sample size n, provide a complete summary of the data.

The simplest situation of sampling from a categorical population is when the population consists of only two categories of individuals. Any categorical population can be reduced to two categories of interest (for example, *tennis* and *other*), so this is not a severe restriction. Since the proportions in the two categories must add to one, the proportion in one particular category provides a complete summary of the proportions (for example, *tennis* 14% implies *other* 86%).

Now consider how a library might respond to an administrative directive to place texts over 25 years old in a special archival collection. First, the library wishes to estimate the proportion of its collection that is in the 'archive' category, so that a suitable space can be planned for the archival collection.

The librarian takes a random sample of 100 titles from the catalogue to provide some data. Each sampled item is examined to determine its publication date and it is recorded as either 'current' or 'archive'. It turns out that 25 titles (25% of the sample) belong to the 'archive' category. Since the entire collection is known to consist of approximately 480 000 titles, the librarian estimates that the archival collection will consist of approximately 120 000 titles.

When the librarian reports this finding, she is questioned on the basis of it, and the ensuing discussion results in her realisation that the sample could be quite misleading. While 25% of the sample is 'archive' type, this is only an *estimate* of the corresponding percentage in the entire collection. What is needed is some theory that will help us to gauge the variability of this sample proportion.

Fortunately, the theory of section 8.4 can also be applied to sample proportions, even though we do not have quantitative data to work with. The 'trick' that allows us to use the theory developed for quantitative data is to assign a code of '1' to one category and '0' to the other category.[10] We can then treat the data as *numerical*.

In this coding, the sample of 100 titles might look like this:

1	0	1	0	0	0	0	0	0	1	0	0	1	0	0
1	0	0	0	1	1	0	0	0	0	1	0	1	0	0
0	0	0	0	1	0	1	1	0	0	0	0	0	0	0
0	0	0	0	0	1	0	0	1	0	0	0	1	0	0
0	0	0	0	0	0	0	0	1	0	0	0	0	0	1
0	0	1	0	0	1	1	1	0	0	0	0	0	1	0
0	1	0	0	0	0	0	0	1	1					

10. This type of 0/1 coding is often called an *indicator* variable.

where '1' indicates a title that is classified as archived. The proportion of 1's in the sample is:

$$p = \frac{25}{100}$$

as mentioned earlier. But what is the sample mean, $\bar{x}$, for this 0/1 data set? The sample mean is:

$$\bar{x} = \frac{\sum x_i}{n} = \frac{25}{100}.$$

The sample mean and the proportion of 1's is the same number in a 0/1 data set!

The central limit theorem (section 8.4) says that, for large samples, sample means, $\bar{X}$, approximately have a normal distribution no matter what the population distribution. (We know the population consists of 0's and 1's in this application and is certainly not normal but, fortunately, this does not matter.) In fact, the central limit theorem says more than this. It says that:

- the average value of $\bar{X}$ is μ, the population mean, and
- the standard deviation of $\bar{X}$ is $\frac{\sigma}{\sqrt{n}}$.

Now, in a 0/1 population, suppose that we denote the proportion of 1's by the Greek letter π (pronounced 'pie').[11] The population mean must also be π; that is, $\mu = \pi$ for this population.

One thing that is not so obvious but certainly true is that, in this 0/1 population:

$$\sigma = \sqrt{\pi(1-\pi)}.$$

In other words, our sample proportion, $p = \frac{25}{100}$ is a single value selected from a distribution that is approximately normal and has mean and standard deviation:

$$\mu_p = \pi$$

$$\sigma_p = \frac{\sqrt{\pi(1-\pi)}}{\sqrt{n}} = \sqrt{\frac{\pi(1-\pi)}{n}}.$$

In the library example, replacing the unknown parameter π by the sample proportion p in these formulae, we get the approximation:

$$\mu_p \approx 0.25$$

$$\sigma_p \approx \sqrt{\frac{0.25(1-0.25)}{100}} = 0.0433.$$

11. The value of π is usually unknown and can be any constant between 0 and 1; its value is *not* the same as the mathematical constant π.

This approximate normal distribution for the sample proportion, p, is shown in the diagram below.

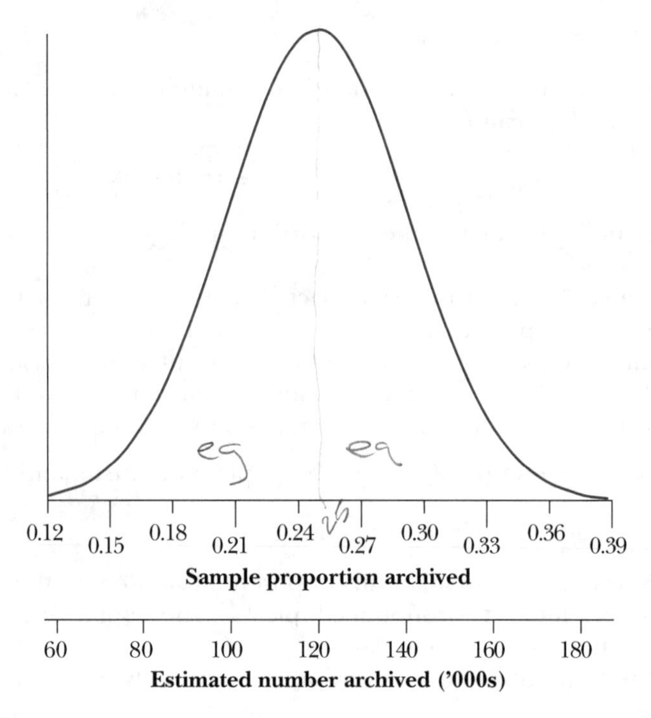

The distribution of the estimated total number of titles that need to be archived (480 000 times the sample proportion) is therefore found by rescaling this distribution (as shown by the second scale below the normal histogram).

Thus the variability of the estimate that the librarian must depend on is quite large; we expect the total number of titles that require archiving to be between 60 000 and 180 000. The librarian may wonder whether a larger sample would have been worthwhile![12]

In summary, the central limit theorem, which was originally presented as a useful result for describing the variability in sample means, can also be applied to sample proportions. We will use this fact for the inferential techniques of chapter 9.

EXERCISE 8.5

1. When is a sample mean also a sample proportion?
2. How do we apply the central limit theorem to describe the variability of sample proportions?
3. In a region, 25% of schoolchildren are asthmatic. What proportion of asthmatics is likely to be found in a class of 30 students?

12. We will show in chapter 9 that a sample of 900 titles would reduce the range of:

$$180\,000 - 60\,000 = 120\,000$$

to a more useful 40 000.

Problems

1. A researcher is about to survey 100 farms to determine their commercial viability. Commercial viability is decided according to certain definitions, and the result of the assessment is that the farm is declared to be 'viable' or 'not viable'. The researcher is told by a statistician that with 100 farms out of the total list of about 5000 farms, the proportion viable obtained from one sample could be quite different from that of the 5000 farms of interest.
   (a) If half of the population of 5000 farms were viable, show that the standard deviation of the sample proportion would be 0.05. If a different proportion of the population was viable, could this standard deviation be larger? (You can either try a range of different population proportions or use a mathematical proof to answer this question.)
   (b) Show that if the unknown population proportion of viable farms was 0.5, the proportion viable in the sample could easily be as low as 0.40 or as high as 0.60.
   (c) What sample size would be needed to reduce the standard deviation of the sample proportion from 0.05 to 0.025?
   (d) What accuracy would be obtained with a sample size of 5000 if sampling is without replacement?

2. The amount of sleep obtained by students on weekday nights, per 24 hours, is about 7.0 hours with a standard deviation of about one hour. Each student in a statistics class is required to undertake a project which involves selecting a random sample of 25 students from the large student body and ascertaining how much sleep each sampled student had in the previous 24 hours.
   (a) Assuming that each student properly selects and assesses a random sample, how much variation can be expected in the sample averages obtained for the projects?
   (b) If the results from 20 students in the class are 'pooled', how much will the standard deviation of the mean sleep time be reduced, compared to a single student's results?
   (c) What are likely to be the practical problems with conducting such a survey?

3. A person weighs himself three times, within a period of a minute, with the results: 91.1 kg, 91.3 kg, 91.0 kg. Is it reasonable to describe these three measurements as a random sample from a population? Discuss.

4. A motor vehicle inspection scheme is to be implemented in which a few owners of registered vehicles are mailed notices about the need to have their car safety inspected by a government agency. Out of a registry of 10 000 vehicles, a random sample of 100 are chosen to be safety inspected. It is estimated that it costs the government $25 to inspect a vehicle when it

passes, and $35 when it fails (because of a re-inspection). It is believed that 70% of the 10 000 registered vehicles would potentially pass the inspection.

(a) What is the approximate distribution of the proportion failing in the sample?

(b) What is the cost of the inspection program? (The cost is clearly a random quantity, so you should express the cost in terms of the proportion, then use this to work out the distribution of the cost.)

(c) How unusual would it be for 80 of the 100 selected vehicles to pass the inspection, while the remainder fail? (*Hint*: See where 0.8 falls in the distribution of sample proportions passing.)

Projects

1. **Sampling M&M's**

This project is intended to give you some experience with random sampling. You will select random samples from a population of 200 or more coloured M&M's, and will observe how the proportion of red M&M's in the samples is affected by the sample size. (Any other population of 200 or more items could be used, provided the physical nature of the items is uniform, except for visual appearance, and some items are distinguishable from the rest. Peanuts could be used with some of them coloured by marker pen.)

Put your collection of M&M's into a pot and draw out 10 of them. Your draw should be done in such a way that any 10 items have the same chance of being drawn. A good way to do this is to mix the items and draw the 10 without looking. This is a 'sample' of size 10 from your population of size 200. Count the number of red M&M's and record this number. Then replace the ten M&M's and repeat the draw for a sample of size 20. Again, record the number of red ones and replace. Continue this process until the draw contains all the items (or 200 of them if your population is larger than 200).

You should be able to generate a table of the form shown below.

Sample size	Number of red M&M's	Proportion of red M&M's
10	2	0.20
20	1	0.05
30	3	0.10
.	.	.
.	.	.
.	.	.
200	20	0.10

Plot the proportion of red M&M's against sample size. Comment on the appearance of the plot, noting any features you think might be reproducible if your project was repeated.

2. **Number of passengers in cars**
   Select some location where there is a steady flow of traffic, such as the entrance to a university car park. Record the number of passengers in each of 10 cars that pass. Wait 20 seconds, then repeat with another group of 10 cars. Repeat until you have observed a total of 20 groups of cars.
   (a) Find the standard deviation of the number of passengers in the combined group of 200 cars that have been observed.
   (b) Evaluate the mean number of passengers in each of the 20 groups of cars and find the standard deviation of these means.
   (c) Use the standard deviations that you have just calculated, and any graphical displays that you think are appropriate, to demonstrate the central limit theorem.

3. **A simulation**
   This exercise is only feasible if you have access to appropriate computer software; a spreadsheet can be used if a statistical program is not available. Your instructor will suggest suitable software.
   Using statistical software, have the computer generate 20 samples from a normal population that has mean 10 and standard deviation 2.
   (a) For each sample, compute the mean. When you have all 20 means, calculate their standard deviation.
   (b) Compare this standard deviation of the 20 means with the standard deviation that you would expect to get from the theory of this chapter.
   (c) Compare the histogram of the 20 means with the normal curve claimed by the theory. Comment.

4. **Paper planes — how many times to try each aeroplane?**
   The number of repetitions of the measurements for a particular aeroplane, in the paper aeroplane projects, was always left to the experimenter; the only guidance was to repeat the trials until the measurement was reliably determined. In this project, you select the number of repetitions to reduce the variability of your measurements to a predetermined level.
   (a) Select a single design and construction for your aeroplane that flies well, and decide how you will measure its performance (for example, time in the air or distance flown).
   (b) Pick a target accuracy; this should be expressed as the standard deviation that you would like for the sample mean from your repeated flights. Remember that your sample mean is likely to be within two of these standard deviations from the population mean measurement.
   (c) Fly your aeroplane 10 times. Find the standard deviation of your performance measurements from these 10 flights, and use this to obtain an estimate of the standard deviation of your *mean* flight time. (You will have to assume that your sample standard deviation is equal to the population standard deviation, σ, here.)
   (d) Determine how big a sample size would be needed to achieve your target accuracy.
   (e) Perform this larger number of flights, and report your results.

CHAPTER 9

From samples to populations — inference

9.1 Introduction
- Inference
- Estimation versus hypothesis testing

9.2 Estimation
- Estimating distributions and parameters
- Point and interval estimates
- Quantifying the confidence level
- Confidence interval for a population mean
- Confidence interval for a population proportion

9.3 Testing hypotheses
- P-values and their properties
- Distribution of p-values when H_A and H_0 hold
- Interpretation of p-values
- P-values as probabilities
- Testing population means
- Testing population proportions

9.4 Comparison of two populations
- Inference for difference between population means
- Inference for difference between population proportions
- Paired data

9.5 Comparison of several population means – analysis of variance
- Testing equality of three or more population means

9.6 Regression inference
- Inference about a relationship

Problems

Projects

Appearances to the mind are of four kinds. Things either are what they appear to be; or they neither are, nor appear to be; or they are, and do not appear to be; or they are not, and yet appear to be. Rightly to aim in all cases is the wise man's task.

Epictetus (c.50–c.120)

9.1 Introduction

In chapter 8, we explained the roles of real and hypothetical populations in describing variation in data. Real populations describe situations like opinion polls or selecting tickets from a hat, whereas hypothetical populations are used to describe variability due to a measurement process, as in reaction times or race results. The unifying concept is that of sampling: elements of the population, whether real or hypothetical, are selected into a sample which we hope will be representative of the population. In all situations, the theory is the same: the sample values that result are assumed to be selected at random from a population.

In both sampling from real and hypothetical populations, the populations are almost always of more interest than the samples themselves. For example, the sampled paint primer quality control data is not really the focus of interest. The main point of the sampling is to learn something about the manufacturing process that produced the paint primer samples. In other words, there is a much larger collection of hypothetical values for the paint thickness that characterise the quality of the paint, and we want to find out about this hypothetical population using the sample data that we have collected.

Whether the data we collect are a sample from a real population or from a hypothetical population, our interest is usually to infer something about a population from the data. The branch of statistics concerned with this process is called *inferential statistics* (or simply *inference*). This is to be contrasted with *descriptive statistics* in which the data themselves, rather than the population from which the data came, are the object of our summary. This chapter is concerned with inferential statistics.

In chapter 8 we were concerned with the sampling process and the variability of the resulting samples. The value of samples is that they provide information based on examination of only a part of the population. However, there is a possible downside to sampling; the information may be misleading! It is important to appreciate that a sample selected at random from a population can be quite unrepresentative of the population sampled, even though it is as 'fair' a sample as can be devised. No sampling scheme will produce perfectly representative samples every time but, with random samples, we can compute the extent to which our sample is likely to be misleading. In practice, we use this knowledge to design samples that are large enough to provide reliable information. Inferential statistics is the theory and method that shows how to make reliable inferences about populations based on sample data.

Two distinct branches of inference address different questions about a population. *Estimation* is the process of summarising information in a sample to describe some characteristic of a population. *Hypothesis testing* uses the sample data to evaluate the credibility of a claim about the population.

For example, a political opinion poll might be conducted to assess voting intentions in an electorate with two candidates. We could use the opinion poll data (random sample of electors) to *estimate* the proportion of the vote that would be received by our favourite candidate, or we could test whether, in the light of the data, it is reasonable to cling to the hope that our favourite candidate will receive the majority of the votes, a *hypothesis test* — similar tasks, but different!

Try to rephrase this distinction in your own words. It is important for understanding this chapter that you understand the difference between these two inferential strategies.

We will begin with a discussion of estimation.

EXERCISE 9.1

1. What is common to, and what is different between, *estimation* and *hypothesis testing*?
2. What is the difference between *inferential statistics* and *descriptive statistics*?
3. Why is sampling such a useful strategy in collecting information about a population?
4. A random sample of 10 of the meat packages on display at a large supermarket is selected and the *E Coli* bacteria concentration is measured. The goal is to determine whether the average *E Coli* concentration from all packages of meat on display is within the limit defined by the government regulations. Identify the sample and the population in this example, and the population characteristic of interest to the investigator.

9.2 Estimation

Estimating distributions

Consider the following problem faced by a property developer. The developer must decide which of three suppliers to use for the standard 5 cm × 10 cm lumber used extensively in his condominium complex. The crucial measurement with this lumber is its length; it is nominally 240 cm long, but variation in the sawmill calibration, then warping of the lumber, and other factors cause the length to vary slightly, complicating the building process and increasing both the labour and material costs of building. The builder visits the yards of the three suppliers and measures the lengths of a random sample of 100 5 × 10 lengths. How should these data be used to choose the best supplier?

Armed with a knowledge of summary measures such as the mean and standard deviation, we might be tempted to base a decision on them. Suppose we have the following data:

	Sample size	Mean (cm)	Standard deviation (cm)
Supplier *A*	100	240.3	0.65
Supplier *B*	100	240.5	0.38
Supplier *C*	100	240.6	0.59

The mean length of supplier *A* is a bit closer to the target of 240 cm than the other suppliers and might, on this basis, be considered the best. On the other hand, the higher mean length for supplier *C* will perhaps reduce the chance of undersized lengths which would be wasted, so perhaps this supplier should be used. Supplier *B* has the smallest variability and this might also be considered to be an important requirement — we need rectangular walls! On the basis of the summary data, we have a tough decision to make. Is there any more information in the data that this summary misses? Let us take a look at the distribution of the sample values.

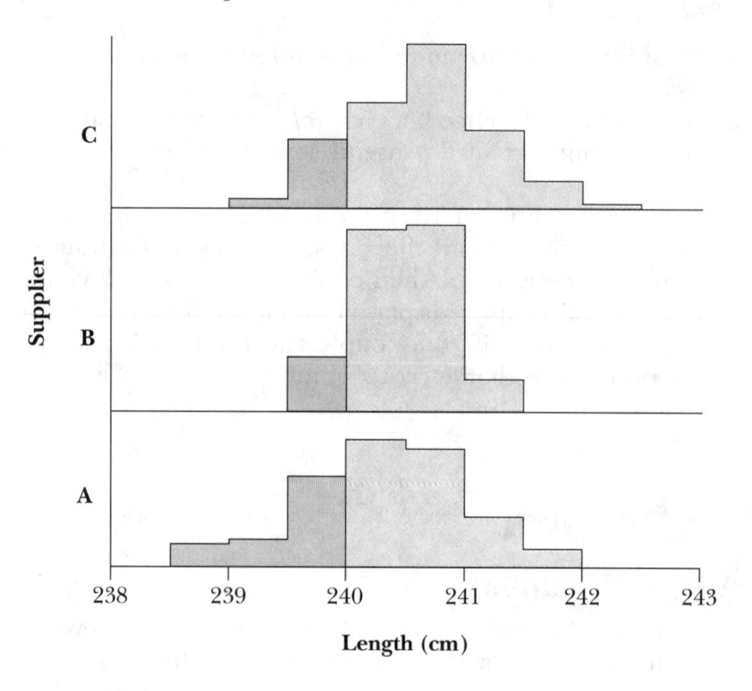

Supplier *B*'s pieces are rarely shorter than the required 240 cm, and this is an advantage over the other two suppliers since a long piece can be made shorter but a short piece must be downgraded to firewood! The decision should clearly be to go with supplier *B*.

If the developer wanted to quantify the above result, he might decide that any pieces shorter than 240 cm were unuseable. The proportion in this category would relate directly to the effective cost of the useable lumber. For supplier *A*, the proportion is $\frac{31}{100}$, for supplier *B* it is $\frac{12}{100}$ and for supplier *C* it is $\frac{17}{100}$. With these data, the developer has a cost-based argument for choosing supplier *B*.

The point of this example is that sometimes means and standard deviations are not directly of interest — what we want from the sample is an estimate of the *whole distribution*. However, there are many situations where population parameters, such as the mean and standard deviation, do contain the important information about a sample, and they are much more concise than the graph of the entire distribution.

Estimating parameters

When a sample is selected at random from a population, we usually want to use that sample to describe the population. Whether the population of interest is real or hypothetical, the task of describing the population based on the sample data is the same — in both cases, the sample values are known but the population distribution is unknown.

In the previous section, we considered a graphical approach to the problem of describing the entire population distribution based on a sample. However, when the real interest in the population is to determine one or more individual parameters, the task becomes more focused. In particular, the mean of numerical populations is often of interest.

Here are two examples of sample data collected for the purpose of estimating the underlying population mean.

- An official of the Ministry of Agriculture wants to estimate the number of ewes available for lambing in the spring. The official knows there are 3535 farms in a region and collects data from a random sample of 100 farms. The mean number of ewes from this sample can then be used to estimate the average for the population of 3535 farms and, from this, a simple multiplication provides an estimate of the total number of ewes on all the farms in the region.

- Two brothers who attend different universities, Roger and David, have their living expenses subsidised by their parents. Roger argues that he should receive a bigger subsidy since living expenses are higher at his university, and the parents ask for evidence of this. By good luck, each university has done a survey of its previous year's students and has determined the annual living expenses of a random sample of 50 students. At Roger's university the sample mean living expense was $8750, whereas it was only $7500 for David's university. These figures do seem to support Roger's claim, but David disputes this and points out that the information is based on samples from their respective populations, and might not be reliable. However, Roger says that the samples were random samples and that there is no good reason to doubt the information.

  The population means would indicate whether there is a tendency for the expenses to be greater at one university or the other and, again, we must use the means of random samples to estimate these population means. However, it is clear that Roger and David need to know more about estimating population means from sample means to be able to sort out their differences.

There are many similar applications of a sample mean being used to estimate the underlying population mean. However, the sample mean is a random quantity and varies from sample to sample, so there is likely to be inaccuracy in its use as an estimate of the population mean. In the next section, we will therefore develop a way to describe the likely estimation errors.

Point and interval estimates

We initially consider the simplest situation where a random sample has been collected from a process that is modelled by a distribution with unknown mean and standard deviation. For example, the 10 values below comprise the first sample of measurements of paint primer thickness that were collected as part of an ongoing process of monitoring the performance of an industrial system. The data were initially examined in section 3.5.

Sample values										Mean	St. devn.
1.30	1.10	1.20	1.25	1.05	0.95	1.10	1.16	1.37	0.98	1.146	0.136

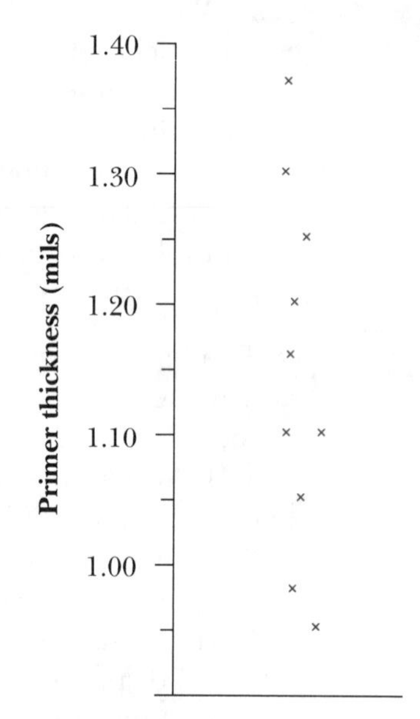

We are interested, among other things, in monitoring the mean paint thickness in the process. On the basis of such a small sample, there is little evidence to support a particular model for the underlying distribution of paint thickness. The jittered dot plot on the right does indicate that the distribution is approximately symmetric, and a normal distribution would therefore be a reasonable model to hypothesise. For the moment, however, we will make no assumption about the shape of the distribution, but will denote its mean by the unknown parameter μ.

We are therefore interested in estimating the value of the parameter μ, which is the mean of this process. An estimate of μ based on this random sample is the sample mean, $\hat{\mu} = \bar{x} = 1.146$.

However, it would be most unusual (indeed surprising!) if the estimate had the same value as the population parameter.[1] This is because the sample mean depends on the particular values selected into the random sample:

$$\bar{X} = \frac{\sum_{i=1}^{n} X_i}{n}.$$

The sample mean therefore varies from sample to sample. Its value in any particular sample is only a guide to the value of the unknown parameter μ. This is illustrated by also considering the subsequent 19 random samples

1. In practice, we could never be surprised by such a surprising occurrence! This is because we could not know that it had happened, since we would not know the *true* value of the parameter!

(each of size $n = 10$) that were recorded twice daily while this process was being monitored. The data and sample means are shown in the table below.

Date	Sample values									Mean
11th a.m.	1.30	1.10	1.20	1.25	1.05	0.95	1.10	1.16	1.37	0.98	1.146
11th p.m.	1.01	1.10	1.15	0.97	1.25	1.12	1.10	0.90	1.04	1.08	1.072
12th a.m.	1.22	1.05	0.93	1.08	1.15	1.27	0.95	1.11	1.12	1.10	1.098
12th p.m.	1.08	1.12	1.11	1.28	1.00	0.95	1.15	1.14	1.28	1.31	1.142
13th a.m.	0.98	1.30	1.31	1.12	1.08	1.10	1.15	1.35	1.12	1.26	1.177
13th p.m.	1.12	1.30	1.01	1.20	1.11	0.93	1.02	1.25	1.05	1.10	1.109
14th a.m.	0.92	1.10	1.13	1.02	0.93	1.17	1.24	0.98	1.34	1.12	1.095
14th p.m.	1.04	1.14	1.18	1.12	1.00	1.02	1.05	1.34	1.12	1.05	1.106
15th a.m.	1.08	0.92	1.14	1.20	1.02	1.04	0.94	1.05	1.12	1.06	1.057
15th p.m.	1.20	1.13	1.19	1.16	1.03	1.25	1.20	1.24	1.10	1.03	1.153
18th a.m.	1.25	0.91	0.96	1.04	0.93	1.08	1.29	1.42	1.10	1.00	1.098
18th p.m.	1.24	1.34	1.40	1.26	1.13	1.15	1.08	1.02	1.05	1.18	1.185
19th a.m.	1.13	1.16	1.12	1.22	1.12	1.07	1.04	1.28	1.12	1.10	1.136
19th p.m.	1.08	1.31	1.12	1.18	1.15	1.17	0.98	1.05	1.00	1.26	1.130
21st a.m.	1.08	1.26	1.13	0.94	1.30	1.15	1.07	1.02	1.22	1.18	1.135
21st p.m.	1.14	1.02	1.14	0.94	1.30	1.08	0.94	1.12	1.15	1.36	1.119
22nd a.m.	1.06	1.12	0.98	1.12	1.20	1.02	1.19	1.03	1.02	1.09	1.083
22nd p.m.	1.14	1.22	1.18	1.27	1.17	1.26	1.15	1.07	1.02	1.36	1.184
23rd a.m.	1.07	1.05	0.97	1.05	1.16	1.02	1.02	1.14	1.07	1.00	1.055
23rd p.m.	1.13	0.90	1.12	1.04	1.40	1.12	1.15	1.01	1.30	1.14	1.131

Because of the sample-to-sample variability in sample means, a single value or *point estimate* from a sample is usually regarded as an inadequate inference about the unknown parameter. The point estimate may be the best guess available, but it is almost certainly wrong.

It is therefore desirable to replace the single value by an interval of values. In effect, this is a way of acknowledging variability in the estimator. Instead of only saying:

> the best estimate of mean primer thickness, μ, based on the first sample of 10 values, is 1.146

it is more appropriate to add a rider along the lines of:

> ... and based on statistical information from the sample, it is reasonable[2] to infer that the true (but unknown) value of μ is some number in the interval 1.05 to 1.24.

An interval of this form is called an *interval estimate* of the parameter μ.

2. The statement 'it is reasonable' needs to be quantified probabilistically, based on information (usually to be obtained from the data) about variability in the process.

Quantifying the confidence level

In view of the spread of values in the jittered dot plot of the first sample of 10 values of paint primer thickness, how confident would you be of each of the following statements?

- μ is exactly 1.146.
- μ is between 1.141 and 1.151.
- μ is between 1.10 and 1.18.
- μ is between 1.05 and 1.24.
- μ is between 0.95 and 1.35.
- μ is between 0.0 and 2.2.

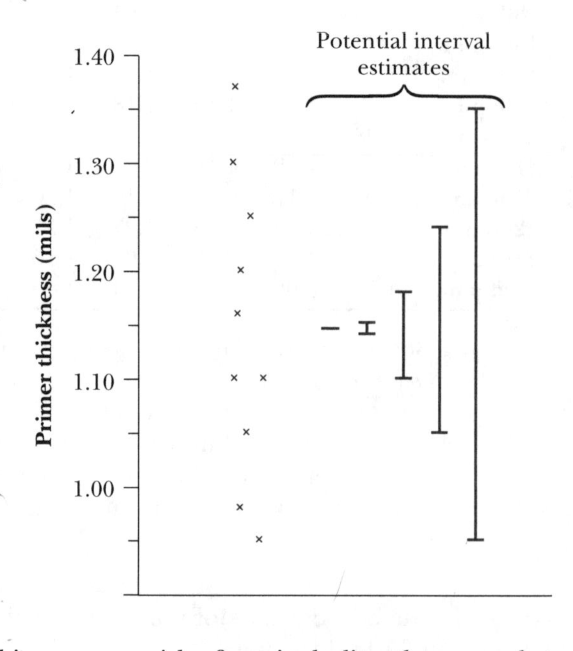

All intervals are centred on the point estimate of μ, 1.146, but the widths of the intervals are different. Our confidence in the first statement should be zero as there is negligible probability of a sample mean being *exactly* equal to μ. Our confidence in the accuracy of the statements should increase as the width of the interval increases. If the interval is wide enough, we should be very confident indeed (almost certain, or '100% confident') in the statement.

Any interval estimate that is narrow enough to be of practical use (for example, excluding such intervals as '0.0 to 2.2' in the paint thickness example) will incur some risk of not including the true value of the parameter. How do we quantify our confidence that any particular interval is wide enough to contain the unknown mean primer thickness in the whole population?

The width of a confidence interval should in some way relate to the variability in the sample and the sample size. In fact:

- the greater the variability in the sample, the less information each value provides about the population mean and, hence, the wider the confidence interval should be made
- as the sample size, n, increases, the sample mean, $\bar{x}$, becomes more stable and, hence, a more accurate estimate of μ; the width of the confidence interval should therefore decrease as n increases.

The confidence interval will have these properties if its width is proportional to the standard deviation of the sample mean. But as discussed in section 8.4, we know that:

$$\sigma_{\bar{X}} = \frac{\sigma}{\sqrt{n}}.$$

In practice, σ is also an unknown parameter, but a reasonable procedure is to replace it by the sample standard deviation, *s*. We therefore initially consider an interval of the form:[3]

$$\left(\bar{x} - \frac{s}{\sqrt{n}}, \bar{x} + \frac{s}{\sqrt{n}}\right).$$

To demonstrate the properties of confidence intervals of this form, we will consider the 20 random samples of paint primer thickness listed at the start of this section. For each sample, an interval has been calculated according to the above formula. The intervals are listed in the table below and are also displayed graphically on page 292. They show what happens in each of the 20 instances in which a sample of size 10 is used to estimate μ.

			Interval estimate	
Sample date	**Mean, $\bar{x}$**	**Standard deviation**	**Lower limit**	**Upper limit**
11th a.m.	1.146	0.136	1.103	1.189
11th p.m.	1.072	0.098	1.041	1.103
12th a.m.	1.098	0.106	1.065	1.131
12th p.m.	1.142	0.120	1.104	1.180
13th a.m.	1.177	0.121	1.139	1.215
13th p.m.	1.109	0.115	1.073	1.145
14th a.m.	1.095	0.136	1.052	1.138
14th p.m.	1.106	0.101	1.074	1.138
15th a.m.	1.057	0.086	1.030	1.084
15th p.m.	1.153	0.079	1.128	1.178
18th a.m.	1.098	0.170	1.044	1.152
18th p.m.	1.185	0.125	1.146	1.224
19th a.m.	1.136	0.070	1.114	1.158
19th p.m.	1.130	0.107	1.096	1.164
21st a.m.	1.135	0.111	1.100	1.170
21st p.m.	1.119	0.137	1.076	1.162
22nd a.m.	1.083	0.074	1.059	1.107
22nd p.m.	1.184	0.099	1.153	1.215
23rd a.m.	1.055	0.059	1.036	1.074
23rd p.m.	1.131	0.141	1.087	1.175

3. We centre the interval on $\bar{x}$ because the central limit theorem states that the distribution of the sample mean will be (approximately) a symmetric distribution centred on μ.

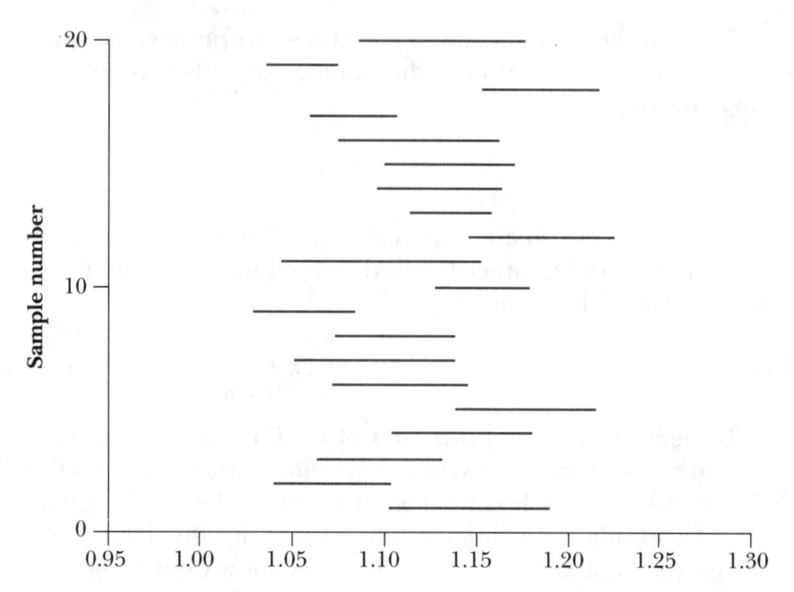

These interval estimates for μ vary greatly from sample to sample and they clearly cannot all contain the true (unknown) value of the parameter, since many of the interval estimates do not even overlap!

However, for a confidence interval to be useful, it does not have to contain the population mean for every possible random sample; it is enough for 'most' random samples to include the parameter value. We do not know the exact value of μ, so we cannot determine the proportion of intervals that include it here but, for the sake of this illustration, let us assume that the true value is $\mu = 1.12$. (We have reason to believe that this will be approximately correct.) Let us see how many of the confidence intervals contain the population mean value 1.12. The diagram below highlights the intervals that include this value. It can be seen that 60%, or 12 out of 20, of the intervals include 1.12.

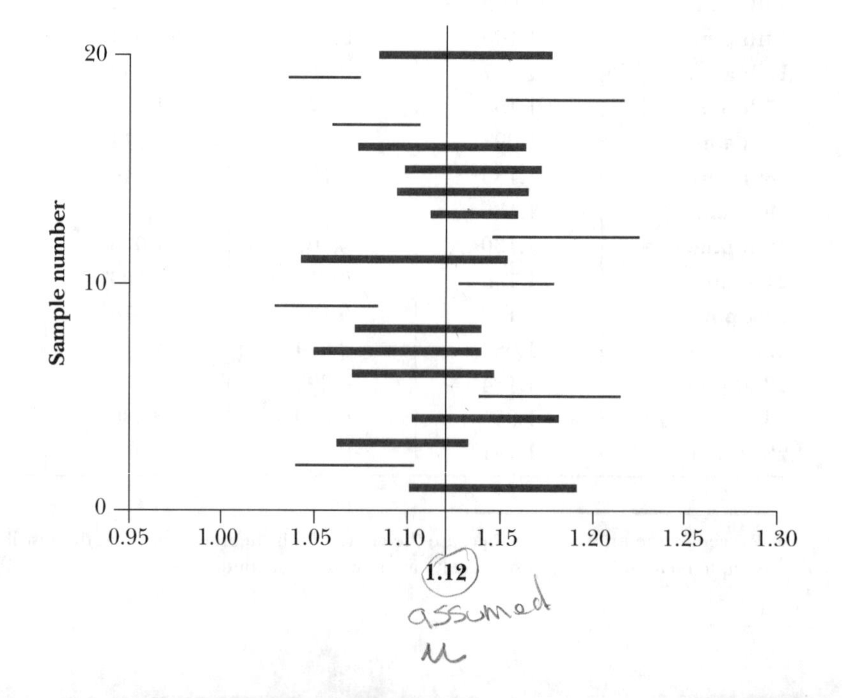

If we were to repeat the whole process of collecting the 20 samples of size 10, and constructing the previous graph again, the proportion of confidence intervals containing the true mean could be different from 60%. But if we put hundreds of these samples of size 10 together, a predictable proportion of them will contain the population mean.

When a fixed procedure, using a formula such as:

$$\left(\bar{x} - \frac{s}{\sqrt{n}}, \bar{x} + \frac{s}{\sqrt{n}}\right)$$

is used to calculate confidence intervals, each will have some chance of including the unknown value of μ. If an indefinitely large number of such intervals is calculated from independent random samples, there will be a limiting proportion that will include μ and this proportion is called the *confidence level* of the procedure. We have observed above that the confidence level of the confidence intervals we found for the paint primer data is approximately 60%. More advanced statistical theory (with some additional assumptions about the population) proves that the confidence level is 65.7%.

In practice, a higher confidence level than 65.7% is usually required and, by convention, confidence intervals are usually reported at a 95% confidence level;[4] these are called *95% confidence intervals.* To increase the confidence level, the interval must be widened. The adjustment required for a 95% confidence interval for a population mean is to use an interval of the form:

$$\left(\bar{x} - t_{n-1} \times \frac{s}{\sqrt{n}},\ \bar{x} + t_{n-1} \times \frac{s}{\sqrt{n}}\right)$$

where t_{n-1} is a constant that may be read from the table below.[5]

v	t_v	v	t_v	v	t_v	v	t_v
1	12.71	**11**	2.201	**21**	2.080	**40**	2.021
2	4.303	**12**	2.179	**22**	2.074	**50**	2.009
3	3.182	**13**	2.160	**23**	2.069	**60**	2.000
4	2.776	**14**	2.145	**24**	2.064	**80**	1.990
5	2.571	**15**	2.131	**25**	2.060	**100**	1.984
6	2.447	**16**	2.120	**26**	2.056	∞	1.960
7	2.365	**17**	2.110	**27**	2.052		
8	2.306	**18**	2.101	**28**	2.048		
9	2.262	**19**	2.093	**29**	2.045		
10	2.228	**20**	2.086	**30**	2.042		

4. Confidence intervals are occasionally reported at 90% or 99% confidence levels. These are obtained in a similar way to the intervals that we describe here; the only difference is in the constant t_{n-1} that is looked up in a different table. See also footnote 5.
5. When the sample size, n, is large, this formula can be explained using the central limit theorem, which states that the sample mean has (approximately) a normal distribution with mean μ and standard deviation, $\frac{\sigma}{\sqrt{n}}$. Using the properties of the normal distribution, $\bar{x}$ will be within 2 standard deviations (that is, $2\frac{\sigma}{\sqrt{n}}$) of μ in 95% of samples. In other words there is a 95% chance that μ will be within $2\frac{\sigma}{\sqrt{n}}$ of $\bar{x}$ — that is, in the interval: $\bar{x} \pm 2\frac{\sigma}{\sqrt{n}}$.

   Noting that t_{n-1} is close to 2 when n is large, and that s will be approximately σ, this is close to the 95% confidence interval given above.

For example, a 95% confidence interval for a sample of 10 paint primer thicknesses would be of the form:

$$\left(\bar{x} - 2.262 \times \frac{s}{\sqrt{10}},\ \bar{x} + 2.262 \times \frac{s}{\sqrt{10}}\right).$$

The diagram below shows the 95% confidence intervals for all 20 paint primer samples.

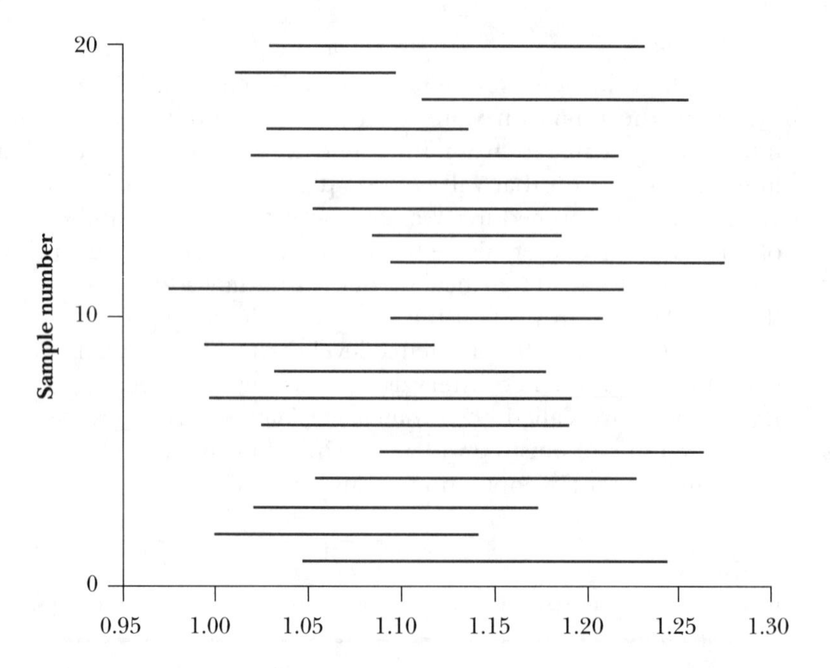

The 95% confidence intervals are wider than the previous intervals, but they are still quite variable; we still have a chance of missing the true (unknown) value of μ with a single sample. We do, however, know that 95% of the confidence intervals obtained using this procedure will include μ, and this is the best we can say with a single sample of $n = 10$ values.

Based just on the first sample of data, our point estimate of the mean thickness of primer, μ, is 1.146 and our 95% confidence interval is (1.048, 1.244), also commonly expressed as 1.146 ± 0.098.

The notion of a confidence interval is easily misunderstood, so be forewarned!

- We are not estimating an interval that includes 95% of the observations in the sample — if we were, we would use an interval such as $(\bar{x} \pm 2 \times s)$ instead of $(\bar{x} \pm t_{n-1} \times \frac{s}{\sqrt{n}})$. It is the population mean we are describing, not the sample data, nor even the entire population distribution.
- Another misunderstanding is that the 95% confidence interval will, with high confidence, contain the sample mean. In fact, any confidence interval for a population mean will *always* include the mean of that sample, since the sample mean is always right in the middle of the confidence interval!

As long as you remember what the confidence interval is estimating, namely the population mean, you will not fall into these traps of misunderstanding.

We will now return to one of the examples that was described at the beginning of this section, and see what interval estimates would produce if sample data were actually collected.

- An official of the Ministry of Agriculture samples 100 farms from the 3535 farms registered in a region and records the number of ewes available for lambing in the spring from each sampled farm. The data from the sample of 100 farms are summarised as follows:

$$n = 100 \quad \bar{x} = 23.5 \quad s = 15.5.$$

The 95% confidence interval for the mean number of ewes per farm among the population of 3535 farms is:

$$\left(\bar{x} \pm t_{99} \times \frac{s}{\sqrt{100}}\right) = \left(23.5 \pm 1.98 \times \frac{15.5}{\sqrt{100}}\right) = (23.5 \pm 3.07)$$

or the interval 20.4 to 26.6.[6]

This implies that the total number of ewes would be somewhere in the interval:

$$20.4 \times 3535 = 72\,114 \quad \text{to} \quad 26.6 \times 3535 = 94\,031$$

In other words, our sample of 100 farms tells us that the total number of ewes on the 3535 farms is likely to be in the range 72 114 to 94 031. This might be useful information, and the fact that we only had to contact 100 farms instead of 3535 is quite helpful. Of course, if we needed a more precise estimate, we should have used a larger sample size. In fact, doubling the sample size can be shown to reduce the width of the confidence interval by a factor $\sqrt{2}$.

The second example involving Roger's and David's living expenses at university will be worked through later in the chapter, since the comparison of two samples requires a more advanced procedure than we have covered so far.

Confidence interval for a population proportion

While quantitative data are most often summarised with means, categorical data are usually summarised with percentages.[7] We will consider populations whose members can each be classified into one of two types, such as female and male, employed and unemployed, or defective and satisfactory. For these examples, we might want to know the proportion of males, employed or defective items, respectively. The population proportion in the category of interest is often estimated by the proportion in that category in a random sample selected from the population.

6. Strictly, the formula we described earlier is only appropriate for sampling from infinite populations or sampling with replacement from finite populations. However, provided the proportion sampled is not too high in sampling without replacement from a finite population, the same formulae can be used as an approximation.
7. Percentages and proportions are used interchangeably in this context — percentages such as 50% or 33% could equally be expressed as proportions 0.50 or 0.33.

As shown in section 8.5, a sample proportion can be regarded as a sample mean if the two categories of interest (such as *male* and *female*), are coded as 0 and 1. The proportion in the category coded as '1' is the same as the mean of the 0's and 1's. For example, the sexes of a random sample of 10 primates from a population with unknown sex ratio might be:

male, male, female, male, female, female, female, male, female, female.

This would be coded as:

0, 0, 1, 0, 1, 1, 1, 0, 1, 1.

The proportion of females, 0.6, is equal to the sample mean of the 0's and 1's.

The properties of sample proportions are therefore very similar to those of sample means. In particular, if the unknown population proportion is π, the central limit theorem gives us the following properties of a sample proportion, p.

- $\mu_p = \pi$.
- $\sigma_p = \dfrac{\sqrt{\pi(1-\pi)}}{\sqrt{n}} = \sqrt{\dfrac{\pi(1-\pi)}{n}}$.
- p is approximately normal when n is large.

We can use these results to obtain a 95% confidence interval for π that is similar in form to that for a population mean μ.

For example, the following table gives results from a public opinion poll conducted in the Palmerston North electorate by Massey University's Department of Marketing shortly before the New Zealand election in October 1996.

Candidate	Number intending to vote for candidate	Percentage
Steve Maharey (Labour)	114	51.4
George Mathew (National)	48	21.6
Gerard Hehir (Alliance)	23	10.4
Trevor Jans (New Zealand First)	16	7.2
Grant Bowater (Christian Coalition)	11	5.0
Val Wilde (ACT)	10	4.5
Other	0	0.0
Total	222	100.0

The table excludes respondents who were undecided, refused to answer the question or indicated that they would not vote. However, for the sake of this example, we will assume that this sample is a random sample of electors from the electorate.

The sample proportion intending to vote for the Labour candidate, $p = 0.514$, is an estimate of the proportion, π, who intend to vote for that candidate in the whole electorate. We know, however, that this sample proportion is random, so a confidence interval should be used to describe the likely values for π.

We saw earlier that a 95% confidence interval of the form:

$$\bar{x} \pm t_{n-1} \times sd(\bar{x})$$

is appropriate for estimating a population mean from the sample mean, where $sd(\bar{x})$ is the best available estimate for the standard deviation of $\bar{x}$, and t_{n-1} is a value obtained from tables. As a sample proportion can be treated as a sample mean (with 0/1 coding of the categories), it might initially *seem* that an interval of the form:

$$p \pm t_{n-1} \times sd(p) = p \pm t_{n-1} \times \sqrt{\frac{p(1-p)}{n}}$$

would be appropriate. However, when the sample size, n, is small, or when the population proportion, π, is close to either 0 or 1, the formula becomes less accurate than it is for most numerical data.[8] We usually therefore restrict use of the formula to situations where n is large and π is not near 0 or 1, and we dispense with the refinement of looking up the value t_{n-1} and simply use the value 2.0.

A 95% confidence interval for a population proportion π is given by:

$$p \pm 2 \times \sqrt{\frac{p(1-p)}{n}}$$

provided the sample size n is reasonably large (say ≥ 25) and p is not so close to either 0 or 1 that the confidence interval includes either 0 or 1.

For example, a 95% confidence interval for the proportion intending to vote for the Labour candidate in the election is:

$$0.514 \pm 2 \times \sqrt{\frac{0.514 \times (1-0.514)}{222}} = 0.514 \pm 0.067 = 0.447 \textit{ to } 0.581.$$

Similarly, a 95% confidence interval for the proportion intending to vote for the National candidate is:

$$0.216 \pm 2 \times \sqrt{\frac{0.216 \times (1-0.216)}{222}} = 0.216 \pm 0.042 = 0.174 \textit{ to } 0.258.$$

Since:

$$2 \times \sqrt{\frac{p(1-p)}{n}} \leq 2 \times \sqrt{\frac{0.5 \times (1-0.5)}{n}} = 0.067$$

for all candidates (and for all values of p), the latter is usually called the *margin of error* of the public opinion poll.

The properties of 95% confidence intervals for a proportion can be illustrated with a similar simulation to that used to explain confidence intervals for means. Consider a large electorate in which 25% support candidate A. We have randomly selected 20 samples of size $n = 200$ from this population. The

8. Since the population of 0's and 1's is so far from a normal distribution, the sample size needs to be considerably larger before p becomes close to normally distributed.

diagram below shows the sample numbers voting for *A* and the 95% confidence intervals for the population proportion that resulted from each of these 20 samples.

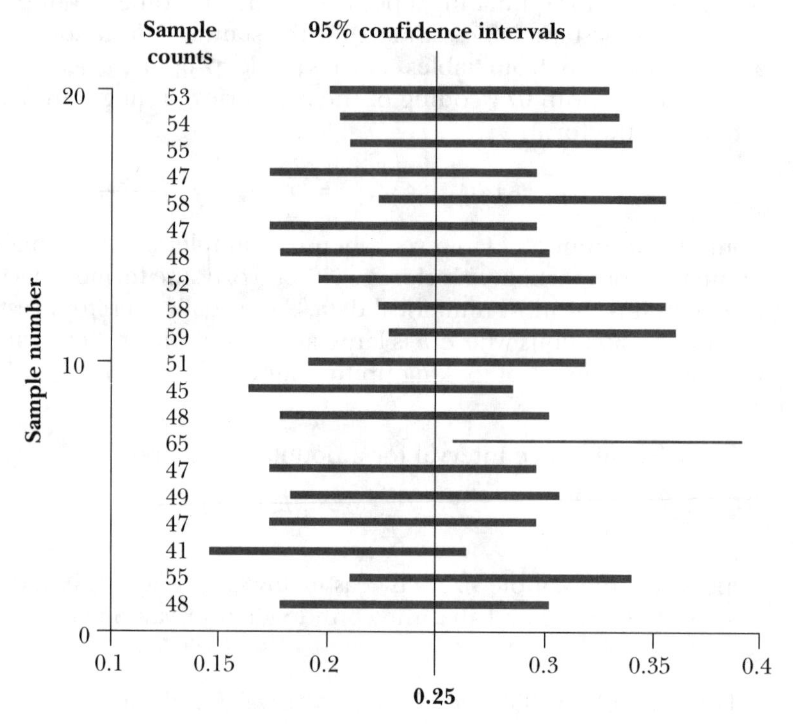

The confidence intervals vary from sample to sample and not all of them contain the true population proportion which is known to be 0.25 in this simulation (but which is unknown in practical applications). However, if we selected enough samples, 95% of them would contain π. (The proportion containing $\pi = 0.25$ happens to be exactly 95% above, but could be slightly different in other simulations of 20 samples.)

The simulation above demonstrates the properties of 95% confidence intervals for a population proportion. In practice, only a *single* random sample is available, and so we only have a *single* 95% confidence interval (such as 0.447 to 0.581 for the Labour candidate). It is, however, a valuable aid to interpreting this confidence interval if you *imagine* that this interval is one of many that *could* have been obtained from repeated random samples. For example, we can imagine the opinion poll being repeated; 95% of the confidence intervals that we would obtain would enclose π, although we do not, of course, know whether or not the actual confidence interval from our data is one of these.

We noted that the formula we have given for a 95% confidence interval for π should only be used when sample size is large and p is not close to 0 or 1. What can be done when the sample size is small? It is still possible to obtain a 95% confidence interval, but the formula is well beyond the scope of this book. Suitable statistical software may, however, be used to obtain the confidence interval.

EXERCISE 9.2

1. In what ways are the purposes of a point estimate and an interval estimate the same, and in what ways are they different?
2. What population parameters are discussed in this section as targets of estimation procedures?
3. The amount of fat in hamburger meat varies from package to package. A sample of ten 1 kg packages from a large shipment is analysed by a laboratory, and the fat contents of these ten packages average 330 g and have a standard deviation of 30 grams.
   (a) Compute a 95% confidence interval for the mean amount of fat in the shipment.
   (b) Would you believe a claim by the shipper that the mean fat content is 300 grams? (This type of question will be addressed more fully in section 9.3, but you should be able to give some assessment based on the confidence interval you have just found.)
4. A food store buys apples from a supplier in '50 kg' boxes. Because of marketing laws, these boxes are required to be at least 50 kg a very large percentage of the time. The effect is to make the average box contain more than 50 kilograms. To monitor the apples that the store receives, the store needs to know what weight of apples is actually supplied, on average. To determine the mean weight of the apples per box, a sample of 16 of the boxes purchased is weighed; their mean is 50.81 kg and their standard deviation is 0.40 kilograms.
   (a) Find a 95% confidence interval for the mean box weight.
   (b) Suppose that another shipment contains a total of 200 boxes. Use the confidence interval in (a) to give an interval estimate of the weight of 'free' apples (over the nominal 50 kg) received in that shipment.
   (c) The exact weight of apples sold from a shipment can be obtained from checkout records and, from this, it is possible to estimate the wastage of apples. The checkout weight of apples sold from the shipment in (b) is 9954 kilograms. How would you estimate the bad apple loss from the shipment?
5. A test of a batch of 10 000 formatted computer disks involved trial writing and reading on a random sample of 100. The disks either operated perfectly or were proclaimed 'defective'. The test resulted in 7 defectives.
   (a) Find a 95% confidence interval for the proportion of defective disks from the supplier.
   (b) What range of defective values might have resulted if the entire 10 000 had been similarly tested?
   (c) Could the same procedure be used if there had only been a single defective in the first sample of 100 disks?
6. Is the following statement true or false? 'A 95% percent confidence interval can be expected to include about 95% of the data values.' (*Hint*: Remember that the 95% confidence interval is an estimate of a population mean.)
7. What features of a distribution that are not revealed by the mean and the standard deviation might be of interest?

9.3 Testing hypotheses

There are various ways of making inferences about parameters on the basis of sample data. In all of them, we are addressing the following question.

- What does the sample tell us about the parameter?

This question can be refined in various ways. The question can be rephrased as:

- What parameter values are consistent with the data?

In this form, we answer the question by deriving a confidence interval for the parameter, as described for population means and proportions in section 9.2. An alternative is to ask:

- Are the data consistent with a particular value of the parameter?

When the question is asked in this form, a confidence interval is not an appropriate answer. The correct way to approach such a question is a form of statistical inference called a *hypothesis test*.

Because they both use sample data to address questions about values of unknown parameters, the ideas and methods of confidence interval estimation and hypothesis testing are interrelated. Whenever sample data are used to address questions about a parameter, either or both of these inferential methods may be used. Very often, in these simple circumstances, confidence intervals provide a more informative inference than do hypothesis tests and should be used in preference. This comment applies especially to inferences about population means and proportions.

In this section, we will consider how a hypothesis test can be used to make an inference about a population parameter, using the data in a single sample from that population.

For example, the complete list of 1994 rolls for all 363 New Zealand high schools was discussed in section 8.3. Whether there is any change in rolls from one year to the next is of interest. As a complete list of school rolls is not published until late in the school year, the school rolls of a random sample of high schools might be obtained by phoning the schools early in the 1995 school year to determine whether the rolls for these schools had changed since the previous year. This information could be used to assess whether the mean school roll in the whole of New Zealand has changed. The table on page 301 shows the change in roll between 1994 and 1995 for a random sample of 20 New Zealand high schools.

School	1994 roll	1995 roll[9]	Increase
Kerikeri High School	849	844	–5
Te Kura Taumata O Panguru	151	144	–7
Marist Sisters College	508	509	1
Mcauley High School	557	560	3
Takapuna Grammar	1228	1271	43
Tangaroa College	589	546	–43
Waitakere College	1453	1427	–26
John Paul College	727	785	58
Taradale High School	875	793	–82
Wairoa College	687	684	–3
Dannevirke High School	613	607	–6
Tararua College	490	481	–9
Turakina Maori Girls' College	114	90	–24
Wellington Girls' College	1102	1114	12
Kaikoura High School	249	257	8
Amuri Area School	243	262	19
Burnside High School	2033	2078	45
Villa Maria College	669	672	3
The Taieri High School	708	704	–4
Menzies College	438	420	–18

If there is no difference in the mean rolls for all high schools, the sample differences will have a distribution centred around zero, whereas if the mean roll has changed, the distribution will be centred around some non-zero value.

A confidence interval could be used to obtain an interval estimate of the mean difference from these data. We could decide instead, however, to test the hypothesis that the mean difference is zero. In this case, we are asking the question, 'Have average rolls changed from 1994 to 1995?' We can observe the change in the above sample (a mean increase of $\bar{x} = -0.64$ per school; that is, a mean decrease of 0.64 per school), but need to infer the change for the underlying population of schools.

P-values

Testing a hypothesis about a parameter is, like estimation, based on a set of data, usually a random sample, and a probability model for the process underlying the data. Two contradictory statements (called *hypotheses*) about the unknown parameter(s) of the model are written down, and our question is expressed by asking which of these two hypotheses is true. One hypothesis is called the *null hypothesis* and is denoted by H_0. This is the more 'conservative' of the hypotheses and is the hypothesis which would tend to be accepted if there were no data. The other hypothesis, called the *alternative hypothesis*, is denoted by H_A.

In the high school example above, we assume that the differences are a random sample from a population with mean μ and standard deviation σ. We

9. *Directory of New Zealand Schools and Tertiary Institutions, 1996*, published for the New Zealand Ministry of Education by Learning Media Ltd, Wellington, 1995, p. 56–66.

then want to test whether or not the parameter μ is zero. The two hypotheses being compared can then be expressed as:

$$H_0: \mu = 0$$
$$H_A: \mu \neq 0.$$

We seek evidence based on a random sample as to whether H_0 or H_A holds; that is, is the sample consistent with H_0 or does it throw doubt on that hypothesis and therefore support H_A? One set of data values might provide clear evidence in support of H_A; a different set of data values might be consistent with H_0.

For example, the hypothetical data sets represented by the two jittered dot plots on the left below are each clearly inconsistent with H_0 and we would strongly infer from either of them that H_A was true. In contrast, the two data sets on the right would not lead us to doubt H_0. However, we require statistical analysis to enable us to assess each of the middle two data sets.

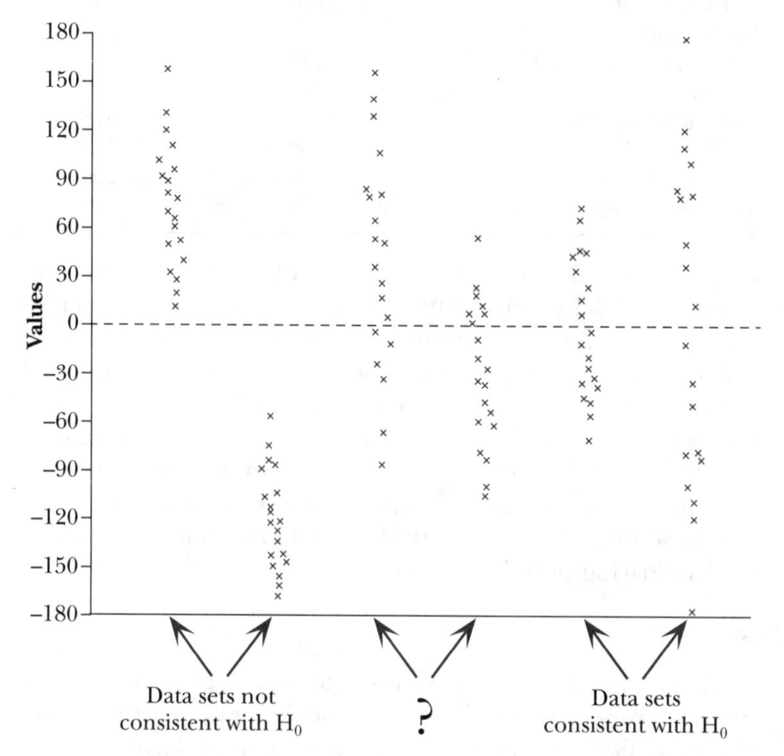

Ideally, we would like to be able to define an index, I, that could be evaluated from our random sample that would distinguish between the two hypotheses:

$$I(\textit{sample data}) = \begin{cases} 1 & \text{if the sample is from a population with} \quad \mu = 0 \\ 0 & \text{if the sample is from a population with} \quad \mu > 0 \end{cases}$$

or, more generally:

$$I(\textit{sample data}) = \begin{cases} 1 & \text{if } H_0 \text{ holds} \\ 0 & \text{if } H_A \text{ holds.} \end{cases}$$

Unfortunately, such a function cannot be found since any set of data values could, however unlikely, be obtained in a random sample from a normal distribution if either H_0 or H_A held. For example, in the school roll example, the data could have been recorded either if $\mu = 0$ (H_0) or if $\mu = 1$ (which is consistent with H_A). The data could even have been observed if $\mu = 150$ (also consistent with H_A), however unlikely that might appear.

We therefore cannot hope for such a clear-cut decision between the two hypotheses. The best separation we can hope for is an index $I(\textit{sample data})$ that will tend to be closer to 0 when $\mu \neq 0$ than when $\mu = 0$, and closer to 1 when $\mu = 0$ than when $\mu \neq 0$. We can therefore interpret values of $I(\textit{sample data})$ close to 0 as giving evidence that H_A holds rather than H_0. Hypothesis testing is conducted using a statistic of this form called a *p-value*.

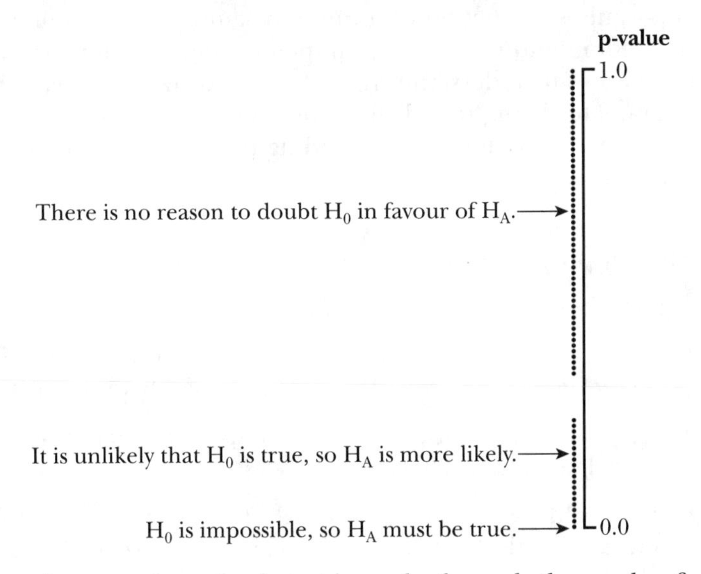

Inference from *any* hypothesis test is made through the p-value for that test. Since computer software will evaluate the p-value, the important skills that you must learn are:

- how to formulate appropriate hypotheses, and
- how to interpret the resulting p-value.

These skills are common to *all* hypothesis tests and will be described in the remainder of this section. Knowing the formula underlying the p-value for any particular test is of lesser importance.[10]

In this book we describe two approaches to understanding p-values. The first approach treats the p-value as an index of the credibility of H_0 against H_A and explains its interpretation through the distributional properties of this index when H_0 and H_A hold. This is the less conventional of the two

10. Many textbooks contain formulae which require tables to be looked up to obtain an approximate p-value. However, for the most commonly used tests such as 't-tests', these tables allow only the roughest approximation to the p-value. It is better to rely on a computer to do the job properly!

approaches, but involves minimal mathematics and is easily generalised. The second approach focuses on the p-value as a probability of the data when H_0 is true, while entertaining the possibility of H_A. This approach is more traditional; it explains how a p-value would be defined for any particular test, but has perplexed generations of students!

You are hereby warned that either approach will require some concentrated effort on your part — this most difficult part of basic statistics will likely require multiple reviews. However, our two approaches are complementary and, for a thorough understanding of hypothesis testing, you should try to master them both.

Properties of p-values

As a p-value is a function of a random sample, it is a summary statistic that is itself random and varies from sample to sample. Interpretation of the p-value depends on an understanding of how its distribution is affected by whether the parameter being tested (μ in the school rolls example) is in H_0 or H_A.

Our first approach to understanding p-values is based on their distributional properties.

The key properties of p-values are:

- a p-value is a random summary statistic
- a p-value has a distribution which can take values between 0 and 1, and each of these values is possible, whichever hypothesis holds
- if H_A holds, a p-value has a distribution in which values near 0 are more likely than values near 1
- if H_0 holds, a p-value has a distribution in which each value between 0 and 1 is equally likely, or values near 1 are more likely than values near 0.

We will explain later how these properties allow us to interpret a p-value, but firstly we will demonstrate the properties for a particular hypothesis test concerning the mean, μ, of a population with unknown mean and standard deviation. Based on a random sample of size $n = 9$, we will test the null and alternative hypotheses:

$$H_0: \mu \geq 10$$
$$H_A: \mu < 10.$$

We will explore the distribution of p-values in this test by repeatedly generating random samples of size $n = 9$ from normal populations with various values of μ.

Distribution of p-values when H_A holds ($\mu < 10$)

We will initially demonstrate that the p-values for this hypothesis test 'tend to be closer to 0 than to 1' when the alternative hypothesis, H_A, holds; that is, when $\mu < 10$ in this example.

As a concrete example, we have generated 100 random samples of size $n = 9$ from a normal population for which $\mu = 8$ and $\sigma = 5$; these random samples are shown in the table below.

Sample number	Sample values									$\bar{x}$	p-value $I(sample)$
1	15.32	9.80	1.13	6.89	3.65	15.09	7.01	0.20	2.11	6.802	0.065
2	16.59	3.43	13.31	19.25	3.13	−0.41	18.30	9.63	6.40	9.958	0.493
3	12.72	6.24	15.56	18.83	5.47	14.14	15.36	4.54	14.34	11.910	0.851
4	6.10	3.01	7.60	4.19	9.03	1.26	4.71	6.98	15.62	6.498	0.018
5	16.84	8.23	13.54	4.23	8.82	9.06	9.11	6.80	4.58	9.024	0.244
6	5.30	5.83	8.49	13.92	11.72	11.16	3.86	9.44	4.28	8.223	0.089
99	7.11	4.44	7.91	−0.42	6.48	6.42	4.86	4.35	8.97	5.570	0.001
100	9.12	4.13	8.81	7.47	9.04	5.01	5.43	−2.45	6.17	5.860	0.004

The two columns on the right show the values of $\bar{x}$ and the p-value I(*sample data*) that were evaluated for each sample. The lower the sample mean, the greater the support for H_A and, hence, the lower we would expect the p-value; this is evident in the table above.

The jittered dot plot and histogram below show the distribution of p-values from the 100 samples.

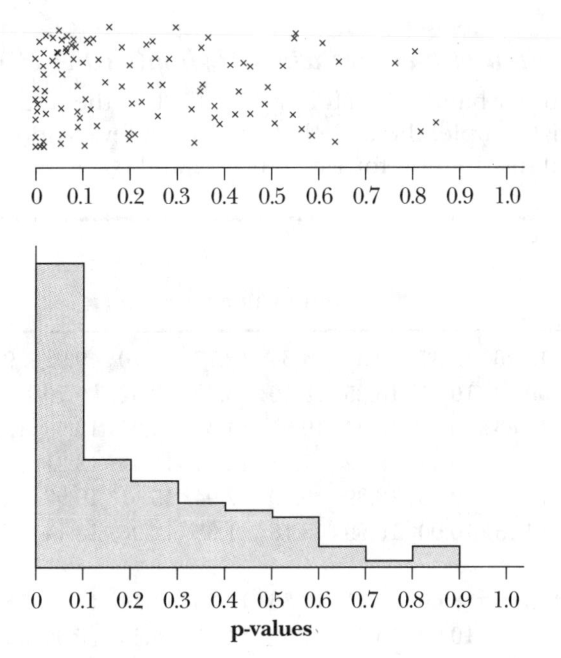

The p-values tend to be closer to 0 than they are to 1. However, even though μ is in H_A ($\mu < 10$), some random samples have p-values near 1 and therefore suggest that μ is in H_0 ($\mu \geq 10$). But, since we simulated the data, we know that these suggestions are wrong!

The lower the true value of μ (the value of μ in the model from which the data are generated), the more concentrated the distribution of the p-values will be around 0. Similarly, the closer μ is to 10 (the closest value consistent with H_0), the closer the distribution of p-values becomes to rectangular. The diagram below shows distributions of p-values for various values of $\mu < 10$.

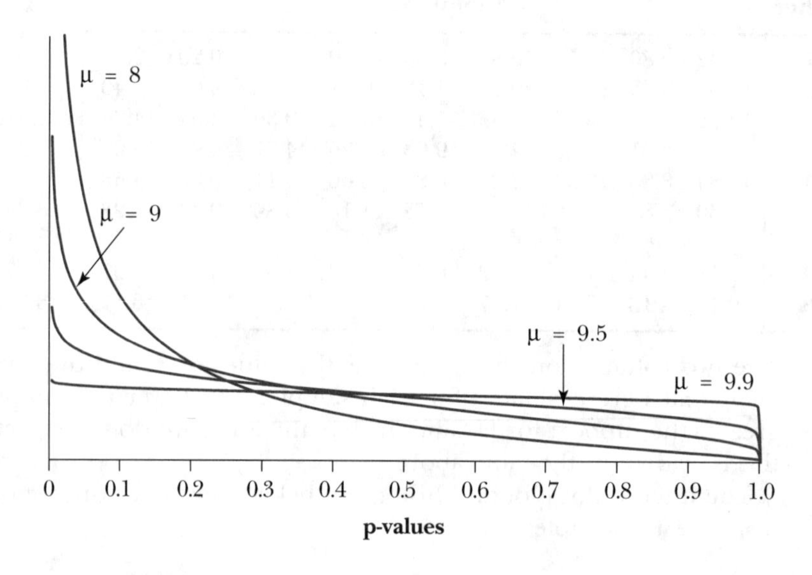

Distribution of p-values when H_0 holds ($\mu \geq 10$)

On the other hand, when μ is in H_0 ($\mu \geq 10$), the p-values will tend to be closer to 1. For example, the table below shows a few random samples of size 9 from a normal distribution for which $\mu = 12$ and $\sigma = 5$.

Sample number	Sample values									$\bar{x}$	p-value *I(sample)*
1	16.40	13.26	12.37	10.39	4.34	12.17	11.70	9.92	9.53	11.122	0.833
2	4.47	9.53	10.65	16.85	8.10	13.30	12.52	9.70	7.68	10.310	0.599
3	6.18	13.83	9.91	13.64	19.95	4.33	14.97	13.58	13.91	12.257	0.904
4	13.03	11.11	9.47	18.92	7.01	12.94	15.43	13.31	13.89	12.789	0.980
5	17.17	13.74	10.15	13.39	15.89	12.94	15.51	19.97	14.53	14.809	1.000
6	3.81	14.15	10.00	16.99	18.18	1.95	12.96	18.14	9.26	11.716	0.793
99	8.40	8.66	8.60	10.41	17.36	13.06	11.14	15.28	18.27	12.356	0.949
100	12.28	9.94	10.09	13.71	11.96	14.45	8.59	6.00	12.44	11.051	0.865

Since we do not know whether H_0 or H_A is true, p-values are obtained from sample data in the same way as in the previous simulation. Again, high values of $\bar{x}$ give p-values near 1 and low values of $\bar{x}$ give p-values near 0.

The diagram below shows the resulting distribution of p-values from the 100 samples.

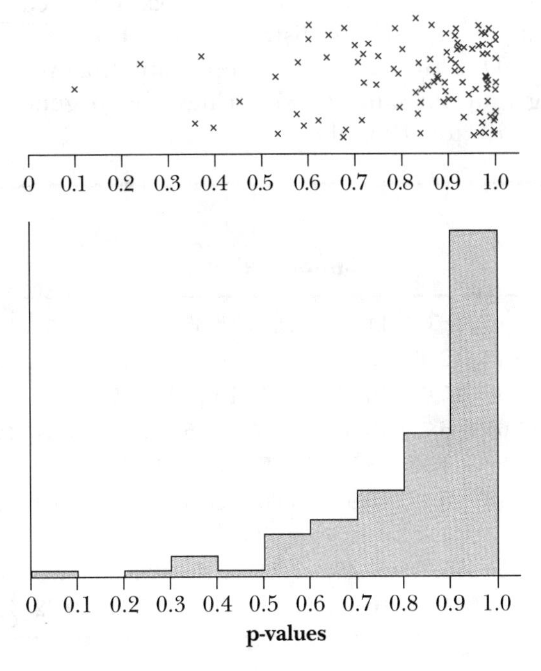

Even though $\mu \geq 10$ (so the null hypothesis H_0 is true), some random samples have p-values near zero and therefore suggest that μ is *not* in H_0. Since we simulated the data, we know that these data sets would lead us to the wrong conclusion! We must accept that, occasionally, samples will be misleading!

The further the true value of the parameter μ from the region specified by the alternative hypothesis, the more concentrated is the distribution of p-values around 1. The diagram below shows some possible distributions for the p-values when μ is in H_0.

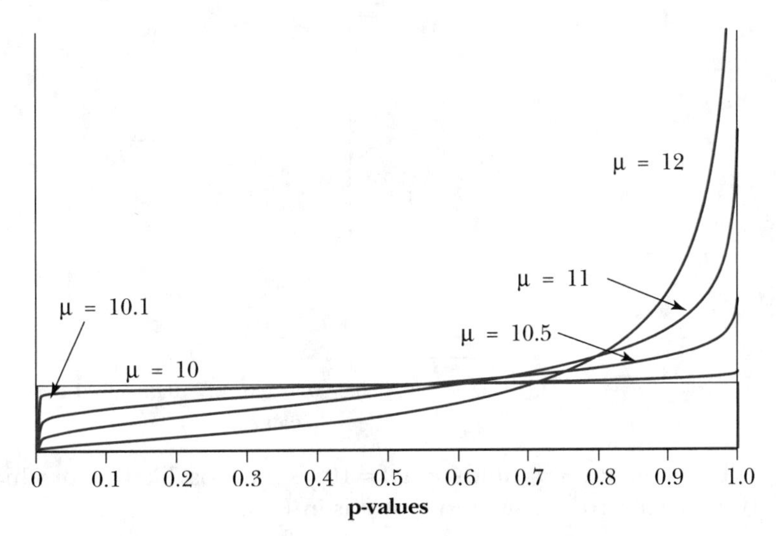

Of these possible distributions, one is particularly important for hypothesis testing. When μ is on the boundary between H_0 and H_A (when $\mu = 10$ in the test we are using as an example), each p-value between 0 and 1 is equally likely (so the distribution of the p-values is *rectangular* between 0 and 1). This situation is so important that we will illustrate it with the following random samples which have been generated from a normal distribution with $\mu = 10$ and $\sigma = 5$.

Sample number	Sample values									$\bar{x}$	p-value $I(sample)$
1	−0.07	11.16	16.05	11.88	12.01	5.35	11.20	4.03	5.39	8.556	0.210
2	5.86	10.43	12.89	3.63	4.32	6.68	11.20	9.56	10.76	8.370	0.089
3	12.22	5.89	15.62	5.30	7.71	16.78	15.10	2.53	9.16	10.036	0.508
4	7.51	11.01	6.75	13.62	8.08	10.55	−1.35	7.07	16.13	8.819	0.248
5	5.09	12.69	4.49	13.98	14.06	17.90	17.47	12.42	9.96	12.006	0.879
6	5.58	2.48	6.49	6.97	13.98	12.75	−1.92	9.64	9.67	7.182	0.062
99	8.70	8.64	15.27	20.40	7.74	2.46	−0.68	16.70	-0.18	8.784	0.320
100	4.92	16.97	15.24	10.94	4.29	9.74	17.13	9.22	12.95	11.268	0.776

The distribution of the p-values is shown below.

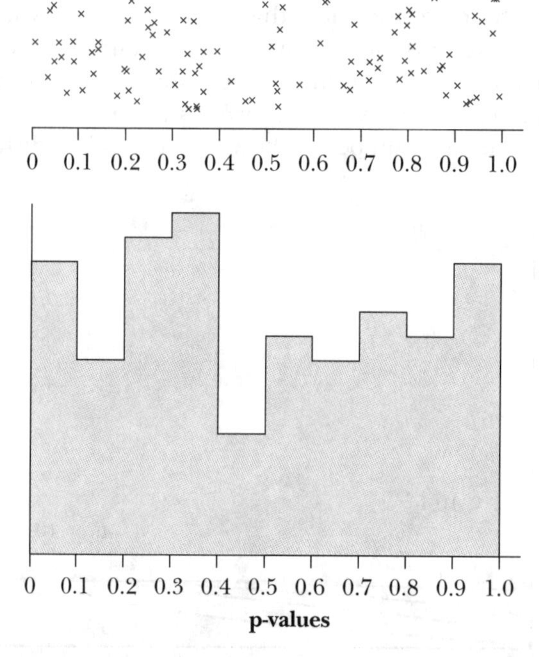

This situation (H_0 holds and $\mu = 10$) is the most likely to produce low p-values (which will wrongly suggest that μ is in H_A).

In some types of hypothesis test, the null hypothesis, H_0, only specifies a single value for the parameter (called a *simple* null hypothesis). For example, we might want to compare the hypotheses:

$$H_0: \mu = 10$$
$$H_A: \mu \neq 10.$$

For this type of test, the p-values *always* have a rectangular distribution between 0 and 1 when H_0 is true.

Interpretation of p-values

We have seen above that a p-value is an index whose distribution is more weighted towards 0 than 1 when the alternative hypothesis holds. When the null hypothesis is true, the distribution of the p-values is more weighted towards 1 than 0; the most heavily weighted towards 0 that is possible when the null hypothesis holds is the rectangular distribution. These results are illustrated schematically by the diagrams below.

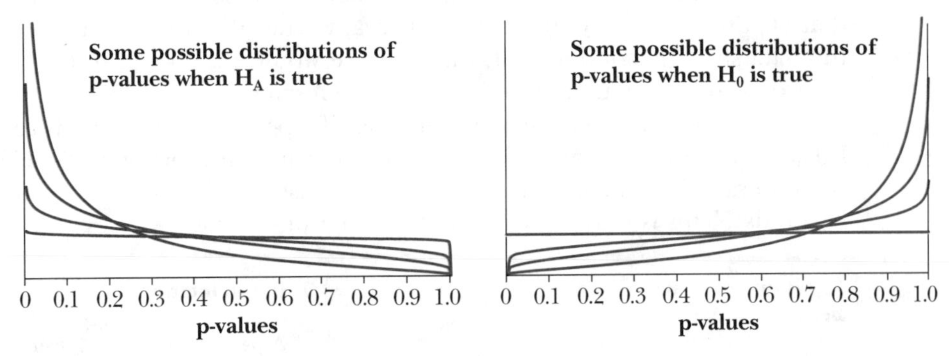

Therefore, p-values close to 0 'favour' the alternative hypothesis, H_A, whereas p-values close to 1 'favour' the null hypothesis, H_0.

The discussion above suggests that p-values less than 0.5 should be interpreted as favouring H_A, whereas values greater than 0.5 should be interpreted as favouring H_0. However, statistical practice does not interpret p-values that are greater than 0.5 in this way.

In statistical practice, there is *not* symmetry between our attitudes towards the two hypotheses. We treat the null hypothesis, H_0, in a fundamentally different way from the alternative hypothesis, H_A.

In statistical hypothesis testing, we accept that H_0 is true unless there is substantial evidence in the data to contradict it. We only accept H_A if the data are inconsistent with H_0. Statistical hypothesis testing challenges H_0 with sample data.

The hypothesis that is labelled the null hypothesis, H_0, should be selected to be one or more of the following:

- the more conservative of the hypotheses
- the hypothesis that you would default to if there were no data.

For example, if an industrial process is monitored, the null hypothesis should be that the process is 'in control'; you should look for evidence to contradict this hypothesis and therefore support the alternative hypothesis of an out-of-control process.

If two groups of children are taught a topic by different methods and both groups are given the same test afterwards, the null hypothesis should be that the mean test results are the same for both teaching methods.

In the New Zealand high school data at the beginning of this section, our null hypothesis should be that there is no change in the mean roll between 1994 and 1995:

$$H_0\text{: } \mu = 0$$
$$H_A\text{: } \mu \neq 0.$$

How does this affect our interpretation of the p-values?

As H_0 is the more 'conservative' of the hypotheses, the p-value should be very small (and certainly less than 0.5) before we should treat it as evidence that H_0 does not hold and that H_A is really true. We should only reject H_0 if the data strongly contradict it; that is, if we would have been unlikely to have recorded such a small p-value if H_0 had been true.

So how strong is the evidence against H_0 if a p-value of, say, 0.05 is recorded? If H_0 is true, a p-value as low as 0.05 will occur in a proportion 0.05 or fewer of random samples (since values near 1 are at least as likely as values near 0 when H_0 holds). This is illustrated in the two diagrams below.

If μ is on boundary between H_0 and H_A

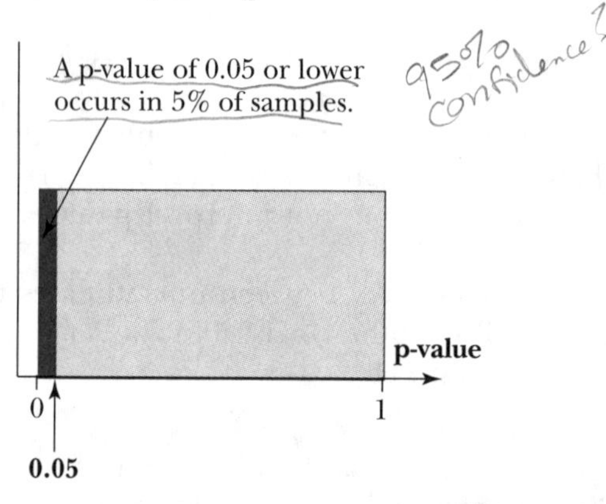

If μ is in interior of H_0

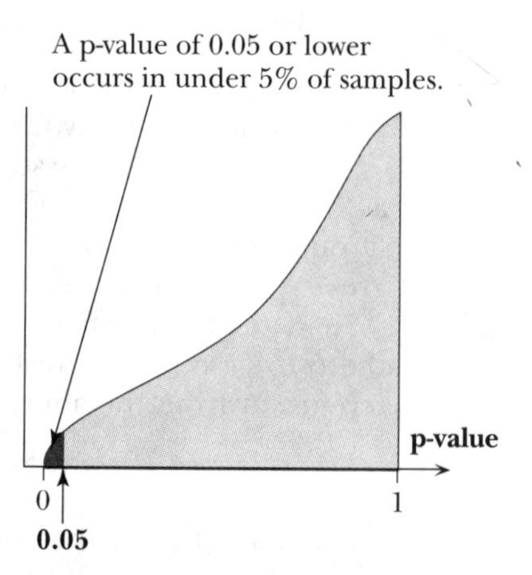

As there is less than 5% chance of getting such a low p-value when H_0 is true, we should conclude that there is moderate evidence that H_0 is not true and, therefore, moderate evidence that H_A is true.

Similarly, consider a p-value of 0.002 which is evaluated from a different data set to compare two hypotheses, H_A and H_0. We should now conclude that there would be less than 0.2% chance of getting such a low value if H_0 were true, so there is strong evidence against H_0 and supporting H_A.

Finally, consider a p-value of 0.3. We might get such a low p-value in 30% of random samples when H_0 is true,[11] so we conclude that the p-value is not unlikely and does not throw any doubt on H_0.

Because of the special role of H_0 in the interpretation of p-values, the p-value can be thought of as a *credibility index* for H_0.

Although it may oversimplify the conclusions that should be drawn from a hypothesis test, the following guidelines are offered for interpreting p-values.

p-value	Interpretation
Greater than 0.10	Minimal evidence against the null hypothesis[12] (H_0)
Between 0.05 and 0.10	Slight evidence that the null hypothesis does not hold
Between 0.01 and 0.05	Moderate evidence that the null hypothesis does not hold
Less than 0.01	Strong evidence that the null hypothesis does not hold

The results of all hypothesis tests are described using p-values. These p-values always have distributional properties of the form described earlier in this section, so they can *always* be interpreted using the guidelines above.

In conclusion, we have seen that the p-value from a single random sample can be used as a *guide* to whether the null hypothesis, H_0, or the alternative hypothesis, H_A, is true. Small p-values are more likely when H_A is true, and large p-values are more likely when H_0 is true.

However, because it is still *possible* to obtain large p-values when H_A is true and it is *possible* to obtain p-values near 0 when H_0 is true, a p-value can only be used as a *guide* to the truth. Sample data is seldom absolutely conclusive, although a p-value close to zero can be treated for practical purposes as conclusive evidence that H_0 should be rejected.

Examples

In the school roll data set discussed at the beginning of this section, the differences were assumed to be a random sample from a population with unknown mean μ. The question as to whether or not there is change in average rolls between 1994 and 1995 can be formulated in terms of the hypotheses:

$$H_0: \mu = 0 \quad \text{no change}$$
$$H_A: \mu \neq 0. \quad \text{change}$$

The sample of 20 differences *could* have come from a normal distribution for which $\mu = 0$, or from one with $\mu \neq 0$, so the best we can hope for is an indication of whether the data support the truth of the null hypothesis.

11. A p-value as low as 0.3 might occur in *less* than 30% of random samples if H_0 is true and the distribution of the p-values is not rectangular. However, as we will only reject H_0 if there is strong evidence against it, it makes sense to compare the p-value against the 'best-case' distribution, the rectangular distribution.
12. More precisely, there is insufficient evidence to reject the null hypothesis as being unreasonable in the light of the information provided by the data.

A 't-test'[13] is appropriate to compare these hypotheses, and most statistical software will evaluate the p-value for this test; the test results in a p-value of 0.803. Since this p-value is so far from zero, we conclude that there is no evidence in the data against the null hypothesis. If H_0 was true, there would be a probability of 0.84 of obtaining such a low p-value; the p-value is not small enough to throw doubt on H_0 and make us think that H_A is likely to hold. There is therefore not enough evidence in the data to support an assertion of a change in average school roll.

Most statistical computer programs can evaluate p-values for a wide variety of hypothesis tests. Even the authors of this book would need to refer to reference material to fully understand the methodology underlying many of these. However, to interpret a p-value, only the null and alternative hypotheses underlying the test need be known[14] and these are often specified in the program's documentation.

For example, we will next describe two hypothesis tests relating to experiments that a scientist, Newcomb, made in 1882; he used his experimental measurements to estimate the speed of light in air.[15] His experimental measurements were the times in nanoseconds (10^{-9} seconds) for light to travel 7442 metres: 24 828, 24 826, ... As the measurements were all close to 24 800, they have been coded:

$$24\,828 - 24\,800 = 28,$$
$$24\,826 - 24\,800 = 26,$$

and so on, so that only 'information-carrying' digits are given. (See section 6.3.)

28	26	33	24	34	−44	27	16	40	−2	29	22
24	21	25	30	23	29	31	19	24	20	36	32
36	28	25	21	28	29	37	25	28	26	30	32
36	26	30	22	36	23	27	27	28	27	31	27
26	33	26	32	32	24	39	28	24	25	32	25
29	27	28	29	16	23						

We hope that these values can be treated as a random sample from a hypothetical population whose mean is μ. If Newcomb's experimental procedure was correct, the true speed of light can be calculated from μ by simple arithmetic. Although μ is of most interest, the following tests relate other aspects of the data.

13. The details of how the p-value is found from a random sample are covered later. In practice, the details of how a p-value is determined from a random sample are of little importance since most statistical computer software will calculate its value for you. The important skill to learn is how to interpret such p-values and *when* to compute one.
14. Of course, we still need to assume that the sample on which the p-value is based is actually a random sample from the intended population.
15. S. Newcomb, 'Measures of the Velocity of Light Made Under the Direction of the Secretary of the Navy During the Years 1880–1882', *Astronomical Papers*, 2, US Nautical Almanac Office, 1891, pp. 107–230.

The diagram below shows a histogram of Newcomb's data. There are two outliers in the data. Were they mistakes in Newcomb's experimental procedure and should they therefore be deleted from the data set?

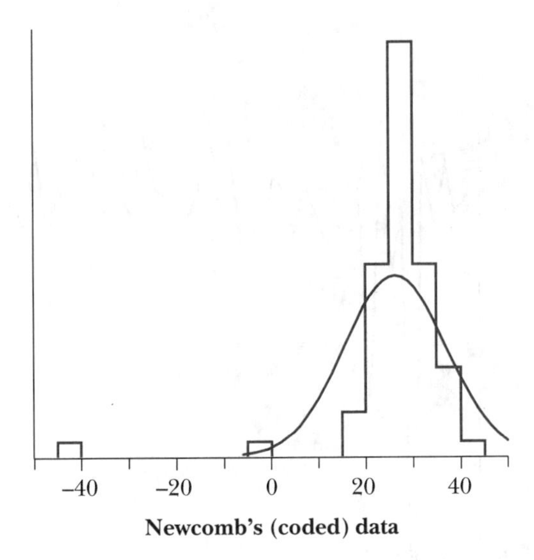

From experience, it has been found that experimental measurements of this type often have a symmetric normal distribution. The diagram above superimposes the best-fitting normal histogram on the data. One way to assess whether or not the two extreme measurements should be treated as outliers is to ask whether they are likely to have arisen as part of a random sample from a normal population.

The statistical program JMP can easily perform a hypothesis test (called the Shapiro-Wilkes W test) to assess whether the data could have come from a normal population. This test compares the hypotheses:

H_0: $x_1, x_2, \ldots, x_n$ are a random sample from a normal distribution.

H_A: $x_1, x_2, \ldots, x_n$ are a random sample from a non-normal distribution.

When this test is applied to Newcomb's data, JMP reports a p-value '0.0000'. As JMP rounds p-values to four decimal places, this really means that $p < 0.000\,05$. We therefore conclude that a random sample so extremely non-normal would occur in less than 0.000 05 of random samples from a normal distribution. As this probability is so low, we conclude that there is extremely strong evidence that the data are not normal.

In contrast, if the two outliers are omitted, JMP reports a p-value of 0.6167. As a p-value this small will often arise (in a proportion 0.6167 of random samples) when the null hypothesis holds, we conclude that the test provides no evidence that the data are not normal. The distribution underlying the data without the outliers may not be normal but, if so, this data set (and test) can find no evidence of it. The test therefore lends support to the assertion that the two outliers resulted from errors in Newcomb's experimental procedures.

As a final example, we note that the Newcomb data are listed (reading across rows) in the order in which the experiments were conducted. The diagram below displays the data in this order.

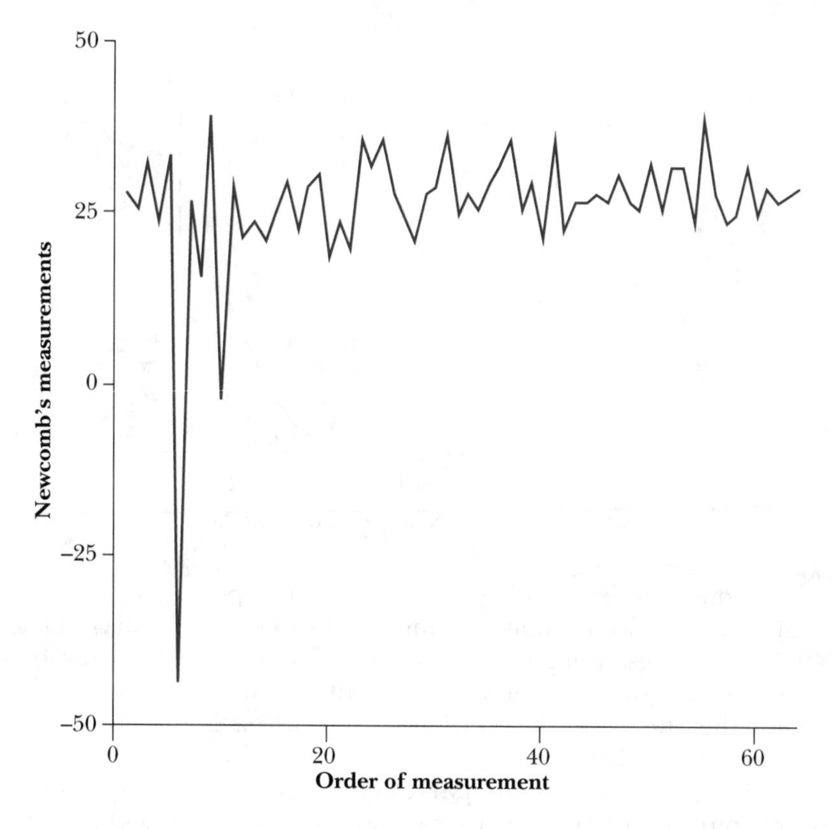

The statistical program Systat provides a hypothesis test (the Wald-Wolfowitz runs test) that can be used to assess whether or not the successive observations are independent of each other. This test is sensitive both to trends (the experimental mean measurement changing during the course of the experiments) and to correlation between adjacent measurements (such as high measurements tending to be followed by further high measurements).

> H_0: $X_1, X_2, \ldots, X_n$ is a random sample (the measurements are independent and all have distributions with the same mean).
>
> H_A: $X_1, X_2, \ldots, X_n$ are either not independent or have distributions whose means change over time.

For the full data set (including outliers), Systat reports a p-value of 0.086. The probability of the test returning a p-value as small as this is 0.086 if the data are independent with the same distribution; this is small, but not small enough to throw serious doubt on the null hypothesis, so we conclude that there is only slight evidence that the data are not a random sample. When the two outliers are omitted, Systat reports a p-value of 0.078 which leads to the same conclusion.

P-values as probabilities

We noted earlier that when a simple H_0 holds, or when the parameter is on the boundary between a composite H_0 and H_A, the p-values have a rectangular distribution between 0 and 1, so that all values in that range are equally likely. As a result, a p-value of k or less will be recorded in a proportion k of samples for any k. We therefore say that there is *probability k* of obtaining such a low p-value when H_0 is true.[16]

For example, consider a set of sample data from which the p-value for a test with a simple H_0 is 0.02; we know that if a simple H_0 holds, then a value of 0.02 or lower would arise in a proportion 0.02 of repeated samples. Because low values of the p-value support H_A, we can therefore state that 2% of samples will support H_A as strongly as the recorded value. Generalising this:

> The p-value for a hypothesis test with a simple H_0 is the probability of obtaining a random sample supporting H_A as strongly as the actual data collected, assuming that H_0 is true.

This result not only gives us another interpretation of the p-value of a test, but is the basis of the formulae that underlie p-values. Note that, as the probability is evaluated assuming H_0 is true, the p-value can again be interpreted as a *credibility* index for H_0.

> In this book, we expect you to use computers to evaluate p-values for hypothesis tests, but we will sketch the reasoning behind some common simple hypothesis tests in order to explain further this interpretation of p-values, and also to allow readers who have no access to computers to evaluate them.

Consider a simple random sample, $x_1, x_2, \ldots, x_n$ from some population with mean μ and standard deviation σ. We will assume that the sample size, n, is large. We will derive the p-value for a test of:

$$H_0: \mu = 0$$
$$H_A: \mu \neq 0.$$

The sample mean, $\bar{x}$, provides the best point estimate of the unknown parameter μ, so we would also expect it to hold most information in the sample about whether or not $\mu = 0$. Values of $\bar{x}$ close to zero will support H_0, whereas values of $\bar{x}$ far from zero will support H_A more strongly.

16. We use the term the *probability* of an event to denote the proportion of times the event happens in repeated random samples.

We therefore define the p-value for the test of:

$$H_0: \mu = 0$$
$$H_A: \mu \neq 0$$

to be:

p-value = probability of getting a sample mean as far from zero as $\bar{x}$ when $\mu = 0$.

To use this result to obtain a formula for the p-value, we need to know the distribution of the sample mean when $\mu = 0$. This is given by the central limit theorem, which states that the sample mean has a distribution with mean and standard deviation:

$$\mu_{\bar{X}} = 0$$
$$\sigma_{\bar{X}} = \frac{\sigma}{\sqrt{n}}$$

and is approximately normal. We can use this normal distribution (with σ replaced by the sample standard deviation, s, since the sample size is assumed to be large) to evaluate the probability that a sample mean will be as far from zero as the observed value, $\bar{x}$.

The procedure will be clearer with a numerical example. A batch of wood chips from various species of softwood is analysed for fibre length by a pulp and paper company that has just taken delivery of a large batch. The grade of chips is determined by the average fibre length since the longer fibres make stronger paper. The pulp and paper company has paid a premium price for chips whose mean fibre length (over the batch of chips) is supposed to be at least 2.0 centimetres. A decision to accept or reject the batch of chips must be based on a small random sample of chips. An instrument determines the average fibre length for each of 50 chips, as shown below.

2.62	1.85	2.11	2.24	1.87	1.78	1.60	1.94	2.30	1.64
1.98	1.59	1.82	2.02	1.72	2.10	1.75	1.77	2.08	1.87
2.04	1.90	1.72	2.15	1.75	1.53	2.08	1.85	2.23	2.52
2.08	1.92	1.79	1.98	2.09	1.13	2.16	2.22	2.43	2.02
1.82	1.63	1.60	2.16	1.42	1.72	1.98	2.05	1.63	2.10

The fibre lengths from the 50 chips are summarised by:

$$\bar{x} = 1.927$$
$$s = 0.280.$$

The values are random and will be assumed to be a random sample from a hypothetical population with mean μ and standard deviation σ. We are interested in testing whether the mean fibre length per chip is greater than 2.0; in other words, we want to test whether $H_0: \mu \geq 2.0$ or $H_A: \mu < 2.0$.

If H_0 is true, and $\mu = 2.0$, the sample mean will be approximately normal with mean and standard deviation:

$$\mu_{\bar{X}} = 2.0$$

$$\sigma_{\bar{X}} = \frac{\sigma}{\sqrt{n}} \approx \frac{0.280}{\sqrt{50}} = 0.0396.$$

The histogram for this normal distribution is shown below.

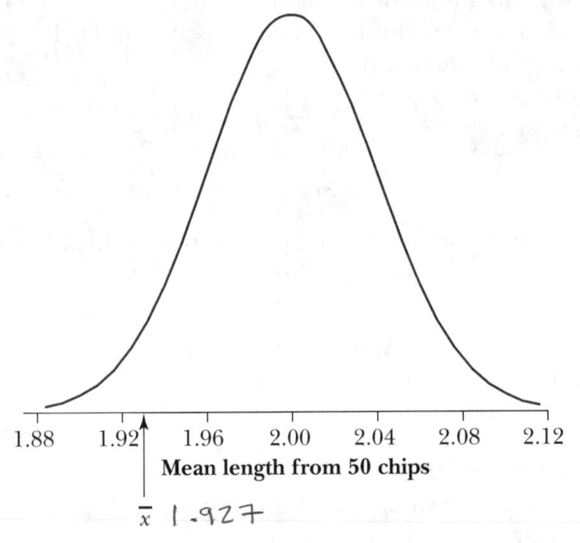

(The distribution will be shifted to the right if $\mu > 2.0$.) From the diagram, we observe that there is very little chance of getting a sample mean as far away from zero as the value in our data, 1.93 (that is, getting a sample mean less than 1.93) when the true underlying population mean is $\mu = 2.0$.

To evaluate this probability, we standardise the sample mean, $\bar{x}$:

$$z = \frac{\bar{x} - \mu_{\bar{X}}}{\sigma_{\bar{X}}} \approx \frac{1.927 - 2.0}{0.0396} = -1.84$$

and look up normal tables (or use a computer) to obtain the probability of recording a standard normal value as low as this. This is:

$$\begin{aligned} \text{p-value} &= \text{probability}(\bar{x} \le 1.927) \\ &= \text{probability}(z \le -1.84) \\ &= 0.033. \end{aligned}$$

Note that an approximation is involved here. We have assumed that $\sigma_{\bar{X}}$ is exactly 0.0396, whereas 0.0396 is only an *estimate*. Provided the sample size, n, is large (say over 30), the estimate will be fairly accurate, so we can ignore the error in the approximation.[17]

17. If n is less than 30, different tables should be used. In practice, you should use a computer to evaluate the p-value — most statistical programs will provide you with the p-value for this and most other tests that we describe here for *any* sample size.

The p-value for the test is therefore 0.033, so we conclude that there is fairly strong evidence that μ is not zero, and therefore that the mean fibre length is less than 2.0.

The p-values for several other common hypothesis tests may be obtained in a similar way. They all follow the following logic.

1. Find an estimator of the parameter whose value is being tested.
2. Find a formula for the standard deviation of the estimator.
3. Evaluate the formula in step 2, assuming H_0 holds (or at least find a good estimate for it).
4. Evaluate the z-score:

$$z = \frac{(\text{Estimate of parameter}) - (\text{Parameter value under } H_0)}{\text{Standard deviation of estimate}}.$$

5. Use the standard normal distribution to find how unusual such a z-score is; this is the p-value for the test.

The general method will be illustrated by the details behind a further test relating to population proportions.

Testing a proportion

A second application of hypothesis testing asks questions about the proportion in some category from a population.

Consider an experiment which investigates whether one subject can telepathically pass information to another about shape. A deck of cards containing equal numbers of cards with circles, squares and crosses is shuffled. One subject selects cards at random and attempts to 'send' the shape on the card to the other subject who is seated behind a screen; this second subject writes the shape imagined for the card. Out of 200 cards, 76 are correctly guessed.

The 'correct/wrong' outcomes each come from a hypothetical population with a proportion π of correct outcomes. If the second subject is simply guessing, the population proportion of correct outcomes will be exactly $\pi = \frac{1}{3}$ since each of the three shapes is equally likely to be turned up. Our hypothesis test therefore compares the hypotheses:

$$H_0\colon \pi = \frac{1}{3}$$

$$H_A\colon \pi > \frac{1}{3}.$$

Following the steps described above:

1. We estimate π with the sample proportion of correct outcomes:

$$p = \frac{76}{200} = 0.38.$$

2. The standard deviation of a sample proportion was given in section 9.2 and is:

$$\sigma_p = \sqrt{\frac{\pi(1-\pi)}{n}}.$$

3. Assuming that H_0 holds, $\pi = \frac{1}{3}$, and this formula is evaluated as:

$$\sigma_p = \sqrt{\frac{0.333 \times 0.667}{200}} = 0.033\ 33.$$

4. The z-score is:

$$z = \frac{0.38 - \frac{1}{3}}{0.033\ 33} = 1.40.$$

5. The p-value is the probability of recording a proportion correct as high as 0.38, when the second subject is guessing, and this is the probability of getting a z-score of 1.40 or higher. From normal tables, this is:

$$\text{p-value} = 1 - 0.9192 = 0.0808.$$

The p-value is not particularly small, so we conclude that there is little evidence against guessing in this experiment.

Note that there are various approximations involved in the above reasoning, and these suggest that the steps should not be followed if the sample size is small or the population proportion is close to either 0 or 1. You will need to use statistical software to evaluate the p-value if either of the above occurs.

EXERCISE 9.3

1. The example of the New Zealand high school rolls was used to introduce the idea of testing a hypothesis. A question of interest was whether the rolls had increased from the previous year, based on information from only a sample of schools. How do we make precise the phrase 'rolls had increased' in this context? Clearly some school rolls will increase and some will decrease, in both our sample and in the full list of schools, so how can we make the question one that has a useful answer?
2. Is it necessary to run the risk of making a mistake when using a p-value to indicate whether the null hypothesis is credible or not?
3. Is it possible to prove that the null hypothesis is true if the p-value is close enough to 1?
4. The diagram at the bottom of page 307 shows the distribution of the p-value in a succession of scenarios for which the population mean gets closer and closer to the alternative values. The boundary between the null hypothesis values of the mean and the alternative values is 10.0 in the example. What is the distribution of the p-value when $\mu = 10$?

5. Bivariate data are often considered to be a random sample from an infinite bivariate population with correlation coefficient ρ. For example, the table below gives two measurements that were collected monthly from a steam plant at a large industrial concern.[18] The two variables are pounds of steam used monthly and average atmospheric temperature (°F).

Steam (pounds)	Temperature (°F)	Steam (pounds)	Temperature (°F)
10.98	35.3	9.57	39.1
11.13	29.7	10.94	46.8
12.51	30.8	9.58	48.5
8.40	58.8	10.09	59.3
9.27	61.4	8.11	70.0
8.73	71.3	6.83	70.0
6.36	74.4	8.88	74.5
8.50	76.7	7.68	72.1
7.82	70.7	8.47	58.1
9.14	57.5	8.86	44.6
8.24	46.4	10.36	33.4
12.19	28.9	11.08	28.6
11.88	28.1		

The question of whether or not steam consumption is related to temperature can be posed as a hypothesis test about the population parameter ρ; if there is no relationship, ρ will be zero. We therefore consider a test of the null and alternative hypotheses:

$$H_0: \rho = 0 \qquad \text{and} \qquad H_A: \rho \neq 0.$$

(a) For a test of this type, why is a small p-value an indication that ρ is not zero?

(b) For the data, the sample correlation coefficient is $r = -0.85$, and the p–value for the above test is reported by computer software as 'under 0.0001'. What is your conclusion?

6. An alumni survey contacted a random sample of graduates who graduated with a major in agriculture during the years 1986–1995. A parallel survey did the same thing for graduates in economics. Of the 25 economics graduates who responded, 80% were in full-time employment in 1996, while in agriculture it was just 64%.

(a) Obtain an interval estimate for the proportion of economics graduates who are employed full-time, based on the information provided.

(b) Test whether a majority of the economics graduates are employed full-time.

(c) To what population do you think these results apply? Discuss. (The answer here is not clear-cut, but you should know what criteria to use.)

The response rates for both surveys were just over 60% but more than 25 were contacted until a response was obtained for 25.

(d) Should you be concerned about non-response error affecting your answers?

(e) Should you be concerned about interviewer error affecting your answers?

18. N. R. Draper and H. Smith, *Applied Regression Analysis*, John Wiley & Sons, New York, 1966, pp. 351–352. Reprinted by permission of John Wiley and Sons Inc.

9.4 Comparison of two populations

Difference between means

Many practical problems require estimation of a population mean from a single random sample. However, there are also many interesting problems that involve comparisons between two separate populations, based on simple random samples from each. The most common questions about the differences between two populations of numerical measurements relate to the difference between the two population means, $\mu_A - \mu_B$.

- Educationalists and psychologists have conducted many studies to try to determine whether the sex of young children affects their mathematical aptitude. The children from a particular class are considered to be random samples of girls and boys of that same age. While the boys and girls will always have test scores that overlap a great deal, the question of interest is whether there is any systematic difference between the groups. One way to approach this question is to attempt to estimate the difference in the average scores of boys and girls, $\mu_G - \mu_B$.

- The ideal pesticide would eliminate crop loss due to pests in a corn crop, but be untraceable in the corn delivered to the market. Supermarkets have to be responsible for the produce they sell, and food distributors must be able to guarantee a certain low level of toxicity in the food they supply.

  An experiment is done, 24 hours after harvest, to compare the toxicity of a new pesticide with the toxicity of an older, widely used pesticide. A random sample of ears of corn is selected from farm areas where each of the two pesticides is used, and the results compared. The toxicity measurements from farms using the old pesticide do vary quite a bit and similarly for the new pesticide. Since the new pesticide is much more expensive than the old one, the researcher needs to describe the relative effectiveness so that a potential user can decide whether the advantage is worth the extra cost. One way to do this is to estimate the difference in the average toxicity in the two pesticides, $\mu_O - \mu_N$.

Any inference based on a comparison of sample means must allow for the facts that the sample means are variable from one sample to the next and different samples can therefore suggest different inferences about the population means. A confidence interval estimate for the difference in population means deals with this problem by summarising in an interval the credible differences in population means, given the sample data.

The following data set shows yearly maxima of wind gusts (miles per hour) in two Scottish cities over a period of approximately 50 years.[19]

---

19. A. J. B. Anderson, *Interpreting Data*, Chapman and Hall, New York, 1989, pp. 29–30.

City	Yearly maxima of wind gusts (mph)									
Edinburgh	78	81	66	79	76	74	75	85	73	81
1918–1967	81	86	87	76	66	69	65	85	74	87
	66	74	82	73	78	69	75	68	76	68
	76	81	75	75	79	74	85	87	69	67
	88	73	71	68	78	83	79	88	71	73
Paisley	66	64	63	59	64	63	69	63	59	63
1914–1967	67	67	63	90	68	55	68	64	64	63
	67	73	83	49	70	59	73	62	58	63
	62	55	66	59	69	69	55	72	63	55
	68	66	67	75	49	63	60	70	77	56
	62	62	59	59						

The jittered dot plots and summarising box plots for the two cities are shown on the left below.

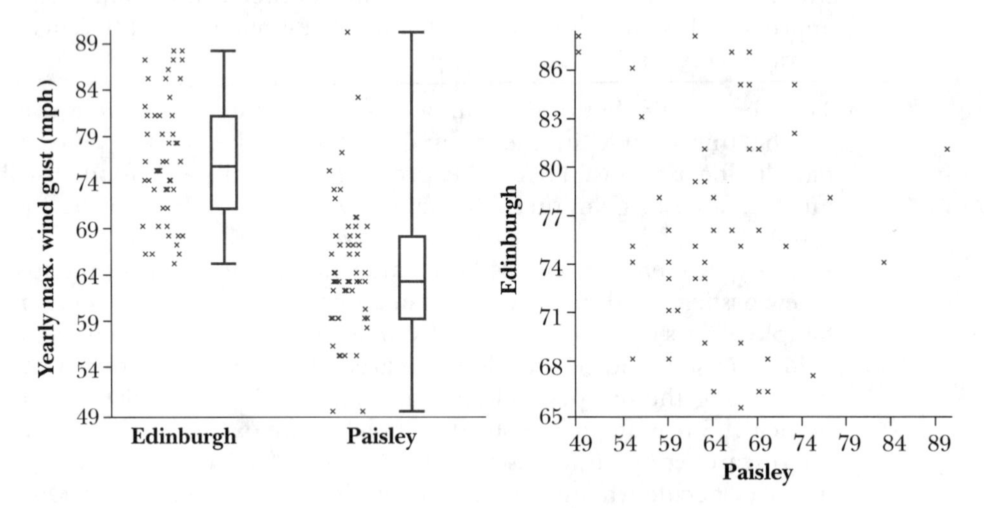

In this section we will not ask whether there is a difference in mean wind speed between the two cities. (We will formally ask questions of this form in the next section, but a difference seems likely from the graphical displays.) In this section, we consider the problem of estimating the difference in mean wind speeds between the cities.

The two cities are approximately 100 kilometres apart, so the two batches of measurements may be regarded as (at least approximately) independent of each other. The scatterplot on the right above plots the Edinburgh maximum wind gust speed against that for Paisley for the years in which measurements were recorded for both cities (1918 to 1967). The random scatter of points supports the independence of the two batches.[20] We will therefore assume that

20. Question 4 at the end of this section (page 329) describes an analysis of these data that can be used even if the assumption of independence of the measurements in the two cities in a year is not tenable. Both analyses lead to essentially the same conclusions.

the two sets of measurements are independent random samples from two populations with population means μ_{Pais} and μ_{Edin}.

We are interested in estimating the difference $(\mu_{Edin} - \mu_{Pais})$, based on these two random samples. The obvious point estimate is the difference between the sample means:

$$\bar{x}_{Edin} - \bar{x}_{Pais} = 76.260 - 64.389 = 11.871.$$

How accurate is this estimate likely to be? Again, a confidence interval is called for.

The following result underlies our approach to two-sample problems such as that above.

If $\bar{X}_1$ is the mean of a random sample of size n_1 from a population with mean μ_1 and standard deviation σ_1, and $\bar{X}_2$ is the mean of an independent random sample of size n_2 from a population with mean μ_2 and standard deviation σ_2, the difference $(\bar{X}_2 - \bar{X}_1)$ has a distribution with mean and standard deviation:

$$\mu_{\bar{X}_2 - \bar{X}_1} = \mu_2 - \mu_1$$

$$\sigma_{\bar{X}_2 - \bar{X}_1} = \sqrt{\frac{\sigma_1^2}{n_1} + \frac{\sigma_2^2}{n_2}}.$$

Using similar logic to the logic underlying our earlier confidence intervals, we reason that *any* random quantity with a reasonably symmetric (and approximately normal) distribution is within two standard deviations of its mean approximately 95% of the time. This would suggest a 95% confidence interval of the form:

$$\bar{x}_2 - \bar{x}_1 \pm 2 \times \sqrt{\frac{\sigma_1^2}{n_1} + \frac{\sigma_2^2}{n_2}}.$$

We must, however, refine this, since σ_1 and σ_2 are unknown.

The appropriate formula for a 95% confidence interval is:

$$\bar{x}_2 - \bar{x}_1 \pm t_v \times \sqrt{\frac{s_1^2}{n_1} + \frac{s_2^2}{n_2}}$$

where s_1 and s_2 are the sample standard deviations in the two groups:

$$v = \min(n_1 - 1,\ n_2 - 1)$$

and t_v is a value looked up in the table on page 324 (the same table we used on page 293 to obtain confidence intervals for a single mean).[21]

21. A larger value of v (and hence a slightly narrower confidence interval) is sometimes suggested, but the difference is usually negligible, so we will stick to the more conservative formula here.

ν	t_ν	ν	t_ν	ν	t_ν	ν	t_ν
1	12.71	**11**	2.201	**21**	2.080	**40**	2.021
2	4.303	**12**	2.179	**22**	2.074	**50**	2.009
3	3.182	**13**	2.160	**23**	2.069	**60**	2.000
4	2.776	**14**	2.145	**24**	2.064	**80**	1.990
5	2.571	**15**	2.131	**25**	2.060	**100**	1.984
6	2.447	**16**	2.120	**26**	2.056	∞	1.960
7	2.365	**17**	2.110	**27**	2.052		
8	2.306	**18**	2.101	**28**	2.048		
9	2.262	**19**	2.093	**29**	2.045		
10	2.228	**20**	2.086	**30**	2.042		

When the sample sizes, n_1 and n_2, are large, t_ν is close to 2, so the confidence interval is close to the earlier formula involving σ_1 and σ_2.

For the wind gust data, we therefore obtain the following 95% confidence interval for the difference $\mu_{\text{Edin}} - \mu_{\text{Pais}}$:

$$\bar{x}_{\text{Edin}} - \bar{x}_{\text{Pais}} \pm t_{49} \times \sqrt{\frac{s_{\text{Edin}}^2}{n_{\text{Edin}}} + \frac{s_{\text{Pais}}^2}{n_{\text{Pais}}}} = 11.871 \pm 2.010 \times \sqrt{\frac{6.642^2}{50} + \frac{7.337^2}{54}}$$

$$= 11.871 \pm 2.755$$

$$= 9.12 \text{ to } 14.63.$$

So, we are 95% confident that the average yearly maximum wind speed in Edinburgh is between 9.12 and 14.63 mph faster than in Paisley.[22]

Difference between proportions

In a similar way, we give a 95% confidence interval for the difference between two population proportions, based on independent random samples from the two populations.

If a random sample of size n_1 is selected from a second population containing a proportion π_1 in some category of interest, and a similar independent random sample of size n_2 is selected from a population containing a proportion π_2 in the same category, the best estimate of the difference $\pi_2 - \pi_1$ is the difference between the sample proportions, $p_2 - p_1$, and a 95% confidence interval for this difference is:

$$p_2 - p_1 \pm 2 \times \sqrt{\frac{p_1(1-p_1)}{n_1} + \frac{p_2(1-p_2)}{n_2}}.$$

22. The expression '95% confident' applied to interval estimates is just a shorthand way of saying that the confidence level of the interval computed is 95%.

For example, in 1988, out of 947 males tried in District Courts in New Zealand, 642 were convicted; out of 121 females tried, 78 were convicted.[23] A 95% confidence interval for the difference between the underlying conviction rates for males and females is therefore:

$$\frac{642}{947} - \frac{78}{121} \pm 2 \times \sqrt{\frac{\frac{642}{947}\left(1 - \frac{642}{947}\right)}{947} + \frac{\frac{78}{121}\left(1 - \frac{78}{121}\right)}{121}} = 0.033 \pm 0.092.$$

We are therefore 95% confident that the male conviction rate is between 5.9% lower and 12.5% higher than that for females.

The width of this confidence interval means that the point estimate (that the male conviction rate is 3.3% higher than the female conviction rate) should not be relied upon to indicate whether males or females have the higher conviction rate. There is simply not enough data to allow the difference to be accurately estimated.

Testing for existence of differences

A confidence interval provides sufficient inference about many problems involving two populations. However, sometimes we want to test whether there is any difference between the populations. When the populations consist of numerical measurements, the most common test is of whether or not the two population means are equal.

We will illustrate this test with the Scottish wind speed data where we will test whether the mean wind speed is the same in Paisley and Edinburgh.

$$H_0\text{: } \mu_2 - \mu_1 = 0$$
$$H_A\text{: } \mu_2 - \mu_1 \neq 0$$

A p-value for the test is found by the following steps, which follow the framework presented at the end of section 9.3.

1. We estimate $\mu_2 - \mu_1$ with the difference between the sample means:
$$\bar{x}_2 - \bar{x}_1 = 76.260 - 64.389 = 11.871.$$
2. The formula for the standard deviation of the difference between two sample means was given on page 323, and is:
$$\sigma_{\bar{X}_2 - \bar{X}_1} = \sqrt{\frac{\sigma_1^2}{n_1} + \frac{\sigma_2^2}{n_2}}$$
where n_1 and n_2 are the sample sizes from the two populations.
3. We do not know the values of σ_1 or σ_2, but we can replace them with the sample standard deviations, s_1 and s_2 in the formula to get the approximation:
$$\sigma_{\bar{X}_2 - \bar{X}_1} \approx \sqrt{\frac{s_1^2}{n_1} + \frac{s_2^2}{n_2}} = \sqrt{\frac{6.642^2}{50} + \frac{7.337^2}{54}} = 1.371.$$

23. Department of Statistics, *New Zealand Official 1990 Yearbook*, 1990, Wellington, p. 300.

4. The z-score is:

$$z = \frac{11.871 - 0}{1.371} = 8.66.$$

5. The p-value is the probability of recording a difference in sample wind speeds as far from zero as 11.871, when the population means are the same, and this is the probability of getting a z-score as far from zero as 8.66. Even without looking up normal tables, we can state that this is virtually impossible, so we can conclude that there is almost certainly a difference in mean wind speeds in the two cities.

The approximations involved in step 3 mean that the method should not be used without modification when the sample sizes from the two populations are less than 30. Computer software will give you the correct p-value.

Similar reasoning can again be applied to comparisons of proportions in two populations. The appropriate formula to use in step 2 is given in section 9.2 and is:

$$\sigma_{p_2 - p_1} = \sqrt{\frac{\pi_1(1-\pi_1)}{n_1} + \frac{\pi_2(1-\pi_2)}{n_2}}.$$

This can be approximated in step 3 by:[24]

$$\sigma_{p_2 - p_1} \approx \sqrt{\frac{p_1(1-p_1)}{n_1} + \frac{p_2(1-p_2)}{n_2}}.$$

This equation also explains the confidence interval formula at the bottom of page 324.

In practice, you should normally use computers to evaluate p-values, but the steps behind their evaluation should help you to understand their properties.

Paired data

The confidence intervals and hypothesis tests above are appropriate when two populations are compared on the basis of independent random samples from each. There is, however, another data structure that initially seems similar, but which should be analysed in a different way.

Consider two experimenters, *A* and *B*, who are interested in testing whether or not doses of sucrose affect the pulse rate of white mice. Experimenter *A* uses 20 white mice for the experiment; 10 of these are randomly selected as controls and the other 10 are given a dose of sucrose.

---

24. Some authors prefer to use the approximation:

$$\sigma_{p_2 - p_1} \approx \sqrt{p(1-p)\left(\frac{1}{n_1} + \frac{1}{n_2}\right)}$$

in this situation, where:

$$p = \frac{n_1 p_1 + n_2 p_2}{n_1 + n_2}$$

is the overall proportion in the combination of the two samples, but the difference between the two formulae is of no practical importance.

Experimenter *B* uses a different strategy; only 10 white mice are used, but two measurements of pulse rate are made from each mouse — one before the dose of sucrose and one after. The data collected by *A* and *B* are shown in the table below.

Experimenter *A* (Independent samples)		**Experimenter *B* (Paired data)**	
Control	**Treatment**	**Control**	**Treatment**
204.93	175.63	178.77	178.77
192.36	233.39	183.33	186.25
201.54	197.57	169.05	173.22
168.34	182.14	154.42	160.55
162.17	182.78	213.83	215.26
175.69	187.36	208.05	228.51
186.01	179.51	181.12	195.38
221.15	185.87	169.49	179.64
198.19	236.4	197.72	220.25
186.03	201.55	194.15	203.09

The summaries from the two experiments are similar, as shown in the following table.

	Experimenter *A*	**Experimenter *B***
Control mean	189.641	184.993
Treatment mean	199.22	194.092
Difference in means	9.579	9.099
Control standard deviation	17.88	18.54
Treatment standard deviation	21.84	22.27

We might therefore think that both experimenters would come to similar conclusions from their data. However, their conclusions would be very different.

Experimenter *A* has two independent random samples and would correctly analyse the data using the methods described earlier in this section. Computer software would give a p-value of 0.31 for a test of the hypotheses:

$$H_0\text{: } \mu_2 - \mu_1 = 0$$
$$H_A\text{: } \mu_2 - \mu_1 \neq 0.$$

The conclusion would be reached that there was no evidence that the dose of sucrose had any effect on pulse rate.

Experimenter *B*, however, does not have independent samples — the pair of measurements from each mouse are positively correlated (if a mouse has a high pulse rate before being given the sucrose, its pulse rate tends to be high after, also). To analyse the data, the difference in means is rewritten as follows:

$$\begin{aligned}\bar{y}_2 - \bar{y}_1 &= \frac{y_{2,1} + y_{2,2} + \ldots + y_{2,10}}{10} - \frac{y_{1,1} + y_{1,2} + \ldots + y_{1,10}}{10} \\ &= \frac{(y_{2,1} - y_{1,1}) + (y_{2,2} - y_{1,2}) + \ldots + (y_{2,10} - y_{1,10})}{10} \\ &= \frac{d_1 + d_2 + \ldots + d_{10}}{10} \\ &= \bar{d}\end{aligned}$$

where d_1 is the difference between the treatment and control measurements from the first mouse, and so on. In words, the difference between the means is equal to the mean of the differences. When written in this way, our hypotheses are equivalent to testing whether the population mean of the d_i is zero, and this is a one-sample problem that can be answered using the methods of section 9.3. Experimenter *B*'s analysis should therefore be based on the final column of differences in the table below.

Control	Treatment	Difference, d_i
178.77	178.77	0.00
183.33	186.25	2.92
169.05	173.22	4.17
154.42	160.55	6.13
213.83	215.26	1.43
208.05	228.51	20.46
181.12	195.38	14.26
169.49	179.64	10.15
197.72	220.25	22.53
194.15	203.09	8.94

A single-sample analysis of the d_i gives a p-value of 0.005 for this test — strong evidence that the sucrose dose has affected pulse rates.

You should note carefully the following points.

- If experimenter *B* had ignored information about the pairing, a two-sample analysis would have (incorrectly) given a p-value of 0.35, and information in the data about differences would have been missed.

- Paired data provide more information than two-sample data when there are large differences between individuals. Taking differences between the two values for a single individual removes this large 'person-effect' and therefore highlights the effect of the treatment.

- Paired data should usually be collected in preference to two-sample data when both are possible.

EXERCISE 9.4

1. Recall question 5 from the exercise in section 9.2 where a test of a batch of 10 000 formatted computer diskettes found seven defectives in a random sample of 100. In a second batch of 10 000, a larger sample of 500 is tested. In this sample, 20 defectives are found. Use a 95% confidence interval to estimate the improvement in the proportion defective, over the first batch. Interpret the meaning of the result in the context of the problem.
2. Recall the example at the start of section 9.2 where two brothers, Roger and David, compared living expenses at two universities. The mean annual living expense in a random sample of 50 students at Roger's university was \$8750, whereas the corresponding figure from a random sample of 50 from David's university was \$7500. To assess whether the difference is greater than would be expected by chance, we need to know the standard deviations in the two samples; at Roger's university it was \$770, whereas at David's it was \$660.
   (a) How much higher do you think living expenses are at Roger's university?
   (b) How big would the sample differences in the two studies have to be to support Roger's contention that his university peers really did have higher expenses on average than David's?
3. Recall the alumni survey from question 6 in the exercise after section 9.3, where random samples of graduates majoring in agriculture and economics during the years 1986–1995 were surveyed.
   (a) Of the 25 economics graduates who responded, 80% were in full-time employment in 1996, while in agriculture it was just 64%. A higher proportion of the economics graduates were employed than of agriculture graduates *in the sample.* Does this establish that a larger proportion out of all economics graduates are employed full-time than for agriculture? Obtain the p-value for the appropriate two-sample test.
   (b) The average income of the economics majors who were employed full-time in 1996 was \$21 103 and in agriculture it was \$25 228. The standard deviation of these full-time salaries was \$2925 for economics and \$3143 for agriculture. Of those in the survey populations who are employed full-time, do agriculture graduates have higher salaries, on average, than economics graduates?
4. Recall the wind gust data at the start of this section. For this question, ignore the data from Paisley between 1914 and 1917 (the values 66, 64, 63 and 59).
   (a) Treating the remaining data as paired observations, test whether there is any difference between Edinburgh and Paisley.
   (b) Again, treating the data as paired, find a 95% confidence interval for the difference between the mean wind gust speeds in the two cities.
   (c) Explain why you would expect to gain little from treating the data as paired in this example.

9.5 COMPARISON OF SEVERAL POPULATION MEANS — ANALYSIS OF VARIANCE

In section 9.3 the comparison of a single population mean with a hypothesised value was discussed, and in section 9.4 we extended this to a comparison of two population means. In this section we examine the comparison of three or more population means.

The most common comparison of several populations is to test if their means are equal, but when there are three or more population means to compare, we have a new problem. To compare two means we simply look at their difference, but with three means there are three differences, and with four means there are six differences. Clearly we need a new approach.

For example, the table below gives the lengths (in millimetres) of cuckoo eggs found in the nests of birds of three other species.[25] The cuckoo (*Cuculus canorus*) is a European bird that lays its eggs in the nests of other species, so there is interest in whether or not the host species affects egg size (perhaps by the cuckoos laying larger eggs in the nests of larger birds).

Host species	Egg length (mm)							
Hedge	22.0	23.9	20.9	23.8	25.0	23.4	21.7	23.8
sparrow	22.8	23.1	23.1	23.5	23.0	23.0	–	–
Robin	21.8	23.0	23.3	22.4	22.4	23.0	23.0	23.0
	23.9	22.3	22.0	22.6	22.0	22.1	21.1	23.0
Wren	19.8	22.1	21.5	20.9	22.0	21.0	22.3	21.0
	20.3	20.9	22.0	20.0	20.8	21.2	21.0	–

These data are displayed graphically in the jittered dot plot, opposite.

We could use the techniques of the previous section to compare the host species two at a time, but we really want to simultaneously compare all three hosts. More critically, if there were a large number of groups, there is a reasonable chance that at least one pair of samples would show evidence of a difference (say a p-value under 0.05) even if the population means in all groups were really the same. The appropriate test compares the hypotheses:

$$H_0: \mu_1 = \mu_2 = \ldots = \mu_k$$
$$H_A: \text{at least one } \mu_i \text{ differs from the rest}$$

where it is assumed that the data are independent random samples from k normally distributed populations with means $\mu_1, \mu_2, \ldots, \mu_k$ and common standard deviation σ.

25. E. Kreyszig, *Introductory Mathematical Statistics: Principles and Methods*, John Wiley & Sons, New York, 1970, p. 269.

Note that we are making the assumptions that:

- the different samples are collected independently of each other
- the standard deviation is the same in each population.

These should be checked before the test is performed.[26]

In the cuckoo example, there will be no dependence between eggs laid in different nests, so the first assumption is satisfied. (If each cuckoo had been trained to lay exactly three eggs, with one egg laid in a nest of each species, the measurements in the three samples would have been related, and a different analysis would have been required.) The jittered dot plot does not throw doubt on the assumption of a common standard deviation in each group.

Most computer software will provide a p-value for this test. For the cuckoo data, the p-value is less than 0.001, so we conclude that there is very strong evidence of a real difference in mean cuckoo egg sizes in host nests of different species of bird. (The sample means would have been extremely unlikely to have been so different if the population means had been the same.)

The details of this hypothesis test are more complicated than those of the tests that we described in earlier sections. The test is often called an *analysis of variance* test or an *F-test*, and is evaluated by hand in steps using an *analysis of variance table*.[27] You should, however, rely on computer software to obtain the p-value for the test.

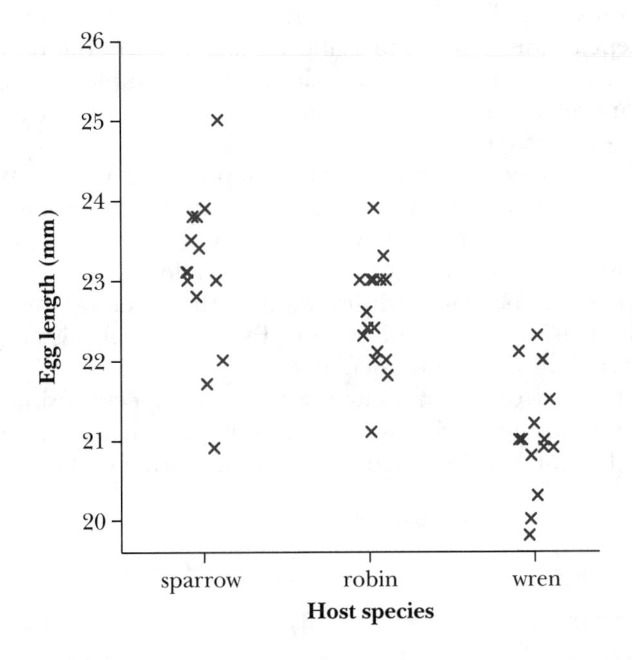

26. The assumption of a normal distribution in each group is the less important of the two assumptions, since the central limit theorem implies that the sample means will be approximately normal.
27. The analysis of variance table is really a computational method and most of the numbers in it do not provide useful descriptive statistics. However, although the p-value for the test and one or two other descriptive statistics would suffice, the full table is often printed in reports.

EXERCISE 9.5

The following questions relate to the cuckoo data. The output below is part of that produced by the computer program MINITAB when asked to test whether the egg lengths in the three host species are different.

```
One-Way Analysis of Variance
Analysis of Variance on egg size
Source        DF        SS         MS          F          p
species        2    30.094     15.047      22.15      0.000
error         42    28.532      0.679
Total         44    58.626
   Level       N      Mean      StDev
       1      14    23.071      1.022
       2      16    22.556      0.682
       3      15    21.120      0.754
Pooled StDev = 0.824
```

1. Above a horizontal axis, mark the three means. Indicate the standard deviations of each host species by drawing a horizontal bar two standard deviations on either side of the mean. Summarise in words what you see.
2. Repeat question 1, but using the standard deviation of the sample means instead of the standard deviation of the sample values. How do these three bars relate to 95% confidence intervals for the mean egg lengths for the three host species?
3. The 'pooled' standard deviation provided in the MINITAB output is a single estimate of the standard deviation of egg lengths, made under the assumption that the standard deviation of egg lengths is the same for each host nest. Does its value seem reasonable?
4. Calculate the standard deviation of the three sample means (whose values are 23.071, 22.556 and 21.120). Explain why this is larger than the standard deviations that were calculated in question 2.
5. What is wrong with analysing the three species using three t-tests (wren v. robin, wren v. hedge sparrow, and robin v. hedge sparrow)?

Note: The analysis of variance table in the MINITAB output is:

Source	DF	SS	MS	F	p
species	2	30.094	15.047	22.15	0.000
error	42	28.532	0.679		
Total	44	58.626			

The only values in this table that you will be able to interpret are the p-value (0.000) and the mean error sum of squares (0.679) whose square root is the pooled standard deviation (0.824).

9.6 Regression inference

In chapter 4, we examined data sets which contained pairs of measurements on each of a collection of individuals. As another example of bivariate data, consider the following data that were collected by biologists from 15 lakes in central Ontario to assess how zinc concentrations in the aquatic plant *Eriocaulon septangulare* were related to zinc concentrations in the lake sediment and therefore to predict plant concentrations of zinc from analysis of sediment samples. The table below gives data that were collected in the study.[28] A scatterplot of the data is shown on page 334.

Lake	Sediment Zn (μg/g)	Concentration of Zn in plant (μg/g dry weight)
Clearwater	37.5	15.9
Crosson	72.5	42.7
Dickie	85.2	85.7
Fawn	76.5	52.6
Gullfeather	64.5	49.1
Hannah	86.8	59.0
Harp	90.8	53.6
Heney	105.8	77.8
Leech	85.8	63.2
Leonard	87.9	62.3
Lohi	53.9	22.7
McKay	102.3	66.1
Moot	90.7	47.4
Plastic	86.0	59.4
Ril	79.0	50.9

When the relationship between the two variables is linear, as seems reasonable in this example, a least squares line is a useful summary of the relationship and the correlation coefficient is a useful summary of its strength. The least squares line fitted to the above data is:

$$(\textit{Plant zinc}) = -14.9 + 0.857 \times (\textit{Sediment zinc}).$$

This is also shown in the diagram on page 334.

In this example, the slope of the least squares line is of most interest — it describes how much extra zinc (in μg per gram dry weight) accumulates in the plants for each one μg per g zinc concentration in their sediment. If it could be assumed that the relationship was causal, we would estimate that a proportion 0.857 of the extra zinc in the sediment would be absorbed into the plants. In view of the wide scatter of points around the least squares line, we must wonder how accurately this estimate has been determined.

28. P. Reimer and H. C. Duthie, 'Concentrations of zinc and chromium in aquatic macrophytes from the Sudbury and Muskoka regions of Ontario, Canada', *Environmental Pollution*, 79, 1993, pp. 261–265. Copyright 1993, with kind permission from Elsevier Science Ltd, The Boulevard, Langford Lane, Kidlington, OX51GB, UK.

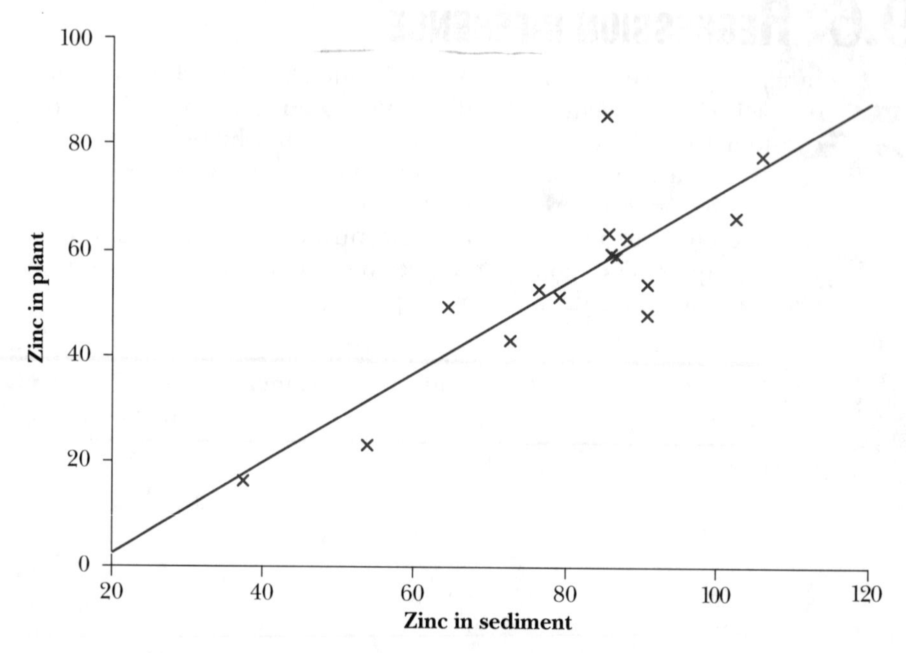

In order to examine the relationship further, we assume that the data are a random sample from a larger population of pairs of measurements (a larger population of sites for analysis of plant and sediment), and that the mean plant zinc concentration in this population is linearly related to the sedimentary zinc concentration.

The general structure of the model that we assume is satisfied by the data is described next.

> We assume that in the population, for any value of the explanatory variable, x, the response, y, has a normal distribution with mean and standard deviation:
>
> $$\mu_y = \alpha + \beta x$$
> $$\sigma_y = \sigma.$$
>
> *Note*: We do not need to make any assumptions about the distribution of the explanatory variable, x.

For these assumptions to hold, the relationship must be *linear* and the spread of the response must be the same for all values of x. The model is called a *regression* model. We will give some guidelines to help evaluate the adequacy of this model later in the section.

The unknown population parameter β is the real focus of attention in this model for our zinc concentrations and the least squares estimate $b = 0.857$ is our best estimate from the available data. If the assumptions above can be made about the underlying population, we can use a 95% confidence interval to describe the accuracy of this estimate. A 95% confidence interval for the slope, β, is given by the formula:

$$(b - t_{n-2} \times sd(b),\ b\bar{x} + t_{n-2} \times sd(b))$$

where b is the least squares estimate of the slope, n is the number of data points, t_{n-2} is a value looked up in the same table provided on pages 293, 324 and in the Appendix, and $sd(b)$ is an estimate of the standard deviation of b. You will usually obtain the values of b and $sd(b)$ from a statistical program but, for completeness, we give the formula for $sd(b)$ in a footnote below;[29] the formula for the least squares slope, b, was given in chapter 4.

For the zinc concentrations, $b = 0.856$, $sd(b) = 0.1531$, and $t_{13} = 2.160$, so the confidence interval for the slope is:

$$0.856 \pm 2.16 \times 0.153 = 0.53 \text{ } to \text{ } 1.19.$$

The best we can say is therefore that we have 95% confidence that an increase of one µg per gram of zinc in the sediment is associated with an increase of between 0.53 and 1.19 µg per gram dry weight of zinc in this type of aquatic plant.

Regression models are not only useful when the relationship between the two variables is linear. In chapter 4, we examined a data set which contained average lifespans and metabolic rates (measured by oxygen consumption) of a selection of mammals. The scatterplot below shows the relationship between the variables, which is clearly not linear.

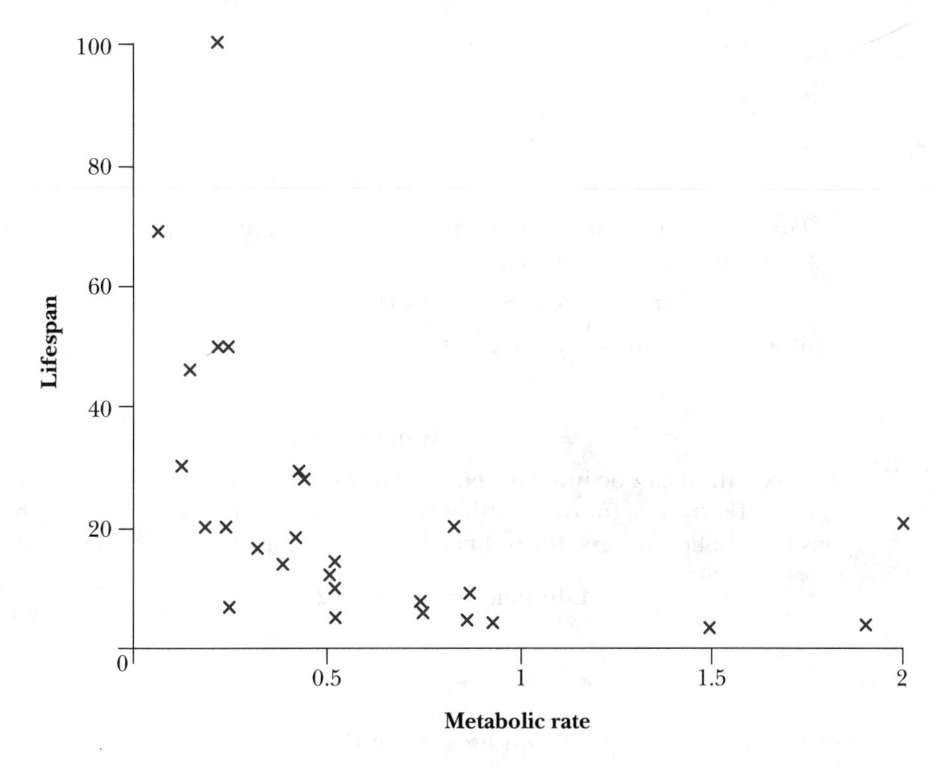

29. The standard deviation of b is:

$$sd(b) = \sqrt{\frac{\sum_{i=1}^{n} (y_i - \bar{y})^2 - b^2 \sum_{i=1}^{n} (x_i - \bar{x})^2}{(n-2) \sum_{i=1}^{n} (x_i - \bar{x})^2}}.$$

However, taking logarithms of both variables in this example seems to linearise the relationship adequately, as shown below.

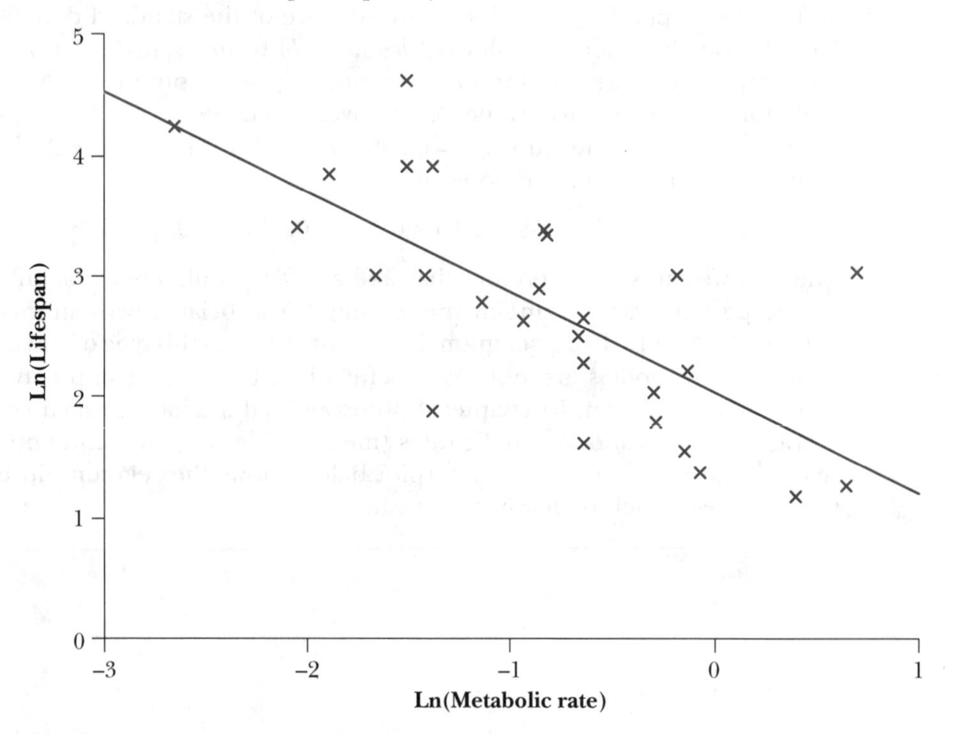

The least squares line for predicting Ln(Lifespan) from Ln(Metabolic rate), fitted to the transformed data, is:

$$\text{Ln(Lifespan)} = 2.029 - 0.828 \times \text{Ln(Metabolic rate)}.$$

This equation can be rewritten in terms of the original variables as:

$$\begin{aligned}\text{Lifespan} &= \text{Exp}(2.029 - 0.828 \times \text{Ln(Metabolic rate)}) \\ &= 7.610 \times (\text{Metabolic rate})^{-0.828}.\end{aligned}$$

This equation can be interpreted by considering two mammals whose metabolic rates differ by a factor of 2, so that the first has metabolic rate m and the second has metabolic rate $2m$. It predicts the lifespan of the second mammal to be:

$$\begin{aligned}\text{Lifespan}_2 &= 7.610 \times (2m)^{-0.828} \\ &= 7.610 \times m^{-0.828} \times 2^{-0.828} \\ &= \text{Lifespan}_1 \times 0.563\end{aligned}$$

or 0.563 times the predicted lifespan of the first mammal. Doubling the metabolic rate is therefore associated with a change in lifespan by a factor of 0.563.

Although we can interpret the relationship well in terms of the original measurements, it is convenient to base further analysis on the log-transformed variables.

The relationship evident in the scatterplot does not seem strong. If one or two crosses at the extremes of the scatterplot were moved, as illustrated in the following diagram, we might doubt whether the variables were related at all.

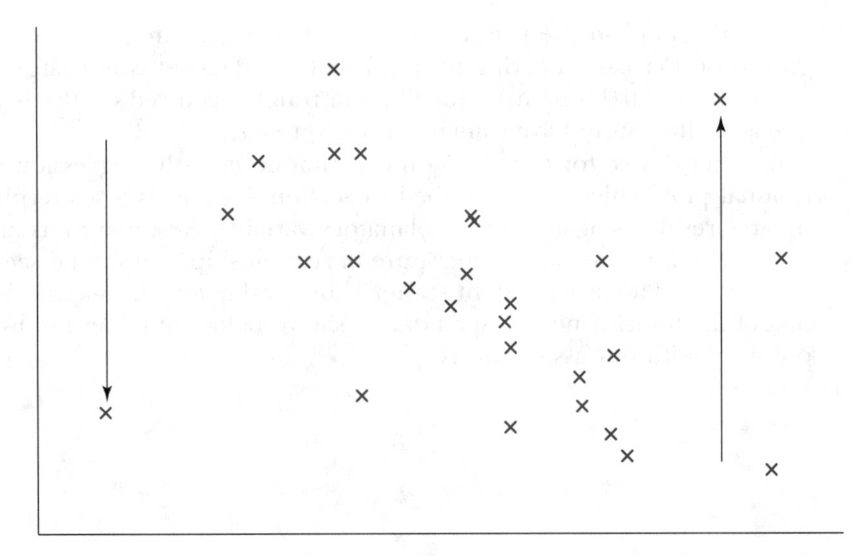

In order to ask whether the variables are related, we again assume that the mammals described by our data are a random sample from a larger population of mammals, and that the mean Ln(Lifespan) in this population is linearly related to Ln(Metabolic rate) with a regression model.

A good guide to whether or not the assumptions underlying the regression model are supported by a particular data set is to draw onto the scatterplot the least squares line and parallel lines at the same distance on each side so that the central band contains approximately $\frac{2}{3}$ of the data. If the scatterplot is split with vertical lines into three or four equally sized groups of points, the distribution of the response, y, in each vertical band should be (roughly) centred on the least squares line and should have similar spread around it.

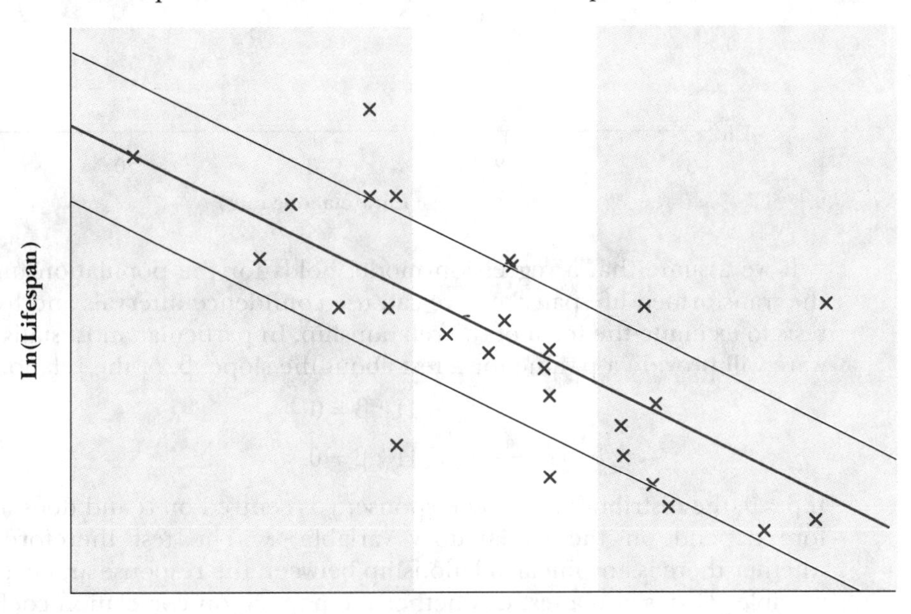

The diagram on the previous page splits the Ln(Lifespan) data into three groups of 9 crosses. Bearing in mind that the data set is not large, the three bands in the Ln(Lifespan) data all seem roughly centred on the least squares line and all seem to have similar vertical spreads.

Another device for examining the assumptions of the regression model is a residual plot, which was described in section 4.4 and is a scatterplot of least squares residuals against the explanatory variable. Residual plots are particularly useful for detection of curvature in relationships; we should see no trends or cycles, and a fairly constant scatter above and below the x-axis. The residual plot of the transformed lifespan data is shown below and does not indicate any problems with our assumptions.

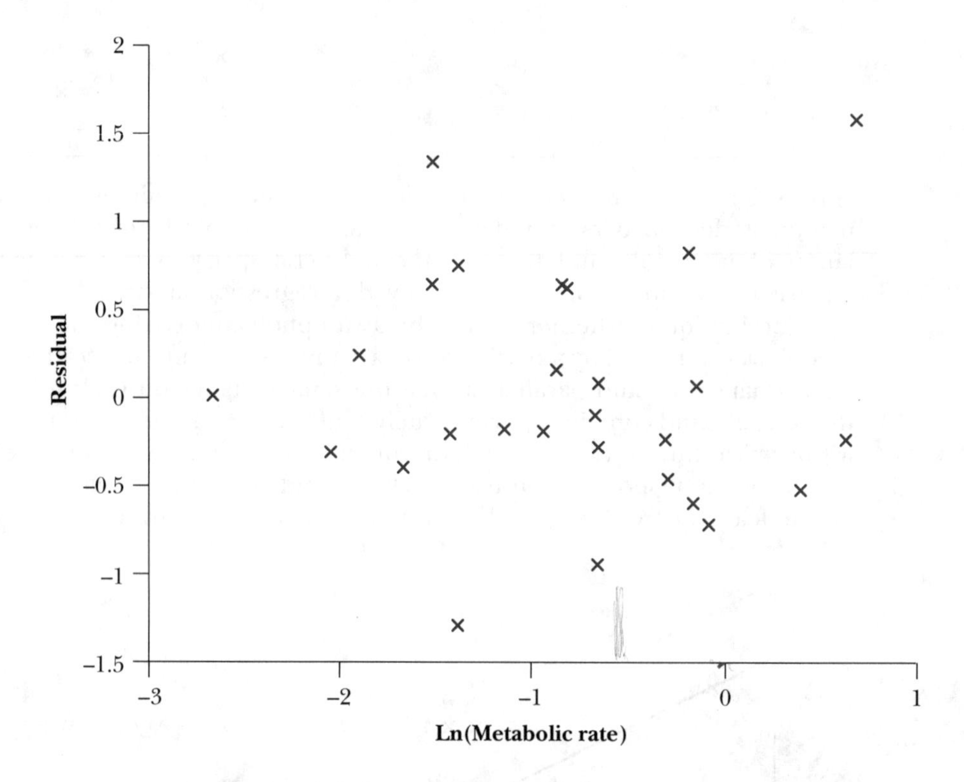

If we assume that a regression model holds for the population underlying the transformed lifespan data, we can use confidence intervals and hypothesis tests to examine the form of the relationship. In particular, most statistical software will provide a p-value for a test about the slope, β, of the relationship:

$$H_0: \beta = 0$$

$$H_A: \beta \neq 0.$$

If $\beta = 0$, the distribution of the response, y, is centred on α and does not therefore depend on the explanatory variable, x. The test therefore assesses whether there is any linear relationship between the response and explanatory variable. This is also a test of whether the population correlation coefficient is zero.

For the Ln(Lifespan) data, the p-value for the test is reported by MINITAB as '0.000', which really means 'under 0.0005'. This is extremely strong evidence that Ln(Lifespan) is related to Ln(Metabolic rate) in mammals.

A 95% confidence interval for the slope, β, obtained in the same way as described for the zinc concentration example, is given by:

$$-0.8279 \pm 2.06 \times 0.1619 = -1.16 \text{ to } -0.49.$$

Since the relationship:

$$\text{Ln(Lifespan)} = \alpha + \beta \times \text{Ln(Metabolic rate)}$$

is equivalent to:

$$\text{Lifespan} = e^{\alpha} \times (\text{Metabolic rate})^{\beta}$$

the confidence interval means that we have 95% confidence that, if the mammals in the data were a random sample from the population of all mammals, doubling the metabolic rate of mammals is related to lifespan changing by a factor between:

$$2^{-1.16} = 0.45 \quad \text{and} \quad 2^{-0.49} = 0.71.$$

EXERCISE 9.6

1. Examine the wind gust data from section 9.4. Can regression analysis be used to predict Paisley wind gusts from Edinburgh wind gusts, for a given year?
2. The regression worked out in this section computes a predictor for animal lifespan based on metabolic rate. Using only the graph of the regression line, find a rough prediction of the lifespan of an animal with a metabolic rate of 0.6. (*Hint*: Remember that the regression line shown is in transformed units.)
3. Check your graphical answer in question 2 by using the regression equation.
4. Refer to the graph on page 336. A rough indicator of the error of prediction is the vertical distance of the points from the regression line, since the height of the regression line at any value of metabolic rate is the predicted response and the height of the point is the actual response. In the units of the Ln(Lifespan), how big are typical prediction errors, roughly?
5. If you compute exp(typical prediction error) using your answer from question 4, how do you interpret the result in terms of the typical prediction error in the original units of the data?
6. The computer program MINITAB produces the following output for the regression of Ln(Lifespan) against Ln(Metabolic rate).

```
The regression equation is
Ln(Life) = 2.03 - 0.828 Ln(Met)
Predictor          Coef        Stdev     t-ratio          p
Constant         2.0294       0.1854       10.95      0.000
Ln(Met)         -0.8279       0.1619       -5.12      0.000

s = 0.6727       R-sq = 51.1%         R-sq(adj) = 49.2%
```

Find the numbers mentioned in the text for the regression line and the estimated standard deviation of the slope of the regression line. Note the line:

```
s = 0.6727    R-sq = 51.1%.
```

This value 0.6727 is the estimated standard deviation of the prediction errors and R-sq is the value mentioned in section 4.4 in connection with least squares line fits. What do these two numbers mean in the context of the lifespan data?

7. MINITAB output can easily be generated for the regression of Ln(Metabolic rate) on Ln(Lifespan), in which the variables have the reverse role of those already described. Here is the output:

```
The regression equation is
Ln(Met) = 0.853 - 0.618 Ln(Life)
Predictor          Coef       Stdev     t-ratio          p
Constant         0.8529      0.3456        2.47      0.021
Ln(life)        -0.6177      0.1208       -5.12      0.000

s = 0.5810         R-sq = 51.1%         R-sq(adj) = 49.2%
```

Do the two regressions produce the same predictive line?

Problems

1. To demonstrate the concept of random sampling, a teacher uses a bowl of approximately 10 000 red and white beads. After mixing the beads well, a random sample of 100 beads is selected from the bowl. The sample contains 60 white beads. The teacher is confused since she believed that the bowl contained an equal number of red and white beads. Has something gone wrong with the teacher's demonstration? Or is this outcome usual in such a sampling situation? Do a helpful calculation as part of your answer.

9.1–9.2

2. A doctor decides to prescribe a new drug to help patients who complain of severe headaches. Over a period of time she alternately prescribes a placebo (with no active ingredient) or the new drug to successive patients with this complaint. After 40 patients have entered the study, 20 have been given the placebo and 20 others have been given the new drug. All patients report the result of the pill treatment (drug or placebo) after two weeks of the prescribed regimen. The result is that 40% of the placebo group report an improvement, whereas 65% of the drug group report an improvement. When the doctor reports these results to her colleagues at a conference, one critic suggests that the difference (40% to 65%) could be due to chance variation and that perhaps the drug has no real effect.
   (a) Compute the probability that the data could favour the drug this much or more, even when there is no real drug effect (that is, a p-value).
   (b) Do you think the difference could be due to chance variation? Give a reason for your answer.
   (c) Previous studies have shown that about 35% of patients will report positive results from a placebo treatment. Use this information to re-test whether the drug has a positive effect.

3. Two fertilisers, *A* and *B*, are applied to 50 plots on which potatoes are to be planted, with *A* being applied to a randomly selected subset of 25 plots and *B* to the remaining plots. *A* is a standard fertiliser while *B* is a new formula which is expected to be an improvement, meaning that the weight of potatoes produced should be greater. Data from the season's yield are:

	***A* plots**	***B* plots**
Mean	45 kg	50 kg
Standard deviation	10 kg	10 kg
Number of plots	25	25

   (a) Is the evidence in favour of fertiliser *B* producing higher yields very strong?
   (b) 'The theory of statistics includes methods for evaluating the plausibility of a hypothesis against the evidence in a data set.' Explain this quotation with reference to this example.

4. On two successive weeks before an election, a political polling company conducts two polls, each based on a simple random sample of 100 from the electorate. It reports that on the first week 30%, and then on the next week 25% of the electorate favour the incumbent party.
   (a) Test the hypothesis that the percentage favouring the incumbent party has declined over the week.
   (b) Which of the following possibilities is more likely?
      (i) the percentage has increased
      (ii) the percentage has decreased
      (iii) the percentage is unchanged

5. A marketing research company wants to find out the total investment by a community on privately owned cricket equipment. A random sample of 100 households from the community of 5250 households is visited by interviewers to determine the new value of the cricket equipment for each household. No cricket equipment was reported in 50 houses, but the remaining 50 in the sample had an average of \$40 worth. The investigator had never heard of the standard deviation but noted that, of the 50 households that had cricket equipment, about two-thirds had values in the range \$30–\$50. Non-reponse was allowed for in determining the sample size and response bias was expected to be negligible.
   (a) Estimate the proportion of households that have cricket equipment. Use a method that indicates the precision of your estimate.
   (b) Which of the following would be the best guess of the standard deviation of the 100 sample values? \$5, \$10, \$15, \$25. Briefly justify your choice.
   (c) Using your choice in part (b), estimate the total value of cricket equipment in the community. Indicate the precision of your estimate.

6. The strength (in kilogram breaking strength) of nylon yarn made from two different production processes is to be compared by testing samples of yarn from the output of each process. One length of yarn is tested from the output of each process during each eight-hour shift for a period, resulting in 100 samples from each process. The 200 measurements may be summarised as follows:

   Process 1 : Mean = 13.5, Standard deviation = 1.0.
   Process 2 : Mean = 13.6, Standard deviation = 1.0.

   In each case, the summary statistics given are based on a sample of $n = 100$ values.
   (a) Do the two processes produce different strengths of nylon?
   (b) The two production processes rely on common raw materials which vary from batch to batch, and daily temperature (also common to both processes) also affects them both. As a result, the differences between each pair of measurements from a shift were evaluated (a total of $n = 100$ differences) and these are summarised by:

      Differences: Mean = 0.1, Standard deviation = 0.3.

      Use these values to re-test for a difference between the two processes.
   (c) Is it possible for a difference to be highly statistically significant and still be negligible? Explain.

7. A supply of 100 experimental guinea pigs is to be used in an experiment to test for the presence of cancerous lesions produced as a result of exposure to certain substances suspected of causing cancer (carcinogens). The examination for the lesions destroys the guinea pig, so comprehensive screening of the animals before exposure cannot be done; however, from past experience with the supplier of the guinea pigs, it is felt that about 20% of the animals would have lesions in the absence of the additional exposure to carcinogenic substances. Lesion rates higher than 20% would therefore provide evidence of carcinogenicity of the experimental substances. If the 100 guinea pigs are exposed to the experimental carcinogens, would the presence of lesions in 35% of the animals offer convincing evidence of carcinogenicity or might the 35% be attributable to random variation? Discuss.

8. A sample survey is being designed for a client and the client wants to estimate the proportion of the target population exhibiting some characteristic within 5% of the true value. After discussion with the survey company, the client agrees that an estimate further than 5% from the correct value would be acceptable one time out of 20. Assuming that the random sample is selected from a large population with zero non-response rate, how large should the sample size be? (*Hint*: Consider the width of a 95% confidence interval based on a sample of size n. Use the value of π in this formula that would be the 'worst case'; that is, for which the confidence interval would be widest.)

9. A sampling survey produces the following data concerning the income level of families in a certain region, based on a random sample of 25 households:

   Mean family income = $40 020, Standard deviation = $9810.

   It was also found that 24% of the sample households had incomes over $40 000.

   (a) How low might the average household income of the entire region reasonably be?
   (b) What can you say about the shape of the distribution of income in the region, based on this information?
   (c) Estimate the proportion of households with incomes over $40 000 in the whole region, allowing for sampling error.

10. A chain of 10 fast food restaurants reports the sales from each restaurant at the end of the day. The daily sales of the 10 stores on a particular day average $3000 and they have a standard deviation of $90. The company statistician claims that average sales of $3100 were predicted for this day, based on historical records and trends. Is the $3000 average figure a cause for concern, or could it be attributed to random variation?

11. Textbooks on display at a bookstore vary in price from about $2.00 to $120.00, with the vast majority of books falling in the range $20.00 to $80.00.
    (a) Guess the standard deviation of the textbook prices and explain the basis of your guess (that is, explain why you think your guess is reasonable).
    (b) Use the standard deviation that you guessed in (a) to determine how many books from the 400 should be sampled to estimate the mean book price to within $5 (in 19 out of 20 samples).

12. An experiment on rats is undertaken to determine if there is a relationship between the amount of a vitamin supplement in their diet and the level of activity they exhibit. A quantitative index of the activity level is worked out based on video tapes. A regression of the activity level on the amount of the vitamin supplement turns out to have an R^2 value of 0.102. However, the test for whether the regression slope is zero results in a p-value of 0.014.

    Assuming that the regression model fits the data adequately and the usual assumptions of linear regression are satisfied, how would you interpret these findings? Justify your answer. (That is, would you find the relationship unreliable because of the low R^2 value or would you focus on the significant p-value?)

13. A university involved in distance teaching does a survey to determine the current students' intentions for the following semester. Four students (let's call them 'surveyors') are hired and each is asked to survey a random sample of 25 students from the undergraduate students enrolled in the previous semester. A list of all such undergraduates is provided by the registrar for this purpose. There are 4000 names on the list. The four surveyors ask the students in their samples how many distance courses they plan to take at this university during the following semester. The four surveyors summarise their results as follows:

		Number of courses intended	
	Respondents	**Mean**	**Standard deviation**
Surveyor 1	25	1.0	1.0
Surveyor 2	25	2.0	1.0
Surveyor 3	25	2.0	1.5
Surveyor 4	25	3.0	2.2

(a) Find an interval estimate for the average number of courses intended, in the population of 4000, based on the sample of surveyor 1.

(b) Pooling all surveyors' data, what is the total number of distance courses intended by all 100 surveyed students?

(c) From the summary information provided by the surveyors, the university estimated the standard deviation of the number of courses intended in the whole population to be 1.51. Based on this, and your answer to (b), find an interval estimate for the average number of courses intended, in the population of 4000.

(d) The mean number of distance courses taken in the previous semester was 1.5. Test whether the mean intended number of courses in the following semester is greater than 1.5.

(e) Does the test in (d) tell the university anything useful? (*Hint*: Are the two semesters' measurements comparable?)

(f) The standard deviation of the four means (1.0, 2.0, 2.0 and 3.0) is approximately 0.8. Is this about what you would expect if the population standard deviation was 1.51? Justify your answer. Does this have any implications for the interpretation of the data?

14. The following table gives weights of adult bears observed in the wild, by sex. Consider the data to be a random sample of bears from this population.

Bear weights (kg)

Males	Females
81	98
111	55
91	63
189	106
210	64
80	126
191	110
96	84
89	144
116	
138	
201	
74	

(a) Are male bears always heavier than female bears?
(b) Are male bears heavier than female bears on average? (Summarise the weight distributions by sex and test for equality of population means.)
(c) Arrange the data in a single column of length 22, 13 males and then 9 females. In another column of length 22, indicate the sex of the animal by 0 for male and 1 for female. Call these columns Y and X respectively. Regress Y against X and compare the result with your analysis in part (b). (*Hint*: Write down what your regression equation collapses to when $X = 0$ or $X = 1$.)

15. In the USA and Canada, scholastic aptitude test scores (SAT) are determined at high school by standardised tests and are used by some colleges and universities to predict academic success and to assist with admission decisions. The computer program MINITAB includes among its sample data sets a set of 100 North-western university students' scores on the Verbal SAT and Mathematics SAT, and the students' subsequent grade point averages (GPA) at university. Possible SAT scores range from 0 to 1000, while the GPA scores range from 0 to 4. The following outputs were produced by MINITAB.

```
Run 1. The regression equation is
GPA = 0.539 + 0.00362 VERBAL
Predictor          Coef       Stdev    t-ratio        p
Constant         0.5386      0.3982       1.35    0.179
VERBAL        0.0036214 0.0006600       5.49    0.000

s = 0.4993        R-sq = 23.5%       R-sq(adj) = 22.7%

Analysis of Variance

SOURCE              DF         SS          MS          F        p
Regression           1     7.5051      7.5051      30.10    0.000
Error               98    24.4313      0.2493
Total               99    31.9364

Run 2. The regression equation is
GPA = 1.44 + 0.00194 MATH

Predictor          Coef       Stdev    t-ratio        p
Constant         1.4402      0.5601       2.57    0.012
MATH          0.0019352 0.0008520       2.27    0.025

s = 0.5564        R-sq = 5.0%        R-sq(adj) = 4.0%

Analysis of Variance

SOURCE              DF         SS          MS          F        p
Regression           1     1.5970      1.5970       5.16    0.025
Error               98    30.3394      0.3096
Total               99    31.9364
```

(a) A regression line having a slope of zero occurs when the predictor variable contributes no information for predicting the response variable. Use the above regression outputs to test whether this is the case.

(b) Are either of the SAT scores good predictors of GPA in this group? Justify your answer.

(c) Do the regression lines produce sensible predictions when the SAT score is zero? Explain.

(d) For SAT scores in the 500–600 range, what is a typical prediction error in predicting GPA?

16. In opinion polls, results are frequently quoted as being 'accurate to within three percentage points, 19 times out of 20'. The results are usually a percentage in favour of a particular party or policy, something like '23.5% favour Party *A*'.
    (a) If the poll is reporting an α% confidence interval for the true population percentage, what is α and what is the standard error of the sample percentage?
    (b) The standard deviation for a sample proportion p in estimating a population proportion π is:

$$sd(p) = \sqrt{\frac{\pi(1-\pi)}{n}}$$

    where n is the sample size. In surveys in which π is to be estimated, it is not known exactly even after the survey is done. Does the formula have practical use?
    (c) Suppose the estimated percentage in favour of party *A* is 23.5%, and the sample size is 900. How would the accuracy of the estimated percentage be reported?
    (d) The formula:

$$sd(p) = \sqrt{\frac{\pi(1-\pi)}{n}}$$

    is largest when $\pi = 0.5$. If you want a poll to report percentages that are accurate to within ± 3 percentage points, 19 times out of 20, how large would the sample have to be?

17. From each batch of 1728 cans of salmon, 100 cans are selected at random and carefully inspected for external imperfections in the can itself. The first batch has 10 rejects among those sampled, while the second batch has five. Is this convincing evidence that the process has improved from the first to the second batch (that is, a reduction in imperfections)?

Projects

1. **Reaction time and handedness**
   This project is related to the reaction time project in chapter 2. Design and carry out a study to test whether a person's hand reaction time depends whether it is the dominant hand. (For example, does a right-handed person have a shorter or longer reaction time with the right hand compared to the left?).

   Report your findings in a report that is no longer than three pages. (The page limitation is to encourage you to make decisions about what a reader would need or want to know, and what displays convey the information most effectively. Review chapter 6 for presentation techniques and chapter 7 for design strategies. The techniques of chapters 8 and 9 will help you to avoid unreproducible claims.)

2. **Service time — quality control**
The objective of this project is to demonstrate the use of control chart techniques on a commercial business.
The time taken for a person to obtain lunch at a cafeteria, the service time, is a measure of the efficiency of the cafeteria. Of course, the customer affects this time by the complexity of selections and other behaviour, but these are not within the control of the cafeteria management. Nevertheless, the average service times of several customers can indicate how smoothly the operation is running.
For this project you are to collect a sequence of 10 service times, on each of 10 days (each weekday for two weeks). These can be sequential service times, but should begin at a particular time of day, perhaps noon. Record the means and standard deviations of the 10 service time measurements from each day.
Estimate the standard deviation of the service time from the service times in the separate days with the formula:

$$s_{\text{pooled}} = \sqrt{\frac{s_1^2 + s_2^2 + \ldots + s_{10}^2}{10}}.$$

Using this value, plot the means on a '3σ' control chart: a chart with out-of-control lines at distance three $\sigma_{\bar{x}}$ above and below the nominal value. Report on your findings, noting in particular any evidence of the process being in control or not in control.

3. **Use of inference in journals**
Obtain an academic journal in a field of interest to you. From the current issue, select three instances of the use of an inferential technique such as a confidence interval or a hypothesis test. (Use more than one issue if necessary). For each example, answer the following questions:
(a) What is the context of the test? What is the research question related to the test?
(b) What were the null and alternative hypotheses for the test? (These are usually only implied — you will probably not see them explicitly defined.)
(c) Comment on the use of confidence levels and/or p-values in the article. Were these strategies properly used?
It is not necessary to describe how the estimates or hypothesis test calculations were done. Be sure to cite the journal and articles used.

4. **Paper aeroplanes — do trim tabs matter?**
Choose one particular design of paper aeroplane and construct two aeroplanes that are identical, except that one aeroplane should have the end of each wing bent up (*trim tabs*). This project will investigate whether or not these trim tabs affect the time aloft for the design.
(a) Can you implement the experiment so that changes in the launching technique do not affect comparisons between the two designs of aeroplane?
(*Hint*: Consider randomisation and 'blind' testing.)

(b) For each aeroplane, time its flight with a standard launching technique 20 times.
(c) Find an interval estimate of the mean time aloft for each aeroplane.
(d) If suitable computer software is available, test whether there is any difference between the mean flying times of the two aeroplanes.
(e) What would have been the benefits of freshly constructing a separate aeroplane for each of the 40 trials in (b)?
(f) Report on your findings.

5. **Rainy days in Wollongong**
In section 2.1, a table was presented containing numbers of rainy days in each month from January 1977 to December 1990 in Wollongong, Australia.
(a) For each month, there is a batch of 14 values. Draw box plots for these 12 batches in a way that allows comparisons of the months to be made easily.
(b) For each month, evaluate a 95% confidence interval for the mean monthly number of rainy days in that month. Draw these confidence intervals on a diagram with *month* on the horizontal axis and *number of rainy days* on the vertical axis. Add the sample means to this graph.
(c) Draw a smooth curve close to the sample means in (b). (Remember that this curve should smoothly link December to January; that is, the pattern should be cyclic.)

6. **World wheat production**
In 1991, the world produced 550 993 million tonnes (t) of wheat. The production for each of 53 wheat-producing countries is listed in the reference, and we record here a systematic sample of 17 of the countries' data.

Country	Wheat production (t)	Area ('000 square miles)
Australia	9 633	2 966
Belgium-Lux	1 620	13
Canada	32 822	3 850
Colombia	94	441
Ecuador	28	105
Finland	431	131
Greece	2 750	51
Iran	8 900	632
Israel	160	8
Korea, North	195	47
Mexico	4 115	756
Netherlands	916	16
Peru	128	496
Romania	5 422	92
Sweden	1 524	174
Turkey	20 400	301
Uruguay	136	68

Source: R. Famighetti (ed.), *The World Almanac and Book of Facts 1995*, Funk & Wagnalls, Mahwah, New Jersey, 1995, p. 141. Reprinted with permission.

Clearly, some of the variation could be explained by the size of the country, and some of it by the suitability of the climate to wheat and to substitute grains. (We will ignore other factors for now.)

(a) Use regression to determine the extent to which a country's wheat production is determined by its size. Summarise your findings. (*Hint*: Use logs.)

(b) Using an atlas or globe, determine a latitude of each country (perhaps the latitude of the capital city). Plot the residuals of the regression in part (a) against the latitude and add to your summary of findings.

(c) Repeat (a) with another systematic sample, shown below. Compare the regression described in part (a) with the similar regression on these data. Is the variation in the slope of the regression line what you would have predicted? Explain.

Country	Wheat production (t)	Area ('000 square miles)
Afghanistan	1 726	252
Austria	1 341	32
Brazil	3 007	3 286
Chile	1 589	292
Czechoslovakia	6 205	30
Egypt	4 483	385
France	34 483	210
Hungary	5 954	36
Iraq	525	168
Italy	9 289	116
Korea, South	1	38
Myanmar	123	261
New Zealand	176	104
Poland	9 269	121
South Africa	2 245	473
Switzerland	574	16
United Kingdom	14 300	94
Former USSR	80 000	8 500 (approx.)

(d) Would your summary of findings about the dependence of the amount of wheat produced on the area and latitude have been the same for each data set?

7. **Market trends**

From a newspaper that includes lists of publicly traded companies and their current stock prices, select 10 companies at random. On two successive trading days, record the closing prices for the 10 companies.

(a) Test for a change in the mean stock price in two ways:
   (i) by considering the data as two independent samples
   (ii) by considering the data as a paired sample.

(b) Select a second set of 10 companies at random. Test for a change in the mean stock price based on these two samples.

(c) Comment on the similarities and differences between the tests performed in parts (a) and (b).

REVIEW PROBLEMS

The review problems have a dual purpose. Since they cover the whole text, they will serve as an aid to review, but there is a more important purpose: a problem based on the material in a particular chapter is invariably solved using the techniques of that chapter. However, in real-world applications, no-one will tell you what chapter a data-based problem comes from so it is important to be able to identify the opportunity for certain statistical tools and strategies without the chapter guide. These problems are intended to give you practice in pulling all the chapters together when answering data-based problems.

1. A tape measure is used to measure the circumference of 25 tree trunks. These trees are selected at random from a large population constituting a particular wood lot. The circumference is found from these measurements to have mean 60 cm and standard deviation 12 cm. However, after the data have been summarised, it is noticed that 4 cm of the tape measure have been cut off the zero end, so all the readings are larger than they should be by 4 cm.
   (a) If possible, calculate the corrected data summaries (that is, mean and standard deviation). If it is not possible, explain why.
   (b) All measurements of circumference are positive. Considering the mean of 60 and the variance of 144, is this a reason to think that the data might have a skewed distribution?
   (c) Estimate the average diameter of trees in the wood lot. Your estimate should be as informative as possible, based on the data.
   (d) Making whatever assumptions are necessary to do so, estimate the percentage of trees in the wood lot that have a trunk circumference of more than 75 cm.

2. An accounting association has membership of 10 000, consisting of 60% full members and 40% associate members (who are still working on their qualifying examinations). One thousand members (full and associate) are selected at random and are mailed a questionnaire. The survey asks for salary information. Of the 500 responses received, 250 are full members and 250 are associate members. Average income for the full members in the sample is $80 000 and the average for the associate members in the sample is $40 000, while the standard deviation of incomes in each sample group is $10 000.
   (a) There is a clear indication from the sample that the full members have salaries that are about $40 000 higher than the salaries of the associate members. Is there any information available about the likely response bias in this difference (between the average salaries of the two groups)? Elaborate.
   (b) By treating the two samples of 250 responses as if they are random samples from their respective populations (that is, full members or associate members), estimate the difference in the average salaries. Use an interval estimate.

3. Each weekday, the parking lot at the university fills up by about 10.00 a.m. with cars of various market values. A student majoring in business administration wants to estimate the total market value of all these parked cars at one point in time. The technique used is to select a random sample of 25 parking permit holders, obtain the year, make and model of each car, and use standard used-car price lists to value the 25 cars. The 25 car valuations average $8000 and have a standard deviation of $4000. Assume that there are 3000 spaces on campus.
   (a) Estimate the total value of all the cars parked at 10.00 a.m., using the data provided.
   (b) Calculate the accuracy of your estimate in part (a).
   (c) Speculate on any bias that might be present in the estimation procedure used in part (a), and give reasons for your suggestions.

4. A certain size and type of lumber is tested for breaking strength by applying increasing force to it. Since the test is destructive, a sample of 20 boards is selected at random from the sawmill's output on a particular day; each of the 20 boards is tested and the breaking strength of these boards averages 200 kilograms and has a sample standard deviation of 50 kilograms.
   (a) Compute an interval for breaking strength that will include the strengths of approximately 90% (that is, 18 out of 20) of boards.
   (b) Compute an interval that estimates the average breaking strength for the day's production.
   (c) Discuss the assumption of normality with respect to your answers to parts (a) and (b).
   (d) If the day's production were 2000 boards, and the sampling method was to select each hundredth board, starting at the 50th board, would the theory underlying your calculations in parts (a) and (b) apply? Explain.

5. The normal distribution is one of the most useful theoretical distributions for statistical applications.
   (a) What theoretical result explains why the normal distribution is often an appropriate model for observed measurements?
   (b) Why is it often said that normal distributions are useful for describing variation in averages? How is the normal distribution used for describing variation in averages?
   (c) Is it possible to tell if a population distribution is normal by looking at a histogram of data sampled from the population? Discuss this question in some detail, considering the shape of the sample data histogram, the normal curve, presence of more than one sample mode, sampling method, sample size, sampling variation and the presence of outliers.
   (d) If the population is assumed to be normal, what summary or summaries of the sample are necessary in order to estimate the 5th percentile of the population? Explain how this calculation is done.
   (e) What connection is there between the table of normal probabilities and the table of the t-distribution in the Appendix?

6. In a marketing firm, you are asked to analyse the data from a customer survey that was conducted to determine family incomes of purchasers of a particular brand of washing machine. In reporting these data, comment on the appropriateness for this purpose of each of the following techniques for summarising the income distribution:
   (a) mean ± standard deviation of income distribution
   (b) mean ± standard deviation of the sample mean income
   (c) median ± interquartile range of income distribution
   (d) lower quartile and upper quartile
   (e) box plot of income distribution
   (f) dot plot of income distribution.

7. Write a brief note explaining the following statements.
   (a) Survey non-response can cause a bias in the results of an opinion survey.
   (b) Very large samples often give statistically significant results that are unimportant from a practical perspective.
   (c) Residual plots for a regression line indicate whether the predictive model can be improved.
   (d) Estimating variability can be more important than estimating the mean, in some circumstances. (Give an example of a study where this would be true.)
   (e) Linearity of a two-variable relationship is often fallacious, even in situations in which it is appropriate to use linear regression.
   (f) A median is preferred to a mean as a measure of the 'centre' of a distribution when some extreme points are suspected to be subject to measurement mistakes.

8. For each of the following statements, explain why the statement is false. (Each statement has a close relationship to a true statement, but is actually very wrong as is — show that you understand what is wrong with each statement.)
   (a) A regression line based on data for variables X and Y can be used to predict X from Y and Y from X.
   (b) The variance of the integers from 1 to 100 is about 25.
   (c) The p-value in a hypothesis test is the probability that the null hypothesis is true.
   (d) A 95% confidence interval estimates the interval of values in which approximately 95% of the sample data will fall.

9. (a) Draw a graph of a normal distribution model for the number of hours of sleep individuals in your university course get per year. Indicate the mean and standard deviation graphically on your graph. Make sure the axes on your graph are labelled and have numeric scales indicated. Your informal knowledge of sleep patterns should give you enough information to produce a graph with a reasonable mean and standard deviation.
   (b) Mark off on your graph a number of hours that would be exceeded by 84% of students. Show or explain how you determined this value.

10. Five students each sit for an exam in English and an exam in mathematics, and the results are:

Student	**1**	**2**	**3**	**4**	**5**
English	73	88	63	50	81
Maths	65	85	70	55	85

(a) Are the two exam scores correlated? How much?

(b) Predict the maths score for a sixth student who obtained 86 on the English exam.

(c) If the five students have been selected at random from a large class of 200 students, test whether the average scores on the two exams (for all 200 students) are different. Assume normality as needed. Give your conclusion in jargon-free English.

(d) How many of the 200 maths exam scores are 90 or more? Use the data to construct an estimate of this number.

11. A sample of 100 students (from a survey population of 15 000 students) was asked whether they required loans to finance their expenses during the current term. Of the 100 students contacted for this survey, only 64 students replied and 40 of them (62.5%) said that they needed loans.

(a) Is it reasonable to treat the 64 responses as if they were a random sample of size 64 from the 15 000 students? Explain how you would decide whether or not to proceed on this assumption.

(b) Assuming that the 64 can be treated as if they were a random sample from the 15 000, use the data to compute a range of credible values for the proportion requiring loans (that is, a range of values that you believe could be the true proportion requiring loans among the 15 000). You may assume that the survey responses are truthful.

12. A machine produces precision parts for a pump, and these parts must be within a specified range of diameters to be useable. The machine has been adjusted so that it accomplishes this task in 90% of the parts it produces, but the machine occasionally gets out of adjustment and produces a lower proportion of useable parts. When this is a concern, the machine is shut down for maintenance. Each hour, a sample of 80 parts is measured to check that the machine is properly adjusted.

(a) Assuming that the machine is working according to specifications, use a normal approximation to the distribution of the proportion of defectives to find the probability of more than 8 defectives in a sample. Similarly find the probabilities of more than 9, 10, and so on, defectives.

(b) How many non-useable parts among the 80 would suggest to you that maintenance was required? Justify your answer.

13. Nine automobiles of the same make and model are tested for the litres of petrol they consume per 100 kilometres of city driving. The nine consumption values are:

8.5, 8.3, 7.9, 8.6, 8.3, 8.0, 8.2, 8.4, 8.3.

(a) Compute the mean and standard deviation of these data.
(b) Based on this information, estimate an interval that you think would contain 95% of the consumption values if many more cars were tested. Justify your method.
(c) An advertisement claims that typical consumption for this model of car is about 8.5 litres per 1000 kilometres. Based on the available data, do you think the advertisement's claim is credible? Justify your answer.

14. Mr Smith measures the petrol consumption of his car over 25 time intervals and then summarises the overall petrol consumption by reporting the mean and standard deviation of these measurements. His summaries are 8.5 litres per 100 kilometres, and 0.3 litres per 100 kilometres respectively.
    (a) Based on this information, what range of values do you think would contain 95% of the 25 measurements of petrol consumption? Explain your choice.
    (b) An advertisement claims that typical consumption for this model of car is about 8.2 litres per 1000 kilometres. Based on the available data, do you think the advertisement claim is valid for Mr Smith's car? Do a calculation on which to base your answer. (Ignore variation due to driver.)
    (c) The petrol consumption data is to be re-expressed as gallons per mile, where 1 litre per 100 kilometres is equivalent to approximately 0.0043 gallons per mile. If possible, compute the mean and standard deviation of the petrol consumption data in gallons per mile. If it is not possible, explain why.

15. For each assertion below, write a short paragraph explaining the assertion. (That is, to the extent possible in a short paragraph, justify the assertion.)
    (a) It is convenient for statistical analysis to consider a population consisting of two categories of things as a population of 0's and 1's, for then averages and proportions are the same thing and the central limit theorem for sample averages can be applied to sample proportions.
    (b) A scatter diagram displays the distribution of two-variable data, but in a different way from how a histogram displays the density of one-variable data.
    (c) An observational study (for example, a survey) can never prove a causal link, but is sometimes useful in suggesting causal links.
    (d) In simple linear regression involving two variables X and Y, the line used to predict Y from X is not the same as the line used to predict X from Y.
    (e) A control chart for means is simply a way of comparing between-sample variation relative to within-sample variation. It allows us to detect changes in mean over time.
    (f) The correlation coefficient measures the association between two variables, but only does a good job of this when the association is linear.
    (g) Using standard units for normal data allows us to relate all normal distributions to the particular one tabulated in the Appendix.

(h) Simpson's paradox is a warning that confounding factors in an observational study can lead to erroneous conclusions.

(i) Consider the two events for rolls of a fair die:
- *A*: 60 rolls turn up ten 1's.
- *B*: 6 rolls turn up one 1.

B is more likely than *A*. However, for 60 rolls, the proportion of 1's is likely to be closer to $\frac{1}{6}$ than for 6 rolls.

(j) A p-value is often interpreted as the probability that the null hypothesis is true, but this is a fallacious interpretation.

16. Crime data for fifteen cities are to be analysed, and data are available from 1980 and 1990. The number of violent crimes per 100 000 population is recorded, and the following data are noted:

City	1	2	3	4	5	6	7	8	9	10	11	12	13	14	15
1980	5.2	10.5	2.8	6.3	13.1	4.4	3.3	16.8	7.5	8.4	2.3	7.3	2.5	3.0	9.8
1990	4.9	10.1	3.2	6.1	12.9	4.4	3.5	16.2	7.4	8.3	2.7	5.9	2.9	3.2	9.5

(a) By considering the rate changes for each city, test whether the crime rate in 1990 is less than in 1980.

(b) Describe in words what you think the data suggest about the crime rates in the 15 cities. (*Hint*: Look carefully at the data.)

17. Thirty men and 30 women agree to participate in an investigation of the effect of coffee on reaction time. Reaction time is measured by dropping a time-scaled ruler through fingers until the fingers close on the ruler. The reaction time is measured before and after drinking coffee, and the 'before minus after' difference is observed to have mean 0.01 seconds and standard deviation 0.05 seconds for the men and mean 0.04 seconds and standard deviation 0.05 seconds for the women. In other words, there is an apparent tendency for the reaction time to decrease after drinking coffee, in both men and women.

(a) Is this an experiment or an observational study? Explain.

(b) What is the relevance of the question in part (a) to the interpretation of the data?

(c) Is the reaction time difference greater for women than for men?

(d) What is the relevance of the question in part (c) to the interpretation of the data?

(e) Separately for men and women, test whether the average reaction time difference is significantly greater than zero.

(f) The average reaction time for the whole group of 60 subjects is 0.20 seconds before and 0.17 seconds after the coffee, and the standard deviation of the 60 reaction times, both before and after, turned out to be 0.11. Using the data from this part only, test the significance of the decline in reaction time from 0.20 to 0.17. Explain why you do, or do not, get the same result about the reaction time difference using these data as you would get using the data provided for part (e) (in the preamble to the questions).

18. A population of 2000 union members is to be sampled concerning the size of the mortgages they hold. A random sample of 50 members responds that they hold an average mortgage of \$50 000. The standard deviation is \$10 000.
    (a) One member claims that the average of \$50 000 seems low — he claims that mortgages of \$100 000 are typical and suggests that the sample is unlucky or that the sampling procedure is biased. What other explanation would you suggest?
    (b) If a second sample of 50 union members was selected, how far might its average reasonably be from the original average of \$50 000, assuming both samples were truly random samples from the population of 2000? (*Hint*: You will need the standard deviation of a difference of sample means for this.)

19. Write a short explanation of each of the following statements (a sentence or two should suffice for each topic), based on the material learned in this book.
    (a) In a hypothesis test, the p-value is computed under the assumption that the null hypothesis is true, but the p-value is also used to test that same assumption.
    (b) The histogram of a sample from a normal distribution can look skewed and multimodal, even for samples of 100 or more.
    (c) The central limit theorem provides a theoretical basis for using the normal distribution in statistical inference.
    (d) The central limit theorem applies to sample proportions of categorical data as well as to sample averages of quantitative data.
    (e) The average is often used as a predictor of individual values in a distribution.
    (f) Medians and interquartile ranges have some advantages over means and standard deviations for describing the centre and spread of a distribution.
    (g) The interpretation of a hypothesis test can depend on the sample size and the practical context of the test.

20. A news story about a study of the possible connection between authoritarian parenting and teenage crime includes the following paragraph:

    'The study included 1100 families selected at random from the 25 000 families in Brownsville who had one or more teenage children. The parental styles were assessed based on interviews with the selected families and classified as predominantly "authoritarian" or "non-authoritarian". Police records were accessed to determine the presence or absence of teenagers in the family who had a criminal record at the time of the survey.'

    The following data were reported:

	Criminal record	**No criminal record**
Authoritarian	15	85
Non-authoritarian	50	950

(a) Summarise the data to describe whether there is a link between parental style and the presence of teenagers in the family with criminal records.
(b) Test formally whether there is a link as hypothesised.
(c) Discuss the statement, 'The strength of the association is not the same thing as the strength of the evidence to support an association' with reference to this study. (*Hint*: What does the p-value measure?)
(d) Can a causal link be inferred from this study? Justify your answer.
(e) If the ages of teenagers in a family were recorded at the interview, would this information be of any relevance to the research hypothesis? Explain why or why not.
(f) Under what circumstances would you conclude that a relationship between parental style and teenage criminality was definitely refuted by the data?

21. A stand of 2500 Douglas fir trees planted in 1950 are of various heights due to random fluctuations in growing conditions. A random sample of five trees is selected and cut down and their heights and base diameters (at two metres off the ground) are measured accurately. The data for the five selected trees are:

Tree number	1	2	3	4	5
Height (metres)	14.5	14.3	15.2	15.6	14.9
Diameter (metres)	0.45	0.44	0.49	0.47	0.45

(a) Use your knowledge of statistical distributions to estimate (or guess) the maximum height of trees in the stand of 2500 trees.
(b) Graph the data. Your graph should be self-explanatory.
(c) Compute the correlation between the tree diameters and their heights, based on the sample data.
(d) If a tree among the 2500 has diameter 0.50 metres, estimate its height.
(e) What proportion of trees has diameters greater than 0.50 metres? (Compute an estimate of this proportion.)
(f) What practical use of the sample data shown might be of interest to the owner of this stand of Douglas fir?

22. A study is done which compares two groups of Year 6 students at a particular primary school. Group *A* consists of those children whose mother works full-time outside the home, while group *B* consists of children whose mother works outside the home either part-time or not at all. Two tests have been administered to these students, and scores have been obtained. The two tests are (a) an IQ test, which purports to measure general intelligence, and (b) an achievement test (AT), which measures learned skills that Year 6 students are supposed to have. So, for each student (in group *A* or in group *B*), there is an IQ score and an AT score. The aim of the study is to see if the home environment in which a mother

works full-time outside the home leads to lower achievement in Year 6 students; that is, group *A* children fare worse than group *B* children.
(a) Is this study an experiment? Explain why or why not.
(b) While analysing the data for the investigator, you notice that the AT scores of the group *A* students are higher on average than the AT scores of the group *B* students, and the test for the significance of this difference has a p-value of 0.021. What would you infer from this finding? (Express in non-technical terms.)
(c) How would you use a scatterplot of AT v. IQ with different symbols for students in the two groups to examine whether the group has any additional effect on AT score after taking account of IQ?

23. Records obtained from a recent random sample of 50 students graduating from private schools shows their average IQ is 115. A comprehensive survey of all private school graduates done 20 years ago shows that the average IQ was 110. Does this recent survey establish that IQs are higher now? You may assume that the tests on which the IQ determinations are based are comparable in both surveys. The standard deviation of IQs in the population tested is known to be 15.

24. The following data have been collected about the weights of pigs before and after they have been kept for four weeks in a feeding facility.

Pig	**1**	**2**	**3**	**4**	**5**	**6**	**7**	**8**	**9**	**10**
Before (kg)	45	40	45	35	40	45	35	50	45	40
After (kg)	55	45	55	50	50	60	45	55	55	50

(a) Compute an estimate of the increase in weight of the pigs during the feeding program and estimate the largest and smallest values that the average increase might reasonably be. Assume that the 10 pigs are selected at random from the population of interest.
(b) A mischievous clerk switches the before and after data of a random selection of five of the 10 pigs. Simulate this process by selecting five of the pigs at random and switching the before and after data, and then test whether the resulting weight gain (based on the data from all 10 pigs) is significantly different from zero. (Be sure to show the detail of how you select the five pigs.)

25. Write a paragraph explaining each of the following statements:
(a) Scatter diagrams are useful for identifying outliers in two-variable data.
(b) A residual plot can reveal failures of the usual assumptions for regression.
(c) A 95% confidence interval usually contains less than 95% of the observations on which it was based.
(d) The sample size in a sampling survey depends on the precision required in the estimates that will be based on the survey data.
(e) Qualitative data may be summarised quantitatively through the use of percentages.

26. For each of the following statements, write a paragraph explaining why it is true.
    (a) The median and quartiles are sometimes preferred to the mean and standard deviation for describing data that have a strongly skewed distribution.
    (b) The one-way analysis of variance is the logical extension of the t-test to more than two independent comparison groups.
    (c) Control charts provide a way to do informal tests of a hypothesis on a process that is being monitored over time.

SOLUTIONS TO EXERCISES

Chapter 2

Exercise 2.1

1. (a), (c), (d), (e), (f), (h)
2. (a), (b), (c), (d), (e)
3. (a), (b), (e) if we consider that 'month' is not really quantitative and conveys information about the season, and (g)
4. (e) It is natural to summarise the data by month. Answer (g) might also be chosen if comparisons between the different aircraft manufacturers are of interest.
5. If the data are sorted, they are much easier to comprehend. If the order in which the data are presented is an important feature of the data, as in a time series, then sorting would not be a good idea. The concept in this exercise is the same as in question 2 — did you recognise it?

Exercise 2.2

1. Most people find the dot plot more effective. The actual numbers are usually less important than the relative position of the dots and the roughly portrayed range of values.
2. We can talk about the centre and spread of the dots, the existence of outliers or clusters and the degree of symmetry of the scatter about the centre of the data.
3. The dot plot is:

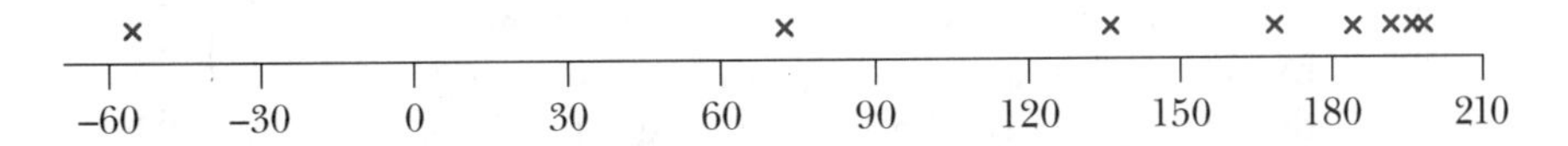

   The spacings double, starting from the right. (If you sorted these data first, as suggested in the previous section, you might have noticed this from the numbers themselves.)

4. The dots in a dot plot have to have some width to be visible, so there is always the problem with continuous data that some dots will overlap each other and therefore be hard to comprehend visually. When this happens in a dot plot, some way of showing the separate dots is needed. Jittering the dots and stacking them are two ways to avoid the problem. However, since a computer screen has the potential for 100 or more separate dots side-by-side, important information is rarely lost by the stacked dot plot. When we want to show each data value precisely, jittering can be used.

Exercise 2.3

1. (a) The basic stem-and-leaf plot is shown on the right.

Stem	Leaf
9	0 0
8	0 1 3 5 8
7	0 1 1 2 3 4 7 8 8 9
6	0 0 3 3 4 4 5 6 7 7 7 8
5	3 6 6 7 8
4	6 9

   (b) We can split the stems to make two bins for each stem, providing a little more detail while still providing an easily grasped display.

Stem	Leaf
9	0 0
8	5 8
8	0 1 3
7	7 8 8 9
7	0 1 1 2 3 4
6	5 6 7 7 7 8
6	0 0 3 3 4 4
5	6 6 7 8
5	3
4	6 9

2. Use only the hundreds digit and higher so the range is from 4 to 116. Then the stem would be the first two digits, from 00 to 11, as shown in the diagram on the right.

   In the 'raw' stem-and-leaf plot, the extremes may reflect aircraft sizes, not their inherent pollution. Standardising by dividing by some index of aircraft size (for example, weight or number of passengers) removes this dependence on size, so the large and small values in the data are more easily interpreted.

Stem unit: 1000
Leaf unit: 100

Stem	Leaf
11	6
10	8
9	
8	1 2
7	3
6	2 2
5	3 6
4	6 8 8
3	2 2
2	0 5 7 8
1	1 2 8
0	4 5 5

3. The only case in which this is warranted is when there is an outlier which is far from the rest of the data; we may need several stems to show the shape of the data where it is dense, and by doing so impose a large number of empty stems in order to portray the outlier.
4. There would be too many leaves on each stem to fit on one page. Alternatives might be to use very small print or use dots or use a roll of paper . . . or, read the next section on histograms!

Exercise 2.4

1. The stem-and-leaf plot has more information in it because the original numbers are still there to see. Furthermore, it is easy to construct by hand, so it is best for a first look. However, the histogram is a more polished graph and conveys the essential information well; it is better for a presentation.
2. (a) The histograms are shown below.

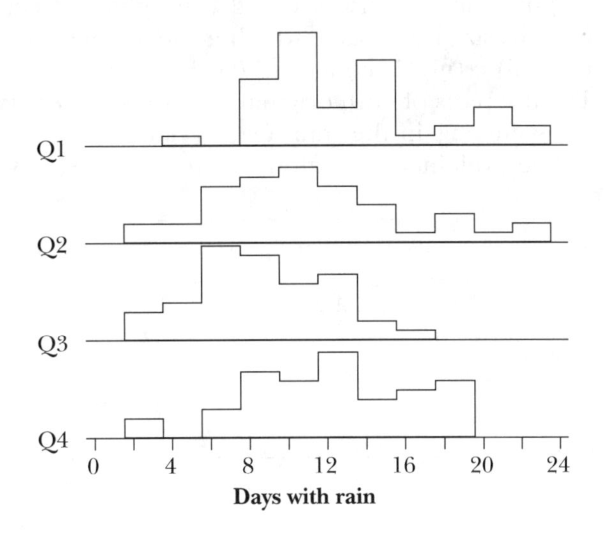

(b) The histograms show that Q1 (first quarter) has the most rainfall, followed by Q4, Q2 and Q3. Note also that the spread of values in the data is greatest for Q2, followed by Q4, Q1 and Q3. It is common in data spreads to find the largest spread for the largest values, but this does not happen here.

3. There are 24 data values, so we need about five cells (the minimum). The range of values is 5 to 394 so a class width of about 80 is needed. Perhaps 0–79.9, 80–159.9, 160–239.9, 240–319.9 and 320–399.9 would be appropriate classes for grouping the data. This would provide frequencies of 12, 5, 3, 2, 2 — a very skew distribution. In describing distributions, the word 'skew' means that the distribution is not symmetric about its centre and has a relatively long tail on one side or the other.

Exercise 2.5

1. The elephant brain weight is 5712 g and this is plotted at a distance 3.121 from the logarithmic origin (which is at $\log_1 = 0$ and corresponds to a mammal weighing 1 g — see 'mole' or 'hamster'). The mouse brain weight is 0.4 g and is plotted at a distance 0.398 below the logarithmic origin. The log scale shows the position of the data after they have been transformed into logarithms, while the original units are also shown because we are more familiar with weights in grams. Both scales reference the plotted points, but only the logarithms are used to actually position the points.
2. Perhaps 100 000 g or even more, since then the log would be 5.0. This brain weight of 100 kg is seen only in science fiction movies. On the small side the log would have to be about −2 to stick out, and this would correspond to a brain weight of 0.01 g, perhaps the weight of an insect's brain.

3. Between 4 and 5. The ratio of highest weight/lowest weight is $5712/0.140 = 40\,800 = 10^{4.5}$. Note that the range of base 10 logarithms in the data is from −0.8 to +3.7 which is a range of 4.5. Each unit difference in the base 10 logarithm corresponds to an order of magnitude in the original data.
4. It spreads out the small values and therefore allows us to detect clusters and outliers in the small animals. Also, log(wt) is a natural transformation of weights since the distance in the dot plot between one animal and another twice its size is the same whether the animal is small or large.
5. (i) e (ii) c (iii) d (iv) a (v) g (vi) b (vii) f
6. The dot plots of both original and transformed data are shown below. Note the symmetry in the transformed data — using symmetry for data analysis will be explained in section 2.9.

(i)

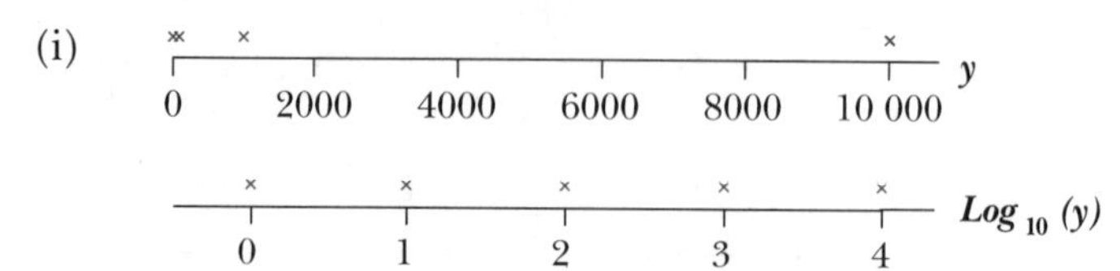

(ii)

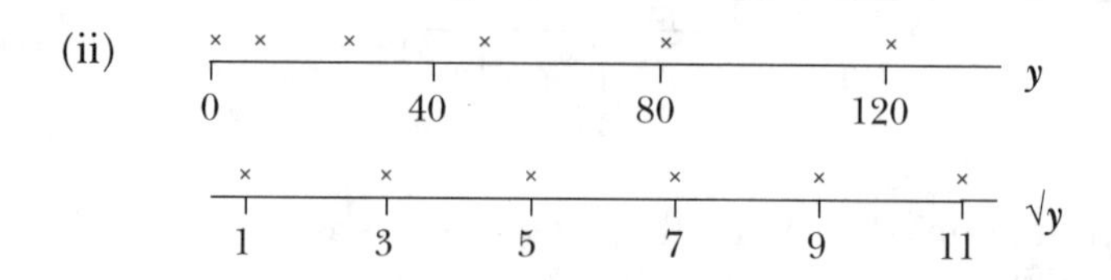

(iii)

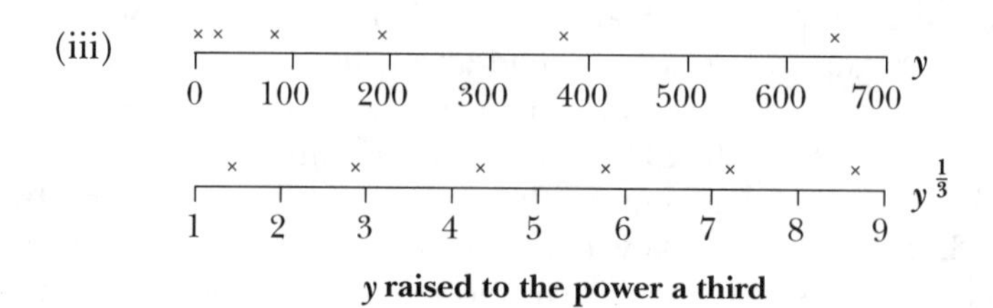

(iv)

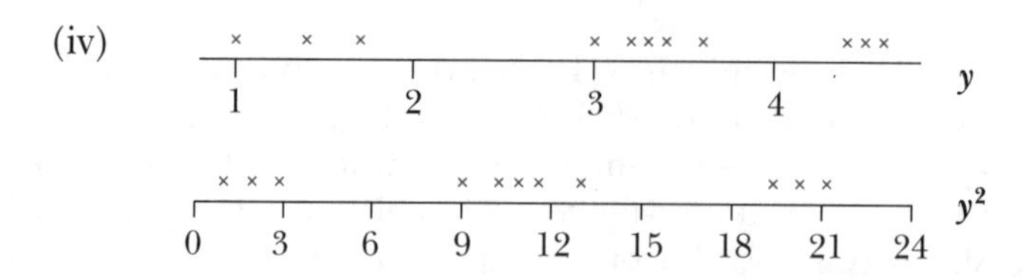

(v)

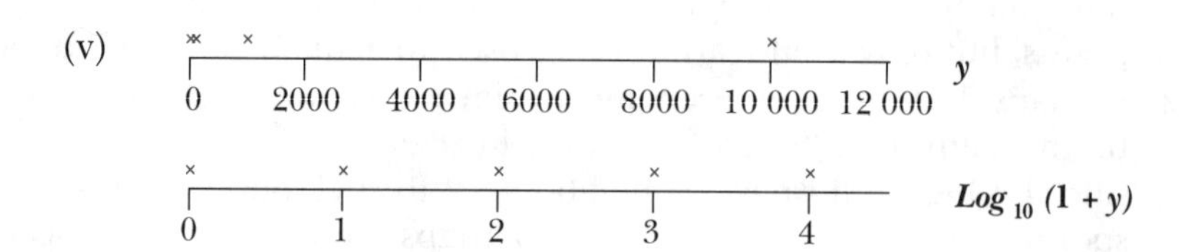

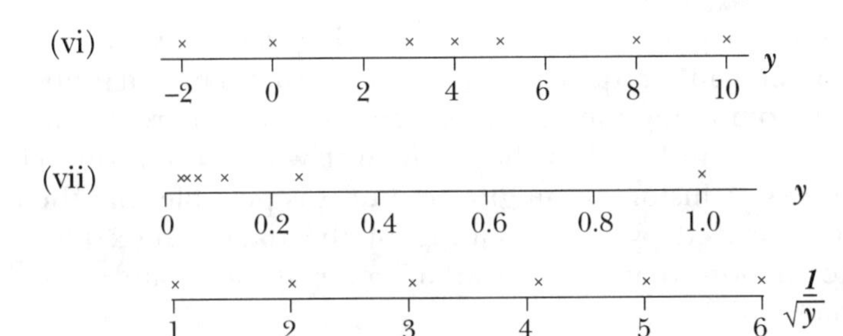

7. No. From the plot in this section, you can see that the 26 largest points would form a very skewed distribution, with most values being close to the lower populations.

Exercise 2.6

1. A bar chart displays the frequency of each discrete data value. A dot plot displays values that are close to each other as a vertical column of dots. In the case of discrete data, the only 'close' values are identical values, so the frequencies plotted by dot plots and bar charts are the same for discrete data.
2. (a) A histogram. There would be a large number of possible ages, so some grouping would be necessary.
   (b) A bar chart. There would be a small number of possible values.
   (c) It would depend on the size of the city. In a big city, there would be many alarms each month, so the range of values in the data would be large and a histogram would be appropriate. In a small town with fewer than 10 alarms in most months, a bar chart would be better.

Exercise 2.7

1. (a) Histograms are an effective way to compare males and females, provided they are scaled to have the same area. This is equivalent to displaying the percentages of males in each group, and similarly with females. We therefore use percentages based on the row totals 123 and 210; for example, the percentage of males aged 60–69 is 18.7% ($100 \times \frac{23}{123}$) whereas for females it is 9.0% ($100 \times \frac{19}{210}$).
   (b) On the other hand, to compare the sex distributions across age groups, use percentages based on 42, 101, 129, 51; for example, the percentage of males in the 60–69 age group is 54.8% ($100 \times \frac{23}{42}$) compared with the percentage of males in the 70–79 age group of 45.5% ($100 \times \frac{46}{101}$).

   The general rule is: to compare columns, use percentages based on row totals and to compare rows use percentages based on column totals.
2. They should add to 50, since each shaded square represents 2% of the males or females in that sub-population. You should get this total for each side of each pyramid.

Exercise 2.8

1. If the data do not have clusters, and this is known either by the appearance of the display or by prior experience with the context of the data, then a display which reveals as much detail as a dot plot may not be necessary. Moreover, the box plot has the advantage that it reveals quantiles graphically, which the dot plot does not do well.

2. Box plots are a graphical way of portraying certain summary measures, such as a median or quartiles, and these summary quantities are very efficient in comparing the broad features of two or more distributions.
3. The dot plot would be best here, since it would reveal the clusters for the two sexes. A histogram might do, but it is possible that the clustering could be obscured by the grouping of the data. A box plot hides the very information that would be of interest here and would not be appropriate.

Exercise 2.9

1. Yes. Consider the following data:

Mean = 0.98

```
 ::.  .   . .          .                      .
 +---------+---------+---------+---------+---------+-----raw
0.00     0.80      1.60      2.40      3.20      4.00
```

In the data set below, the original 4.38 maximum is stretched to 20.0; the mean is 2.54 whereas the second largest value is 2.33.

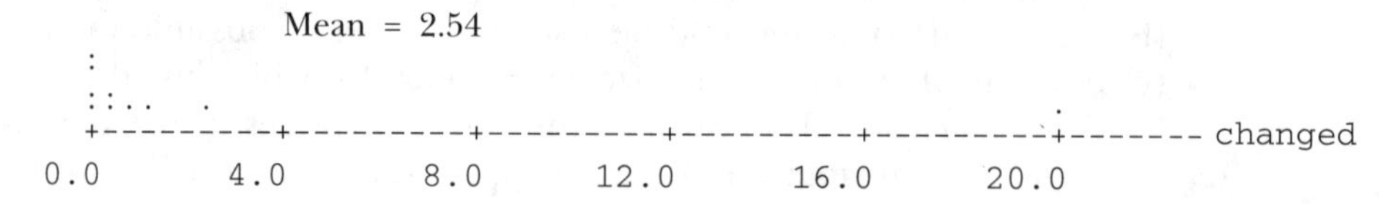

2. When the data are skewed to the right; that is, when the long tail is to the right (*Note*: see question 1).
3. Not exactly! If we take away an extreme value, the median will often change, though usually not much. The relative sizes of *all* the points determine which one is the middle-ranked one. It is true that we can change the values of the maximum and minimum as much as we want (in the positive and negative directions respectively) without changing the median. This is a useful feature of the median, especially when the extremes have dubious validity.
4. No. The mode is quite unreliable for indicating the 'location' of the data. Imagine a distribution that has two or more clusters of values — the cluster containing the mode is not necessarily in the centre of the data. Even with single mode data, if the data set is skew, the mode could be quite far from any number that reasonably represents the 'location' of the data.
5. The range for the first two values (9, 10) is 1. For three values (9, 10, 5) it is 5. It increases as the number of years gets larger so, as a measure of spread of the data, the range will depend on the number of values used. However, if we want to describe how variable the number is from year to year, we should use a summary that is not systematically larger or smaller because of how many years of data we have. We need a measure of spread that does not depend on the number of data values. The range does not have this property.

6. The interquartile range is the easier to describe and to comprehend. The standard deviation is difficult to explain and sensitive to outliers. The standard deviation can be influenced unduly by the skewness of the data — two sets with the same interquartile range can have very different standard deviations if one data set is skew. An advantage of the standard deviation is the 70-95-100 rule associated with it. Another advantage is that while the standard deviation is not always the best descriptive measure of variability, it is certainly the most commonly used and is even better known than the interquartile range.
7. No, because the units are wrong. The variance of ages in years would be in units of square years. Standard deviation is closely related to variance but is in the original units, so it is much preferred to the variance as a descriptive measure. Variance should be considered a technical term — it is widely used in inference strategies in more advanced statistics courses, but is of little use for descriptive statistics.
8. Both ways, the mean is $\frac{40}{25}$ or $1\frac{3}{5}$.
9. The standard deviation is the same for both the number of children per household and household size, since they differ by only a constant two. The standard deviation turns out to be $\sqrt{\frac{40}{24}}$ which is approximately 1.29.

Chapter 3

Exercise 3.1

1. Temperatures, rainfall, days with rain, hurricanes and even the presidents and prime ministers data are time series, even though we treated them as unordered batches in chapter 2. The presidents and prime ministers data are different, though, in that the time intervals associated with each value (of age at death) are unequal, and so the treatment of these data as a time series is a bit more complicated than if the time intervals were all equal.
2. There is a trend to older women! All the plotting methods mentioned should illustrate this trend.

3. (a) (i) and (iii) are discrete time, (ii) and (iv) are continuous time.
   (b) All are discrete. Even (iii) is discrete because monetary values are discretised by the smallest-value coin. The actual inventory of milk

would be continuous but the number of 2-litre cartons is discrete. However, in practice, discrete variables that have a large number of values, and for which specific values have no special importance, are often treated as continuous variables in choosing statistical methods. If you answered that you would treat the monetary data as continuous, you would be on the right track from a practical perspective. In this case, the discreteness of the variable is a mere technicality.

Exercise 3.2

1. For the food workers, the annual pattern is constant except for slight vertical shifts, whereas the airline passengers series has an annual pattern whose amplitude grows with time. Another more subtle point that could be made here is that the trend in the food workers series moves in two directions (down, then up) while the airline passengers series pattern keeps marching upwards.
2. None. The time ordering is essential for these trends.

Exercise 3.3

1. Exponential smoothing as it has been described here is for forecasting a time series and is one-sided in the sense that the smoothed value at time t depends on the raw values at time t, $t-1$, $t-2$ and so on, but not on $t+1$, $t+2$ etc. So, there is only one end where smoothing cannot be done — the left or 'past' end.
2. Values between 0 and 1 are all feasible. A value of a greater than 1 results in a series that is less smooth than the original data! Also, note that the closer a is to 0, the greater the degree of smoothing, while $a = 1$ corresponds to no smoothing at all. In fact, when $a = 0$, the result is that $y_t = y_0$ for all t, a horizontal straight line.
3. Running medians produce a 'smooth' result that usually changes in steps, but they are insensitive to occasional extremes. Running means tend to be smoother but are sensitive to extremes.
4. When the transformed series is plotted, we would like the deviations around the smoothed series to be of similar magnitude throughout the series, and to be fairly symmetric with positive deviations of similar size to the negative deviations.

   A related objective is to arrange that any seasonal component should be of roughly the same size for each year of data, so that this single seasonal effect can be more easily summarised.

   A third objective is to linearise the trend in the time series. If we are very lucky, we can find a transformation that does all these things at once.
5. (a) The smoothed series is the same as the original series {1, 2, 3, 4, 5}.
   (b) The smoothed series is {1, 2, 2, 2, 1}. Repeating the smoothing does not change it further.
   (c) An example is {6, 2, 3, 1, 5, 4, 6, 2}. After a single smoothing, the series becomes {6, 3, 2, 3, 4, 5, 4, 2} and after a second smoothing {6, 3, 3, 3, 4, 4, 4, 2}. Further smoothing does not affect this series.
6. The log(CPI) series looks quite different, with the apparent change in the series coming about 1970 instead of 1980. In fact, the log(CPI) series looks quite linear in 1970–1989 whereas, in the original series, this is the period of most marked nonlinearity.

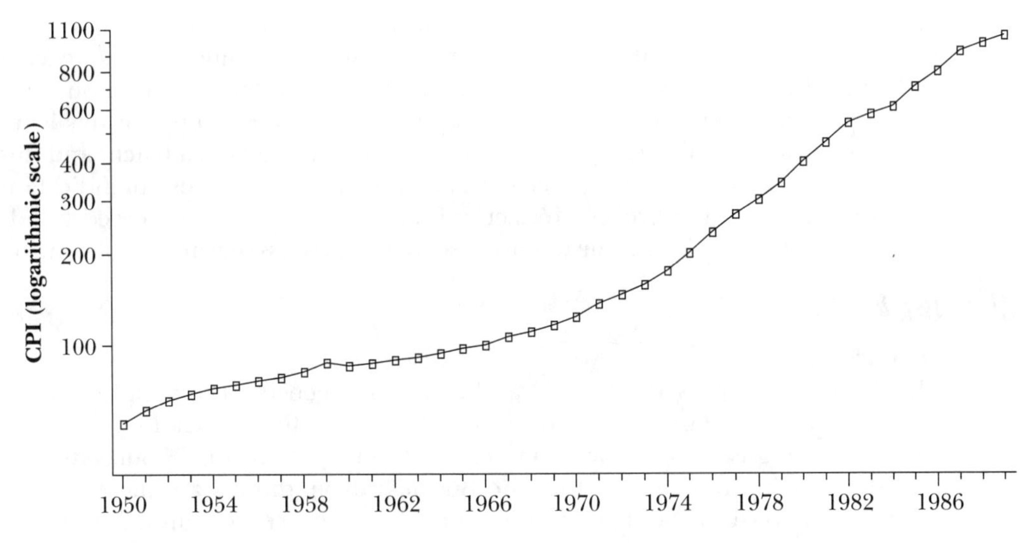

Exercise 3.4

1. Yes, because an additive seasonal component is defined this way. In other time series, the seasonal pattern may well change each year, as is apparent in the untransformed airline passengers data, but the seasonal component, as defined, cannot capture such a changing pattern. Transforming the data so that the seasonal pattern is unchanging simplifies things.
2. No. The seasonal changes are subtracted from the series before the trend is portrayed.
3. It would look very similar, but would not exactly reproduce the original. For example, the summer values in 1974–5 are higher than the reconstructed series would represent. These small differences are computed as the 'residuals'.
4. Certainly. It helps us to verify that the model used to separate the components (additive or multiplicative, for example) does the job properly. The residual time series should show no trend or seasonal pattern. Also, large residuals may indicate by their timing that some other factor may have influenced the data, and this may be useful for either forecasting or control of the time series in future.
5. When seasonal and residual fluctuations increase in magnitude in conjunction with a trend to higher values, a logarithmically transformed series often has a seasonal pattern of constant magnitude. When the seasonal component of the log data is additive, it is multiplicative on the original data. An additive seasonal component for the log data would be the best way to summarise such a seasonal trend.
6. Since the logarithm drops from 1.5 to –0.5, the actual value must drop a proportion equivalent to $\log_{10}(-2)$, or in other words to a $\frac{1}{100}$ proportion of its Q3 value $(10^{-2} = \frac{1}{100})$.

Exercise 3.5

1. Yes. An essential aspect of the control chart is that each measurement, or batch of measurements, is charted as soon as it is available, so the data are collected and plotted in sequence — a time series.

2. Yes. Control charts help to define 'exceptions' in an objective way. Very good or very bad outcomes are worth noting. The timing of these exceptions often reveals some causes of them, other than random variation.
3. In pre-computer days, its ease of computation was important, but this is no longer the case. The range, *R*, is very sensitive to outliers in a batch, and this is useful in quality control since extreme observations are a useful indication of problems in the process. In fact, it is often the timing of observed trends or shifts that provides a clue to the causes of our process veering out of control.

Chapter 4

Exercise 4.1

1. No. We must preserve the paired linkage between the two measurements, either in a table or a scatter diagram, to examine this association.
2. Yes. Sorting is a useful way of improving the readability of a tabular display. After sorting the rows so that the Business expenditures are increasing as we read down the table, then the trend in the Government and Universities expenditures will indicate whether it is positively or negatively associated with the Business expenditure variable; it may even be suggestive of a more complicated relationship when it is present. In this case, the weak positive association will be evident.
3. There is no way the number of cycles to failure could cause a certain amplitude of stress. However, it is conceivable that we might want to predict the stress, given the cycles. For example, the data might be used to predict the stress from the cycles to failure, in a situation where the actual stress was unknown and not easily measurable. Then the cycles to failure variable would be useful as a predictor.

Exercise 4.2

1. (a) Research expenditure data: a weak positive relationship
   (b) Stress loads: a very strong dependence of cycles to failure on the stress applied, with the cycles to failure declining dramatically at first, as stress is increased, and then levelling off
   (c) Animal lifespans: a fairly strong relationship between increasing metabolic rate and declining lifespan — the relationship is nonlinear
   (d) Aircraft emissions: a weak relationship, approximately linear, between the variables. There is some evidence that the NO_x does not increase linearly as the CO approaches levels over 100, but that the NO_x levels are lower than a linear relationship based on the lower values of CO would predict.
2. The relationship between two variables may be different in each cluster. For example, a scatterplot of heights and weights of male and female gymnasts may show different clusters for males and females. The relationship of height to weight in each group would be obscured if the clusters were analysed as one group.

   When clusters of data are observed in a scatterplot, it is wise to consider what attribute of the items plotted might account for the clustering. If the data can be split into meaningful groups corresponding to the clusters, we can study the relationships between the two variables plotted for each cluster separately. But, first, we have to notice the clusters.

3. The association is positive within the main cluster, and there is no information in the single outlier about the association in that cluster. It would certainly be unwise to conclude that the association is negative between the variables, even though a naive look at the scatterplot might suggest this is so. A single measure of association is meaningless if there are distinct clusters.

Exercise 4.3

1. Since conversion from one set of units to another is a linear transformation, the correlation would be exactly the same, 0.6.
2. A correlation that is not in the interval between −1 and +1 is impossible. The reporter or the typesetter must have erred; the value is not reasonable.
3. The correlations are +0.3, –0.9, +0.6 and +0.6 respectively. Notice that it was not necessary to put quantitative scales on the graphs — any scale with the same configuration would produce the same correlation. However, the subjective judgement of correlation can depend on the scaling, at least to the unpractised eye. Did you think the last two graphs had different correlations? If so, you have lots of company. To avoid this pitfall, try to draw your scatterplot so that the variability in each direction is approximately the same.
4. The correlation coefficient summarises the first data set adequately since there are no other patterns in the data (such as clusters, outliers or curvature). To use this summary for the other three data sets would be misleading: for the second and third plots, it would fail to convey the strength of the dependence between the two variables and, for the last two plots, it would fail to indicate the presence of an outlier.

Exercise 4.4

1. By drawing a vertical line at the birth rate 45, and a horizontal line from where it meets the median trace to the 'death rate' axis, the death rate is predicted to be about 17 per 1000. However, a birth rate of 60 per 1000 is out of the range of the data and extrapolation of the median trace could be very inaccurate.
2. No. The smoothed or fitted line is first produced, then the distances of the data points from this fit are computed as residuals.
3. The response variable is usually plotted on the vertical axis, so a vertical residual is a measure of the errors in predicting the response from the explanatory variable. However, if the response is plotted on the horizontal axis, horizontal residuals would have the same interpretation.

   If neither variable can be easily classified as a 'response', and we are simply determining the extent to which the fit represents the point scatter, then perpendicular distances to the fitted line from the points are the appropriate measure of residuals.
4. $\dfrac{Var(y-\hat{y})}{Var(y)} = \left(\dfrac{2}{3}\right)^2 \quad \text{so} \quad r^2 = 1-\left(\dfrac{2}{3}\right)^2 = 0.56$
5. In the original scatterplot of the raw data, the residuals are very small distances, even for the outlier. Therefore, patterns in the residuals and, in particular, the presence of the outlier, cannot easily be detected. To see a graph with the proper scaling makes a big difference to what we can perceive from it, and the residual plot uses a scale that expands the residuals and therefore allows us to see patterns in them more easily.

Exercise 4.5

1. About 23 mm pressure, the same as for boiling point 199.4°F! See the graph in the subsection 'Transforming relationships to linearity', explaining why this is so. The graph shows how misleading extrapolation of a polynomial trend can be.
2. In each, the correlation would be zero. Consider the quadrants centred at the mean of the scatter, and the fact that the contributions of the 1st and 3rd quadrants will exactly balance the contributions of the 2nd and 4th quadrants.
3. Since most students spend time on study when they think it will improve their mark appreciably, few students spend extremely large times studying. Over the *usual* range, the relationship may be *approximately* linear. However, the relationship is likely to flatten out with a large number of hours of study — extra hours will then yield little return. Thus, the relationship will not be exactly linear and, over the whole range of possible hours, a straight line would not represent the true relationship well.

Exercise 4.6

1. The 'unusually cold' cities close to the Great Lakes and highlighted on the scatterplot matrix are further 'north' in the scatterplot than their true geographic position. Similarly, the 'unusually warm' cities on the western coast of the United States are plotted further 'south' than they should be.
2. The surface would slope down towards the north, and would also slope down slightly from west to east. The paper would have a slight lip around the eastern and western coasts, and a fairly high lip on the south, especially in the south-east. The interior 'dip' would be most noticeable around the middle states in the north. If water were poured onto this surface, it would drain out of the middle northern part.
3. The *Setosa* are smaller on all variables. The *Versicolor* are smaller on average than the *Virginica* on all variables but the overlap between *Versicolor* and *Virginica* is considerable. It is necessary to consider the entire four-variable description of these two varieties to properly classify some of these flowers, based on the measurements alone.

Exercise 4.7

1. (a) Average education level and population are both increasing over time, and so would be positively associated.
   (b) Again, there is an increase in most countries of the participation of women in higher education, so this temporal trend would likely induce a positive association.
   (c) If it is true that the people who eat calorie-reduced foods are usually overweight, there would be a positive association.
   (d) Assuming the farmer knows what he is doing, there is likely to be a positive association.
2. (a) There is no sense in which the increase in the population causes the change in the education level, and it is also unlikely that increased level of education has anything to do with increasing population.
   (b) A greater proportion of women coming in will lead to a greater proportion of women graduating, so we can consider this a causal link. (It might be argued that it is not direct, because of the time lag, but we will ignore this argument for now.)

(c) This one is definitely not causal. Reduced calorie foods would not increase a person's weight. The overweight problem would usually have preceded the reduced-calorie diet foods, and so the reverse causal link is impossible in these cases.

(d) This one is likely to be causal. More fertiliser will cause a larger yield.

3. Rush hours tend to involve both cars and buses, so there will be a positive association between these two variables. However, there is no sense in which the number of cars influences ('causes') the number of buses directly. Both numbers are controlled by the sun!
4. Although association can occur when there is no direct causal link, if a direct causal link is present, it will usually be revealed as an association, provided the measurement process is accurate enough. So when biochemists noticed that the accidental occurrence of mould on a Petrie dish was associated with retardation of bacterial growth, they suspected a causal link. This is the way penicillin was discovered. The causal link was later proven by proper experiments, in which the mould was deliberately assigned to certain bacterial culture samples and the reduction in bacterial presence measured.

Chapter 5

Exercise 5.1

1. (a) Categorical
   - (i) type of object: stone, shell etc.
   - (ii) colour
   - (iii) attractive, interesting or both

   (b) Numerical
   - (i) size (diameter)
   - (ii) value to tourist
   - (iii) number of items of this type found previously
2. No! Categorical data are not made numerical by assigning a code to their categories — these numerical codes are just like labels, and have no quantitative connotation. Numerical summaries of this kind cannot be used with categorical data, as if they were numerically meaningful.

Exercise 5.2

(a)

Religious group	Number of adherents (%)
Christians	33.5
Muslims	18.2
Non-religious	16.3
Hindus	13.5
Buddhists	6.0
Atheists	4.3
Chinese folk	2.5
New religionists	2.2
Tribal religionists	1.8
Other	1.6
Total	**100%* (5 575 954)**

*The percentages actually add to 99.9% instead of 100% because of rounding.

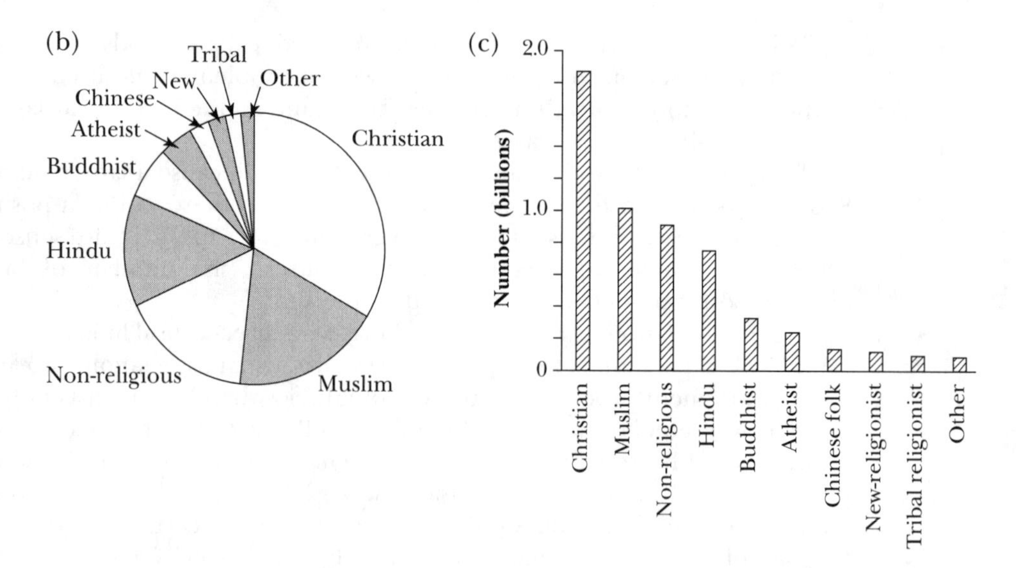

(d) The bar chart facilitates an accurate comparison of any two categories (for example, showing that there are fewer non-religious than Muslims), whereas the pie chart facilitates a visual reading of the actual percentages without referring to a numerical scale (for example, showing that half the world's population are either Christian or Muslim).

Exercise 5.3

1. (a) The two bar charts are shown below.

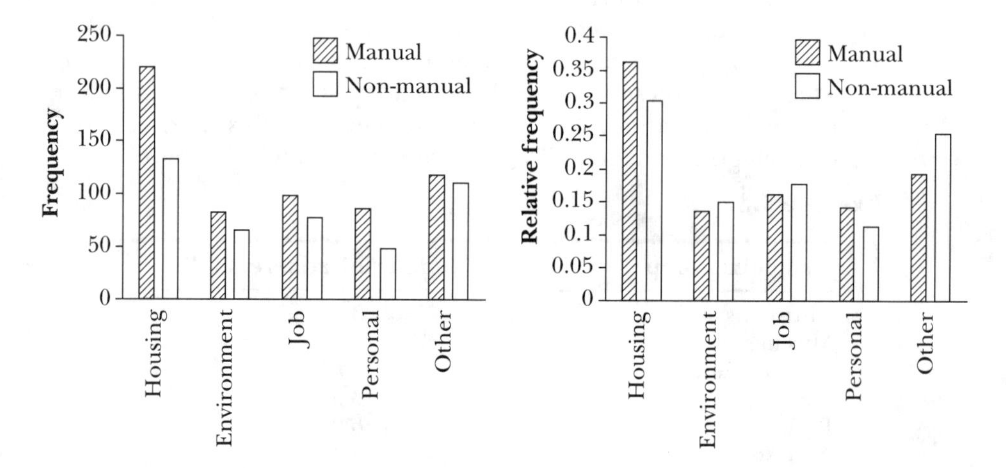

The bar chart on the left shows that there are fewer non-manual than manual households in the survey for each of the reasons. However, the bar chart on the right shows that a bigger proportion of manual than non-manual households moved because of housing and personal reasons and lower proportions moved because of environment, jobs and 'other' reasons.

(b) As we are interested in comparing manual and non-manual households, relative frequencies should be calculated within each of these two groups. It is harder to compare the lengths of the 'middle' categories than in the unstacked bar chart of relative frequencies in question 1, so the stacked bar chart is a poorer display of the data.

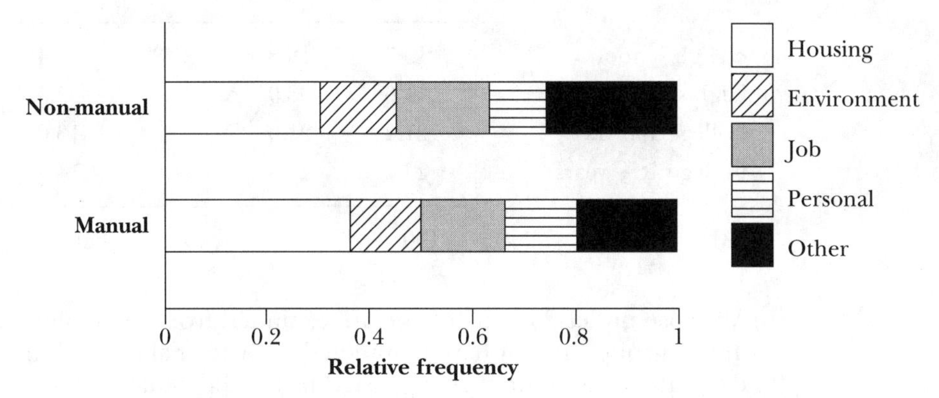

2. The dot chart shows that the number of unreturned cans has remained fairly stable and that the big increase has been in the returned cans. The relative frequency bar chart misses this relevant feature. Both charts have important information that complement each other.
3. By showing the number of cans *not* returned next to the axis in the dot plot, it is easier to display the trend in the number of unreturned cans. The dot plot below shows that the number of unreturned cans has increased only slightly between 1978 and 1989. 1979 stands out as a year in which an unusually large number of cans was not returned.

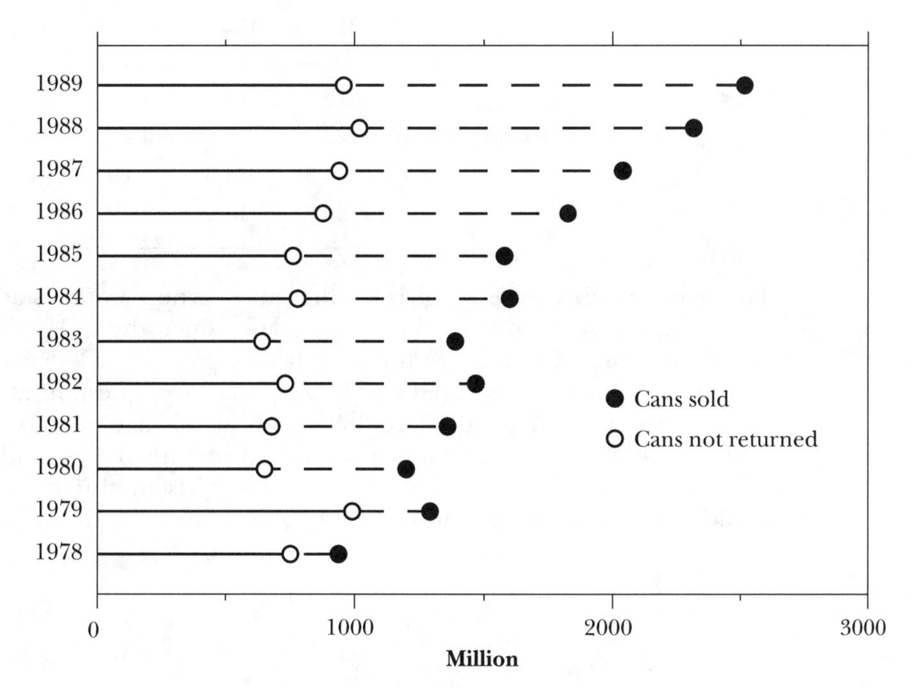

Exercise 5.4

1. (a) Compute percentages based on column totals: that is, $100 \times \frac{5.9}{43.0}$ and so on.

	25–34	35–44	45–54	55–64	≥ 65
< High school	13.7	13.8	21.8	32.4	46.3
High school	41.6	38.0	41.6	39.8	33.0
College 1–3 yrs	21.2	21.0	15.5	13.0	10.2
College ≥ 4 years	23.7	26.8	21.0	14.8	10.5
	100	100	100	100	100

(b) All those under 25 years of age are excluded from the table, since many in this group will not have completed their formal education.

(c) Percentages computed as suggested in (a) show that educational attainment is generally higher for the younger age groups. This reflects the trend over time for people to spend longer these days on their formal education than used to be the case.

(d) To compare the top two rows with the bottom two rows using age group, we need to compute percentages over rows. This can be done either before or after combining the two pairs of two rows, but in either case the percentages will be based on row totals. The answer to the question posed is Yes, based on the table of row percentages:

	25–34	35–44	45–54	55–64	≥ 65	
< High school	16.3	13.3	14.4	19.4	36.6	100
High school	30.4	22.4	16.8	14.6	16.0	100
College 1–3 yrs	35.3	28.3	14.3	10.9	11.2	100
College ≥ 4 years	33.1	30.2	16.2	10.4	9.7	100

2. The age at which 90% of girls in the study group would start to menstruate can only be guessed from the graph, since the girls were not followed over time. The 92.78% for girls in the age range 14 years, 3 months to 14 years, 5 months is based on the response to a question at one point in time, and only 97 girls of the 3898 were of this age. However, the pattern shown by the graph could be smoothed roughly by eye, and we might still guess from the graph that 90% of the girls in this group of 3898 would likely start menstruating by 14 years 4 months, the middle of the interval 14 years, 3 months to 14 years, 5 months.

Exercise 5.5

1. To evaluate the strength of an association between two variables, think about how well we could predict one variable from knowledge of the other variable.
   (a) This is likely to be a very strong association (since men are unlikely to read the magazine).
   (b) This would be moderate (and weakening as the years go by).
   (c) This association is quite strong.
   (d) This would be a very weak association.
   So, a ranking from strongest to weakest would be (a), (c), (b), (d).
2. (a) The sex of the individual will causally affect the magazines read.
   (b) Although there are other factors involved, sex affects society's expectation of occupation, which also affects occupation. The relationship is therefore causal.
   (c) Although there is a relationship, neither variable directly affects the other. Other factors, such as social background and sex are likely to affect both variables, so the relationship is not causal.
   (d) Any relationship (if there is one) is not causal.
3. No. Getting older certainly does not cause a person's educational level to change — at least it certainly does not cause it to decrease! The passage of time affects both measurements.

Chapter 6

Exercise 6.1

1. (a) Exploration only
   (b) Presentation
   (c) Presentation
   (d) For a general audience, exploration only. However, to a sophisticated audience that knows how to interpret it, a box plot can also be a very effective presentation technique.
2. A tabular display is effective when the message in the data can be conveyed to the reader with a small number of values. Tabular displays are therefore often effective summaries for small or simple data sets.
3. While most (63%) of the adolescent girls wanted to lose weight rather than maintain or increase their weight, the tendency was much stronger among white girls (66%) than among African-American girls (47%).
4. Title: *Relationship between January temperature and latitude in the USA.* Caption: There is a general trend for January temperature to be lower in cities further from the equator. However, the cities on, or close to, the West Coast are warmer than would be expected from their latitude, and a group of cities in the centre, near the Canadian border, is colder than would be expected from their latitude. In particular, Juneau (Alaska) is considerably warmer than expected. The two island cities (Honolulu in Hawaii and San Juan in Puerto Rico) have the temperatures you would expect from their latitudes.

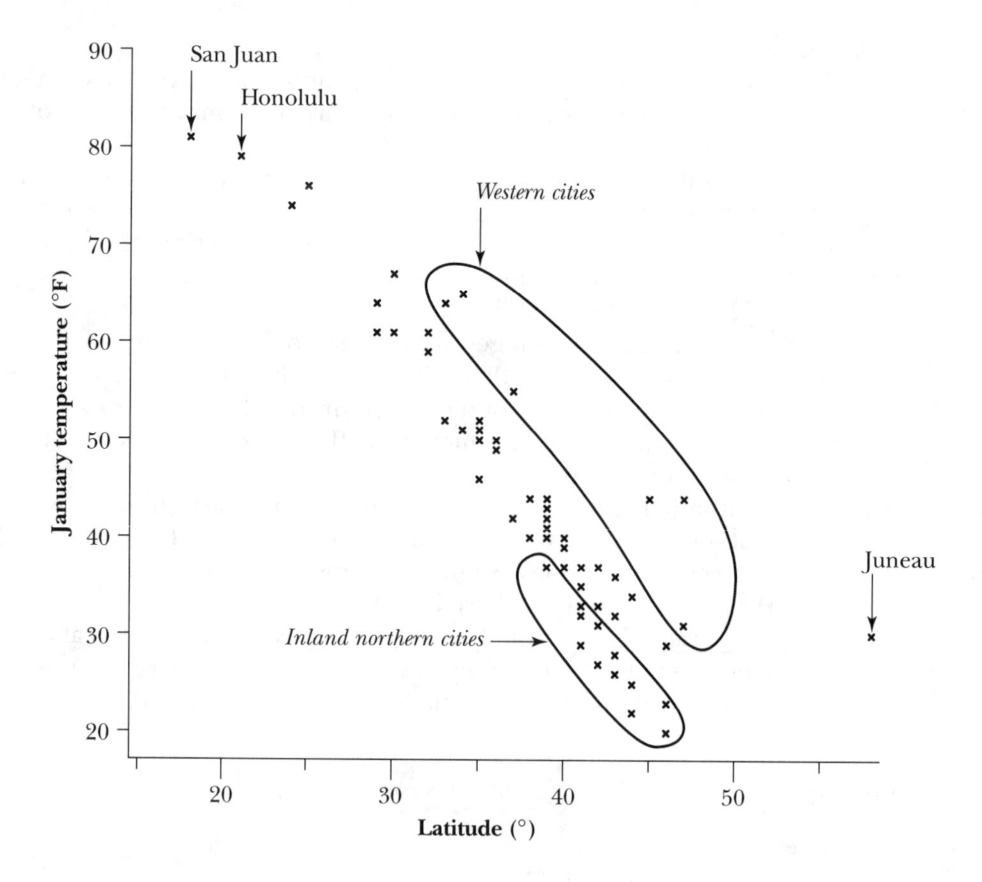

Exercise 6.2

1. Smoothing is based on the assumption that there is a 'smooth' trend in the series over time — the signal in the series. The distances of points from this trend, the residuals, are assumed to be noise. Large residuals (that is, values much higher or lower than the smooth trend) may have a known cause (for example, a strike or war) and may therefore be part of the signal in the data, but are not captured in the smoothed series.
2. Yes, the order of the data values is discarded. (Note that if the order of the data is important, the jittered dot plot will have discarded information that is *not* simply noise.)
3. (a) No. The original ordering of rows and columns had no significance. The actual counts shown were rearranged, not changed.
   (b) If a display uses an arbitrary choice for some feature, this choice can sometimes be exploited to improve the readability of the display.
   (c) A common practice is to order the categories so that the bars increase or decrease as the eye moves across the categories from left to right.
4. The countries/regions are alphabetically ordered, so ordering them by size retains all the information, but allows comparisons to be more easily made.

Country/region	Exports ($ million)
Japan	2660.3
Australia	2608.9
United States	2008.1
Western Europe (excluding UK)	1812.4
Asia (excluding Japan and China)	1738.7
United Kingdom	1036.2
China	820.9
Middle East	427.1
Oceania	413.8
South and Central America	404.9
Eastern Europe (including USSR)	395.7
Canada	261.2
Africa	71.5

Rounding to the nearest $10 000 000 would further clarify the message in the data (see section 6.3). The pie chart below is also clearer from this re-ordering of countries.

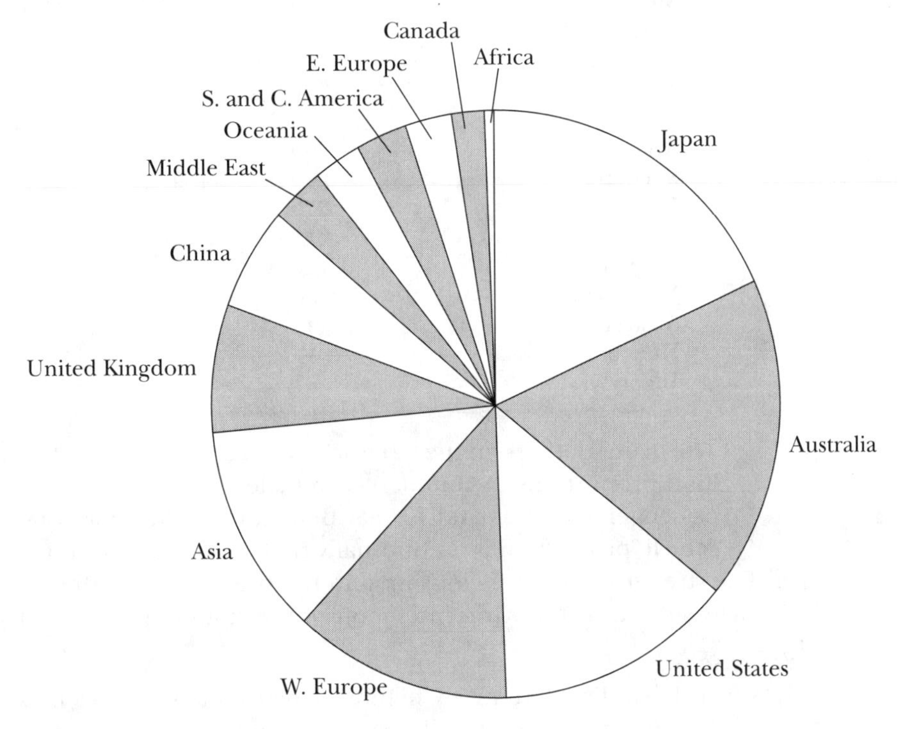

From this pie chart, it is clearer that approximately half of New Zealand's exports went to Japan, Australia and the United States.

Exercise 6.3

1. (a) The number of cars manufactured in a year is unlikely to be known precisely. The figures are unlikely to be accurate to more than the nearest thousand.

(b) The countries should be ordered by their total vehicle production. The commercial figures should then be rounded to the nearest 1000 and the passenger figures to the nearest 10 000.

Country	Passenger vehicles (thousands)	Commercial vehicles (thousands)
Japan	8500	2730
United States	5980	4883
Germany	3750	237
France	2840	319
Canada	1350	889
Korea, South	1590	457
CIS	1210	600
Spain	1510	262
United Kingdom	1380	193
Brazil	1100	288
China	220	1089
Italy	1120	150
Mexico	840	245
Australia	350	131
Belgium	350	57
Taiwan	280	105
India	200	172
Argentina	290	55
Sweden	280	58
Poland	270	38
Czech Rep./Slovakia	210	24
Malaysia	120	0
Netherlands	80	19
Austria	41	4
Hungary	11	5
Yugoslavia	7	1

The main 'outliers' in the table are China and India, which have unusually high proportions of commercial vehicle production (compared to their passenger vehicles) and Korea, Belgium and Malaysia whose passenger vehicle production is an unusually high proportion of total production.

2. The strength of the relationship and the presence of outliers. Both of these factors can affect our interpretation of the resulting smoothed series.

Exercise 6.4

1. Vertical grid lines are rarely helpful. Horizontal grid lines make it easier to read off precise values, but this is rarely important in presentation graphics. However, one or two judiciously placed, fine horizontal lines may help compare values at different parts of a series with several peaks and troughs.
2. It provides a lot of information in a single display — the departure and arrival times at 13 stations for about 30 one-way trips. Also, the graph needs no explanation — its structure is obvious. (We must allow for the fact that the display would be clearer in its original size.)

Exercise 6.5

1. You would need to know the inflation rate and whether there was an increase in the number of students. If the inflation rate was 3% and there was a 2% increase in student numbers, funding *per student* would have *decreased* by 1% in real terms.
2. Although the height doubles in (a), the area of the bin quadruples and, if it is treated as a three-dimensional object, its volume becomes eight times as large. Although the 'volume' of the bin is doubled in (b), its area is less than doubled, so there is some visual ambiguity. Changes should never be represented by rescaling two-dimensional or three-dimensional objects. Although (c) and (d) correctly double the number of bins, the extra distance between them in (d) suggests more than doubling; (c) is the best display. Of course, a standard bar chart would be just as effective!
3. There is a potential for the change to be interpreted by the change in the size of the white region of each cigarette — a greater percentage decrease than the true decrease, as illustrated in the two diagrams below.

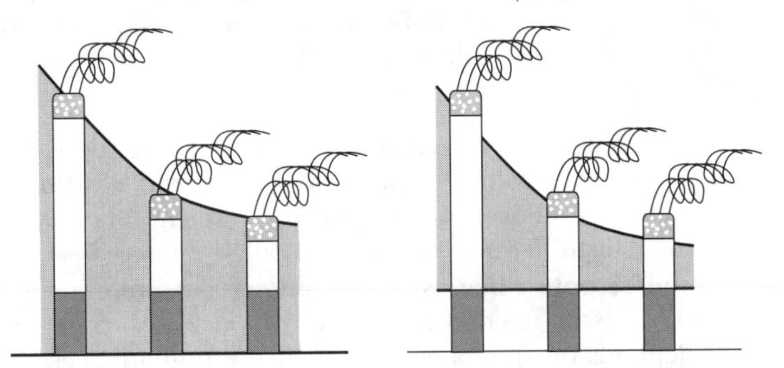

Exercise 6.6

1. The plot is shown below.

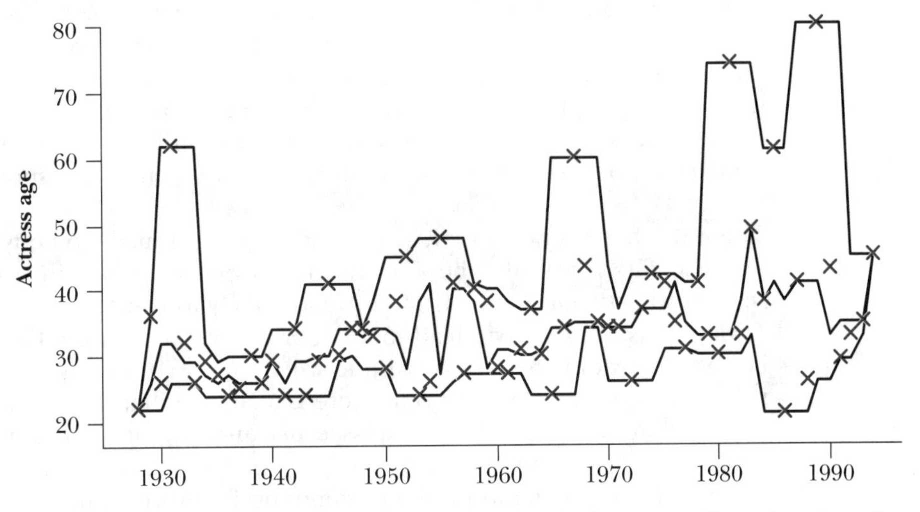

   It accentuates the changing spread of actress ages, as well as showing the increasing trend.
2. Napoleon's troops should have stayed at home. Most of them perished.

Chapter 7

Exercise 7.1

Anecdotal data can suggest hypotheses. (To have an idea of what *might* be true or what some people *think* is true before the idea has been properly tested with reliable data is very valuable. No matter what outcome a formal investigation of the idea is, anecdotal data are of interest.)

Exercise 7.2

1. In an experiment the investigator sets the experimental conditions under which measurements are made whereas, in an observational study, the investigator makes measurements under conditions that have already been set at levels by nature or by the subjects themselves.
2. In an observational study, variables which are correlated should not be interpreted as having a causal link, since many further factors may change along with the observed variables, and any of these may affect both correlated variables. On the other hand, in a well-designed experiment, the investigator can eliminate the influence of extraneous factors, and may then infer causal links reliably when they are present.

Exercise 7.3

1. The people who were asked their voting intentions represented only one sector of the voting population — the more affluent sector. The large sample size reduced sampling error, but still retained a large non-sampling error. This non-sampling error is called *coverage error.*
2. If the response bias is strong enough, a sampling proportion of nearly 100% may be needed to reduce response bias to negligible proportions — it depends on the extent to which the potential response is related to the tendency to respond. If this relationship is strong, the respondents and non-respondents will have very different potential responses, and the respondents alone will present very biased information.
3. The response rate is likely to be lower with a telephone survey, but the sample size will be greater since it is quicker and therefore cheaper to telephone people than to interview them in person. The telephone survey will be more accurate if the non-respondents would have given similar responses to the respondents — that is, if the response bias is low.
4. There will be coverage error, since readers are not likely to be typical of all females. There will also be sizeable non-response error, since only those readers with strong views on the issue are likely to respond.
5. A long questionnaire will lead to a larger non-response rate. Further, it will be costlier, leading to a lower sample size. However, unless a small questionnaire has been well thought out, there is a chance that the analyst will later find that an issue cannot be addressed because relevant questions were not asked.
6. When a list of population units (a sampling frame) cannot be created. For example, it is not possible for a survey company to obtain a complete list of all adults of Chinese descent in a country, or to obtain a complete list of all households without a car.

Exercise 7.4

1. (a) A subjective rating of quality might be given by a tester on a scale of 1 to 10. Alternatively, if the 'lightness' of the cake was important, its height might be used as an objective measure.
   (b) The blocks would be the different batches of 6 cakes that are baked simultaneously.
   (c) *Amount of sugar*: Suggestion in recipe, 20% more and 20% less.
   Baking time: Suggestion in recipe, 10% longer and 10% shorter.
   Temperature: Suggestion in recipe, 10°C hotter and 10°C cooler.
   (d) Each block (that is, batch of 6 cakes) should contain the same number of cakes baked with each level of the factor (for example, two cakes with the amount of ingredient suggested, two with 20% more and two with 20% less).
   (e) The cakes in each block should be randomly positioned within the block — they should be randomly positioned within the oven. If some parts of the block are unexpectedly different from others (for example the top shelf of the oven), randomisation makes it unlikely that a treatment will always be advantaged (or disadvantaged) by its position.
2. Each cake would be marked with a different number. Only the experimenter would be able to tell the treatment received by any particular cake. Neither the person administering the taste test nor the subject assessing a cake would know the treatment, so this knowledge could not influence the cake's rating.
3. The purpose is to make the comparisons of treatments robust to differences in the experimental units and to unintended differences in the application of the treatments. Such differences will tend to occur 'at random' in the treatment groups (that is, the groups of experimental units receiving a particular treatment) and will not systematically bias the comparisons among the treatments.
4. Eight. The combinations of the three factors, denoted *S*, *W* and *B*, are shown below.

	***S* low**		***S* high**	
	***W* low**	***W* high**	***W* low**	***W* high**
***B* low**	×	×	×	×
***B* high**	×	×	×	×

5. (a) Varieties are confounded with differences between the fields, so it would be impossible to tell whether differences between the fields or differences between the varieties would have caused any observed differences in yield. Further, there is no replication, so there is no way to tell how accurate each estimate of yield is.
   (b) Randomly allocate each variety to two of the sub-plots in each field. Differences between variables will not be confounded with differences between fields (since each variety is used the same number of times in each field). Randomisation of the varieties within each field lessens the chance of unforeseen fertility differences between sub-plots being confounded with the varieties (that is, one variety getting consistently better sub-plots). Replication allows assessment of accuracy.
6. 'That's not an experiment, that is an observational study.'

Chapter 8

Exercise 8.1

1. (a) Real. In principle, a list could be compiled of all the potential voters.
   (b) Hypothetical. The measurement from the soil sample will vary due to inaccuracies in the measurement process, so the possible outcomes cannot be considered to exist.
   (c) Real. These leaf areas really exist if we are considering a particular tree. However, if we had referred to leaf areas of a particular species of tree, *in general*, the population would have been hypothetical because of the infinite population of possible leaves.
   (d) Hypothetical. The process of forging the wheels can produce an infinite number of wheel diameters, even though they are designed to be the same. This infinite population can only be imagined.
2. (a) The time that a particular aeroplane will stay aloft is dependent on so many uncontrollable conditions that it would be almost impossible to get the same measurement twice in a row. The thing that would be useful to describe from our trials is not a fixed number, but actually a whole distribution. We want to know what the usual range of times aloft is. The distribution in which we are interested is nevertheless not real, but hypothetical.
   (b) The river is flowing, and there is no fixed real population that is being described. Rather, there is a process going on that we hope to be able to describe. One way is to imagine a hypothetical population that underlies the samples taken, and that provides a way of describing the outcome of the complex process that affects the acidity in the river. In this example, we would likely hypothesise two populations, one relating to the river upstream and one downstream.
   (c) We conceive of the population of yields that would result if we were able to plant the whole batch of seeds in the laboratory (under the same controlled conditions as the 25 were subjected to). We want to claim some knowledge of these yields before actually using the seed. To be able to relate the 25 measured values to the population of interest, we conceive of the population of interest. It is not real since most of the seeds are never going to be planted under the artificial conditions of the laboratory.

Exercise 8.2

1. (a) Suppose that your sample resulted in 11 out of 20 pages containing graphics. If the full text had 400 pages, a commonsense guess of the number of pages among the 400 that have charts would be $400 \times \frac{11}{20}$, or about 220, pages.
   (b) If we are limited to a sample of size 20, there is no point in looking at the same page twice — we would expect to get more information by looking at 20 different pages. So, repeats are avoided to maximise the information we get. This is sampling without replacement if we ignore repeats.
   (c) No. We could not sample pages '111', '323' or other pages with repeated digits. In the terminology of section 7.3, this is *coverage error.*

(d) Yes. From each sampled page, we would record the number of italicised words. The total count from the 20 pages could be scaled up in a similar way. For example, if we detect 7 italicised words, the 400 page book would have approximately $400 \times \frac{7}{20}$, or about 140, italicised words.

Note that the previous question scaled up the proportion of pages with a characteristic, whereas this question scales up the mean number of items per page. However, the principle of scaling up what we get in the random sample is the same.

2. (a) The infinite population of possible filled bottles is useful to model variability in the filling process.
   (b) The mean $\mu = 2010$ millilitres and the standard deviation $\sigma = 5$ millilitres are the parameters.
   (c) We would expect the volume of drink to be in the range $2010 \pm 3 \times 5$ millilitres, or from 1995 to 2025 millilitres, since 99.7% of the values are within three standard deviations of the mean for any normal distributions.
3. (a) normal (b) exponential (c) rectangular
4. The sample maximum is much more variable between the two simulations.

Exercise 8.3

1. (a) The maximum is most variable, the mean is relatively stable and the minimum is least variable. (These features depend on the particular population — in this case, the population lower bound of about 100 limits the variability of the minimum, whereas there is no such bound for the maximum.)
   (b) No. With four smaller and three larger, there is no systematic difference.
   (c) Yes. The range of the mean rolls is 248.0 (572.4 to 820.4) for the seven samples of size 10. This is greater than the range of 115.2 (583.8 to 699.0) for the seven samples of size 40. (The seven mean rolls have standard deviations of 96.9 and 40.8 respectively.)
   (d) No. With three smaller and four larger, there is no systematic difference.
   (e) Yes. The range of 262.6 for the seven samples of size 10 is greater than the range of 127.7 for the seven samples of size 40.
2. Although the simulated samples will have means and standard deviations similar to the real data, the simulated data distributions have lower maxima and minima and they are more symmetric — the real data were skewed to the higher values, with a bunching of the smaller values in the 10 to 26 kg range.
3. It does not mean much. In this section, you were warned about reading too much into the pattern of a sample distribution, and this is an example of how misleading samples can be. In the plot, the two distributions of sample values were actually simulated from a single population defined by a particular normal distribution. So, the apparent gap in the first sample is an accident of the sampling process — it signifies nothing in this context.

Exercise 8.4

1. (a) The central limit theorem states that the standard deviation of the sample means will be $(\sqrt{10})^{-1}$ times the standard deviation of the individual values or, in other words, the standard deviation of the individual values will be $\sqrt{10} = 3.16$ times the standard deviation of the means. In the simulation, this ratio was $\frac{0.1136}{0.0386} = 2.94$, which is close.
   (b) A similar result to the central limit theorem might be expected for other measures of spread, such as the interquartile range (the length of the central box in a box plot). The ratio of the larger to the smaller length is approximately 3.0, showing that this measure of spread is also approximately $\sqrt{10}$ times greater for individual measurements.
   (c) The central limit theorem says that means have a distribution that will be approximately normal, and the normal distribution is symmetric.
2. (a) If you went to a lot of trouble to do this accurately, you would find that the proportions for $n = 1, 2, 4$ and 8 are 0.86, 0.91, 0.96 and 0.99 respectively.
   (b) In each case the mean is exactly 1. This is most clear from the $n = 8$ chart, in which the distribution shown, which is approximately normal, is almost symmetric about 1. The charts are displaying the distribution of the sample mean, and it was stated in the text that this distribution always has mean μ, the population mean.
   (c) It is becoming more concentrated about the mean. The reason is that the standard deviation is the population standard deviation divided by $\sqrt{n}$, and n is getting larger.

Exercise 8.5

1. When it is a mean of data that consists only of 0's and 1's. The mean is then the proportion of 1's.
2. The mean and standard deviation of a sample proportion based on a sample size of n, are:

$$\mu_p = \pi \text{ and } \sigma_p = \sqrt{\frac{\pi(1-\pi)}{n}}.$$

   The central limit theorem states that its distribution will also be approximately normal when n is large. This allows us to use properties of the normal distribution to describe sample proportions.
3. The number of asthmatics in a sample of $n = 30$ children must be an integer, so only certain proportions are possible:

$$\frac{0}{30}, \frac{1}{30}, \ldots, \frac{30}{30}.$$

   However, the normal distribution provides a useful approximation. The sample proportion of asthmatics (in a class of $n = 30$) has approximately a normal distribution with mean and standard deviation:

$$\mu_p = 0.25 \text{ and } \sigma_p = \sqrt{\frac{0.25(1-0.25)}{30}} = 0.079.$$

   We would therefore expect 95% of samples (classes) to have a proportion of asthmatics that lies within two standard deviations of the mean; that is, $0.25 \pm 2 \times 0.079$ or 0.09 to 0.41. In terms of actual numbers of asthmatics, we therefore expect between 2.7 and 12.3, which means in whole people, 3 to 12.

Chapter 9

Exercise 9.1

1. Both techniques are inferential techniques; that is, ways of inferring something about a population based on a sample from that population. Estimation summarises the information in sample data about a population parameter, whereas in hypothesis testing the sample data are used to evaluate the credibility of a claim about the population, usually a claim about the value of some population parameter. (Sometimes non-numerical characteristics such as 'two humps' or 'normal' are of interest, and these would be characteristics that would not be thought of as 'parameters' but for which we might still want to test.)
2. Inferential statistics refers to a collection of strategies for describing a population based on a sample from that population. Descriptive statistics does not require the data to be a random sample from a population, and simply involves describing characteristics of the data. (However, the strategies of descriptive statistics are a useful part of inferential statistics. Sample data should always be examined with numerical and graphical summaries before inferential statistics is used.)
3. It allows us to obtain information about a population without measuring the whole population, and this is usually a great saving of time and money.
4. The sample is the collection of 10 measurements. The population is the collection of measurements for the entire collection of meat packages on display. The investigator is interested in the mean, μ, of this population.

Exercise 9.2

1. Both are intended to express information in a sample about the value of a parameter. The point estimate is a single value which is computed from the sample to be as close as possible to the parameter; it does not, however, give any indication of its accuracy. In contrast, an interval estimate is an interval of values that is intended to enclose the parameter value and its width describes the accuracy of estimation.
2. The mean, μ, and standard deviation, σ, of a numerical population and the proportion, π, in some category of a categorical population.
3. (a) The confidence interval is:

$$\bar{x} \pm t_{n-1}\frac{s}{\sqrt{n}} = 330 \pm t_9\frac{30}{\sqrt{10}} = 330\pm2.262\frac{30}{\sqrt{10}} = 330 \pm 21.46$$

   or 308.5 to 351.5.

   (b) We are 95% confident that the mean fat content in the shipment will be between 308.5 and 351.5, so we should be sceptical about the shipper's claim that the mean is 300 g.
4. (a) The sample data are summarised by:

$$n = 16 \qquad \bar{x} = 50.81 \qquad s = 0.40$$

   so the 95% confidence interval for the average box weight is:

$$\left(\bar{x} \pm t_{15} \times \frac{s}{\sqrt{16}}\right) = \left(50.81 \pm 2.131 \times \frac{0.40}{\sqrt{16}}\right) = (50.81 \pm 0.21)$$

   or 50.60 kg to 51.02 kg.

(b) A 95% confidence interval for the 'free' apples received would be

$$(50.60 - 50) \times 200 = 120 \text{ kg to } (51.02 - 50) \times 200 = 204 \text{ kg.}$$

(c) The bad apple loss from the shipment is the difference between the weight received and the checkout weight, so a 95% confidence interval would be:

$$200 \times 50.60 - 9954 = 166 \text{ kg to } 200 \times 51.02 - 9954 = 250 \text{ kg.}$$

This confidence interval would give the store manager a firm basis for deciding on the amount of effort required to reduce this spoilage.

5. (a) The sample proportion of defectives is:

$$p = \frac{7}{100} = 0.07$$

which is a point estimate of the proportion of defectives from the supplier, π. Its standard deviation is:

$$\sigma_p = \sqrt{\frac{\pi(1-\pi)}{n}} \approx \sqrt{\frac{p(1-p)}{n}} = \sqrt{\frac{0.07 \times 0.93}{100}} = 0.0255$$

so that the 95% confidence interval is:

$$0.07 \pm 2 \times 0.0255$$
$$\text{or } 0.018 \text{ to } 0.122.$$

(b) We estimate the number of defectives in the whole batch to be the number of defectives in our sample of 100 (that is, 7), plus a proportion, π, of the remaining unsampled 9900 disks:

$$7 + 9900 \times \pi.$$

The range of defective disks that might reasonably exist in the full batch is therefore:

$$(7 + 9900 \times 0.018) = 185 \text{ to } (7 + 9900 \times 0.122) = 1215.$$

This is an argument for a big discount!

(c) No. The warning about the use of this approximation when p is too close to 0 or 1 would suggest this technique is not acceptable. The proper method in this case is beyond the scope of this text.

6. False. The 70-95-100 rule says that approximately 95% of data values will be within two standard deviations of the mean. Using the sample mean and sample standard deviation as estimates of the population values, the interval:

$$\bar{x} \pm 2s$$

will therefore include approximately 95% of data values. This does not depend on the sample size.

However, the 95% confidence interval depends on the variability of the sample mean, which is much less than that of individual values, and decreases as sample size becomes bigger.

$$\bar{x} \pm t_{n-1}\frac{s}{\sqrt{n}} \approx \bar{x} \pm \frac{2s}{\sqrt{n}}$$

The confidence interval will therefore usually include fewer than 95% of the population values.

7. The skewness (or asymmetry) of the distribution, or the quartiles or similar cut-off points, or the presence of outliers or clusters of points.

Exercise 9.3

1. The phrase 'rolls had increased' is interpreted to mean that the average change in the rolls is positive, in the population of all the schools. This is a hypothesis that data from a sample of schools will throw light on, so the null hypothesis we test is that the average roll has not changed, with an alternative that it has in fact increased.
2. Yes, virtually any p-value between 0 and 1 is possible when the null hypothesis is true. A p-value as low as 0.001 would be interpreted as strong evidence that H_0 was not true, but such low values will occur in 0.1% of samples even when H_0 is true. Furthermore, if the alternative hypothesis is true, it is also possible to obtain large p-values, giving us the impression that H_0 is true. (However, we can reduce the chance of this latter kind of mistake by increasing the sample's size.)
3. No. We can only evaluate the credibility of the null hypothesis, which is high if the p-value is close to 1. However, in a test comparing the hypotheses H_0: $\mu = 10$ and H_A: $\mu \neq 10$, a mean of $\mu = 10.001$ or 10.000 01 would be practically indistinguishable from $\mu = 10$, so it is never possible to *prove* that the alternative hypothesis is not true. 'High credibility' is not the same as 'proven true'.
4. The rectangular distribution between 0 and 1.
5. (a) When the null hypothesis is true, the p-value will take all possible values (those in the range 0 to 1) equally often; that is, the variation due to sampling error will cause the p-value to have a rectangular distribution. A p-value as small as, say, 0.05 would therefore occur in only 5% of samples; a p-value as small as 0.01 would occur in only 1% of samples and so on. Very small p-values are therefore unlikely when $\rho = 0$, but are more likely when $\rho \neq 0$. Because of this, it is reasonable to use the occurrence of a small p-value to indicate that the alternative hypothesis is true.

   (b) A p-value as low as 0.0001 would only occur in 0.01% of samples if ρ was zero — very unlikely. There is, therefore, extremely strong evidence that ρ is not zero and that steam usage is therefore related to atmospheric temperature at the plant.
6. (a) $p = 0.80$ and the standard deviation of p is estimated as:

$$\sigma_p = \sqrt{\frac{\pi(1-\pi)}{n}} \approx \sqrt{\frac{0.80(1-0.80)}{25}} = 0.080$$

   so a 95% confidence interval is approximately:

$$0.80 \pm 2 \times 0.080 = 0.64 \text{ to } 0.96.$$

(b) If π is the proportion of economics graduates in full-time employment in the population of alumni, the test compares the hypotheses:

$$H_0\colon \pi \geq \frac{1}{2}$$

$$H_A\colon \pi < \frac{1}{2}.$$

We find:

$$\sigma_p = \sqrt{\frac{0.5 \times 0.5}{25}} = 0.1.$$

The z-score for the test is therefore:

$$z = \frac{0.80 - 0.5}{0.1} = 3.0.$$

The p-value for the test is the probability of obtaining a smaller proportion than that observed, 0.80:

$$\text{p-value} = \text{Probability}(z \leq 3.0) = 0.9987.$$

There is no evidence against the null hypothesis that the majority are in full-time employment.

(c) Strictly speaking, they only apply to the population of graduates from which the sample was selected. A generalisation may be valid if other institutions are like the one sampled, but we must not say this broader generalisation is supported by data. This step is more an exercise of judgement.

(d) Yes, those who cannot be contacted are less likely to be in full-time employment and are likely to have lower salaries; those who are in good jobs are more likely to be contactable and are more likely to be willing to respond. As the response rates are different for the two groups, this could also bias comparisons.

(e) Yes, because respondents may well inflate their salaries to the interviewer and may not want to admit being unemployed. However, since this is likely to appear in both groups, comparisons are likely not to be seriously biased.

Exercise 9.4

1. A 95% confidence interval for the difference of two sample proportions is required. Since:

$$p_1 = \frac{20}{500} = 0.04 \text{ and } p_2 = \frac{7}{100} = 0.07$$

we use the point estimate:

$$p_2 - p_1 = 0.07 - 0.04 = 0.03$$

as the centre of our confidence interval. The 95% confidence interval is therefore:

$$p_2 - p_1 \pm 2 \times \sqrt{\frac{p_1(1-p_1)}{n_1} + \frac{p_2(1-p_2)}{n_2}} = 0.03 \pm 2 \times \sqrt{\frac{0.07 \times 0.93}{100} + \frac{0.04 \times 0.96}{500}}$$

$$= 0.03 \pm 0.054$$

or between −0.024 to +0.084.

The interpretation is that while a reduction from 7% defective to 4% is encouraging, it is not convincing evidence that the proportion defective has decreased.

2. (a) A 95% confidence interval for the difference in population means is:

$$\bar{x}_2 - \bar{x}_1 \pm t_v \times \sqrt{\frac{s_1^2}{n_1} + \frac{s_2^2}{n_2}}$$

where $v = \min(50 - 1, 50 - 1) = 49$. The confidence interval is therefore:

$$8750 - 7500 \pm 2.093 \times \sqrt{\frac{770^2}{50} + \frac{660^2}{50}} = 1250 \pm 300.$$

We would therefore be 95% confident that the mean living expenses were between \$950 and \$1550 higher at Roger's university.

(b) One way to answer this question is to determine how much of a difference between the sample means would cause the 95% confidence interval for the difference in the population means to just miss 0. (This shift would also lead to a p-value of 2.5% for a test of $H_0\colon \mu_{Roger} = \mu_{David}$ against $H_A\colon \mu_{Roger} > \mu_{David}$.) If the difference between the sample means is greater than \$300, the distance from the sample mean difference to the edge of the confidence interval, then we would therefore have moderately strong evidence that Roger's expenses are greater than David's.

3. (a) A test is required. The estimated difference of proportions is:

$$p_2 - p_1 = 0.80 - 0.64 = 0.16$$

and its standard deviation is estimated to be:

$$\sqrt{\frac{0.80(1-0.80)}{25} + \frac{0.64(1-0.64)}{25}} = 0.125$$

so the z-score for the test would be:

$$z = \frac{0.16 - 0}{0.125} = 1.28.$$

The chance that $p_1 - p_2$ would be at least as far as 0.16 from zero is the probability of obtaining a standard normal value as far from zero as 1.28. From tables, the p-value is therefore about $2 \times 0.10 = 0.20$, so the answer to the question is no.

(b) The numbers of economics and agriculture graduates in the sample are $n_1 = 0.80 \times 25 = 20$ and $n_2 = 0.64 \times 25 = 16$, respectively. The estimate of the difference in means is:

$$\bar{x}_2 - \bar{x}_1 = 25\,228 - 21\,103 = 4215.$$

The standard deviation of the difference is estimated to be:

$$\sqrt{\frac{2925^2}{20} + \frac{3143^2}{16}} = 1022$$

and the number of standard deviations that 4125 is from zero (its z-score) is:

$$z = \frac{4125}{1022} = 4.0$$

which is much larger than usual for a standard normal distribution. (The p-value is less than 0.0002.) In other words, we conclude that agriculture graduates almost certainly do have higher salaries than economics graduates.

4. (a) The mean of the paired differences, ignoring the first four years at Paisley, is 11.56 and the standard deviation of the differences is 9.75, so the t statistic for the test that the mean difference is zero is:

$$\frac{11.56}{\frac{9.75}{\sqrt{50}}} = 8.39$$

and the p-value associated with this t statistic is 0.0000 to four decimal places. The mean wind gust speed is therefore almost certainly higher in Edinburgh than in Paisley.

(b) The 95% confidence interval is:

$$11.56 \pm 2.01 \times \frac{9.75}{\sqrt{50}}$$

so we are 95% confident that the mean Edinburgh maximum annual gusts are 11.56 ± 2.77 mph higher than those in Paisley.

(c) With the two cities 70 km apart and near opposite coasts of Scotland, with wind gusts at least partly a local phenomenon, we would not expect the pairs of data to be highly correlated and this is confirmed in the scatterplot at the start of the section. The pairing would therefore not be much more sensitive to differences than just comparing the means themselves. Also, city effect is so great that even a crude comparison shows the difference.

Exercise 9.5

1. The graph is shown below.

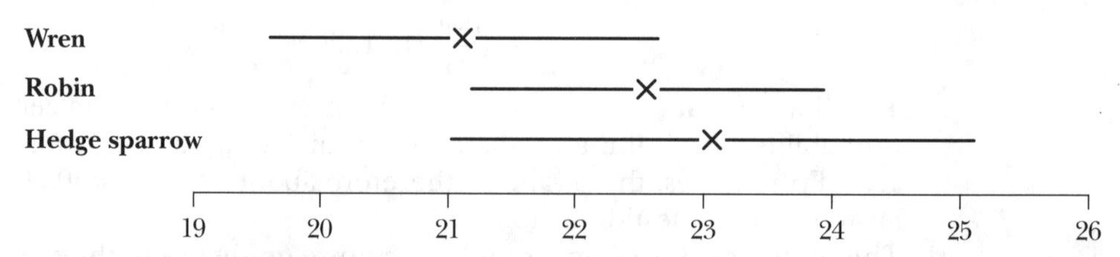

The distributions overlap but the eggs in wren nests do seem quite consistently smaller than those in the other two species. The hedge sparrow and robin nests have distributions which overlap quite a bit although the hedge sparrow's eggs seem slightly larger on average.

2. The standard deviation of the sample mean is estimated to be $\frac{s}{\sqrt{n}}$, so the intervals are narrower, as shown in the diagram below.

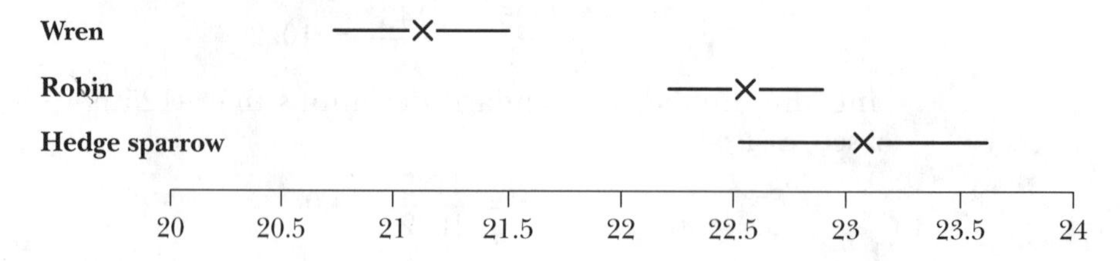

The conclusion is the same except that the focus on the mean, as opposed to the whole distribution, makes the hedge sparrow's mean egg size seem clearly larger than the robin's. The 95% confidence interval for each mean is of the form:

$$\bar{x} \pm t_{n-1} \times \frac{s}{\sqrt{n}}.$$

This has the same form as the intervals drawn above, but with the value 2 replaced by t_{n-1}. As the latter is close to 2 for this example, the intervals are very similar.

3. The pooled standard deviation should be close to the middle of the three standard deviations in the three individual groups and does seem to be a kind of 'average' of the three standard deviations.
4. The standard deviation of the three mean egg sizes is 1.011, whereas those evaluated in question 2 were 0.273, 0.170 and 0.195. The observed value of 1.011 is much larger because the differences between the means are not just caused by egg-to-egg randomness, but are also caused by inter-species variability.
5. If the null hypothesis of equal populations means was correct, the probability of obtaining a p-value less than 0.05 in *at least one* of the three tests would be greater than 0.05. (The probability of getting such a p-value in a *specific* comparison would be exactly 0.05, but there are three such comparisons.) This means that we would be more likely to conclude that there was a difference between at least two of the species than we should be.

Exercise 9.6

1. Yes. Regression analysis must be based on pairs of data values (X, Y), where X is the predictor and Y is the value we want to predict. Such pairs exist for the years 1918–1968. The least squares line fitted to these data can be used to predict Y for values of X not in the data, such as the missing Paisley data for the years 1912–1917.
2. The natural logarithm of 0.6 is –0.51. From the graph, the Ln(Lifespan) appears to be about 2.5, so the predicted lifespan is Antilog(2.5) which is the same as Exp(2.5) = 12.2 years. So, the predicted lifespan is 12.2 years.
3. Using the regression equation:

$$\text{Ln(Lifespan)} = 2.0294 - 0.8279\ \text{Ln(Metabolic rate)}$$

we compute Ln(lifespan) = 2.45 and Exp(2.45) = 11.61 years. Note that the small percentage error in the logarithmic scale becomes a larger percentage error in the original scale.

4. The errors range from approximately –1.5 to +1.5, with most in the range ±0.6. Perhaps ±0.5 would be a typical error in this Ln() scale.
5. Exp (+0.5) = 1.65 and Exp (–0.5) = 0.61 so, in the original units, the prediction is typically in error by a factor of 1.65 high or 0.61 low. Predicting lifespans is not very precise, even with good data on metabolic rate!
6. The equation Ln(Life) = 2.03 – 0.828 Ln(Met) and the line

```
Ln(Met)    -0.8279    0.1619    -5.12    0.000
```

show the numbers mentioned. (The zero p-value is for the test that the slope is zero, and this H_0 is clearly rejected.)

The value $s = 0.6727$ is the number we tried to intuit in question 5 from the graph (we guessed 0.5, which is a bit low as it turns out). It is a measure of the typical prediction error in the Ln scale, but should be transformed to the original scale as in question 5. '`R - sq = 51.1%`' means that 51.1% of the variance in the Ln lifespans of the animals in the data set can be eliminated if the metabolic rate is known.

7. Solving:

$$\text{Ln(Met)} = 0.853 - 0.618\ \text{Ln(Life)}$$

for Ln(Life), we obtain:

$$\text{Ln(Life)} = 1.380 - 1.618\ \text{Ln(Met)}$$

which is quite different from:

$$\text{Ln(Life)} = 2.03 - 0.828\ \text{Ln(Met)}$$

that was obtained from the regression of Ln(Life) on Ln(Met). The lines are different because the least squares lines minimise the prediction errors for different variables (in the vertical direction when the response variable is on the vertical axis).

APPENDIX — TABLES

t Distribution

Quantiles of *t* distribution for 95% confidence intervals

v	t_v	v	t_v	v	t_v	v	t_v
1	12.71	**11**	2.201	**21**	2.080	**40**	2.021
2	4.303	**12**	2.179	**22**	2.074	**50**	2.009
3	3.182	**13**	2.160	**23**	2.069	**60**	2.000
4	2.776	**14**	2.145	**24**	2.064	**80**	1.990
5	2.571	**15**	2.131	**25**	2.060	**100**	1.984
6	2.447	**16**	2.120	**26**	2.056	∞	1.960
7	2.365	**17**	2.110	**27**	2.052		
8	2.306	**18**	2.101	**28**	2.048		
9	2.262	**19**	2.093	**29**	2.045		
10	2.228	**20**	2.086	**30**	2.042		

Standard normal distribution

Cumulative normal probabilities $p(Z \leq z)$ for $z \leq 0$

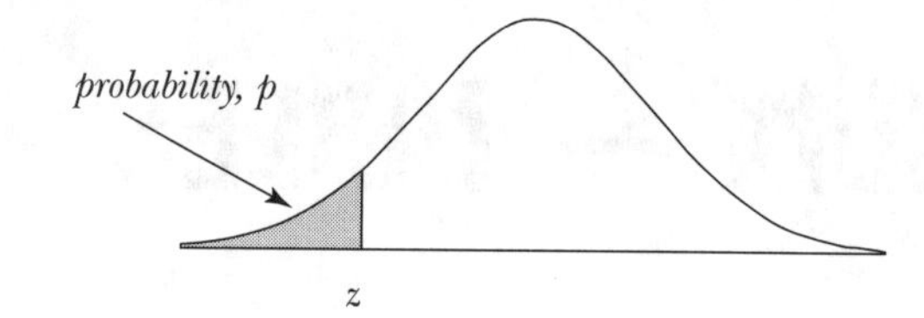

.09	.08	.07	.06	.05	.04	.03	.02	.01	.00	z
									.0002	−3.5
.0002	.0003	.0003	.0003	.0003	.0003	.0003	.0003	.0003	.0003	−3.4
.0003	.0004	.0004	.0004	.0004	.0004	.0004	.0005	.0005	.0005	−3.3
.0005	.0005	.0005	.0006	.0006	.0006	.0006	.0006	.0007	.0007	−3.2
.0007	.0007	.0008	.0008	.0008	.0008	.0009	.0009	.0009	.0010	−3.1
.0010	.0010	.0011	.0011	.0011	.0012	.0012	.0013	.0013	.0013	−3.0
.0014	.0014	.0015	.0015	.0016	.0016	.0017	.0018	.0018	.0019	−2.9
.0019	.0020	.0021	.0021	.0022	.0023	.0023	.0024	.0025	.0026	−2.8
.0026	.0027	.0028	.0029	.0030	.0031	.0032	.0033	.0034	.0035	−2.7
.0036	.0037	.0038	.0039	.0040	.0041	.0043	.0044	.0045	.0047	−2.6
.0048	.0049	.0051	.0052	.0054	.0055	.0057	.0059	.0060	.0062	−2.5
.0064	.0066	.0068	.0069	.0071	.0073	.0075	.0078	.0080	.0082	−2.4
.0084	.0087	.0089	.0091	.0094	.0096	.0099	.0102	.0104	.0107	−2.3
.0110	.0113	.0116	.0119	.0122	.0125	.0129	.0132	.0136	.0139	−2.2
.0143	.0146	.0150	.0154	.0158	.0162	.0166	.0170	.0174	.0179	−2.1
.0183	.0188	.0192	.0197	.0202	.0207	.0212	.0217	.0222	.0228	−2.0
.0233	.0239	.0244	.0250	.0256	.0262	.0268	.0274	.0281	.0287	−1.9
.0294	.0301	.0307	.0314	.0322	.0329	.0336	.0344	.0351	.0359	−1.8
.0367	.0375	.0384	.0392	.0401	.0409	.0418	.0427	.0436	.0446	−1.7
.0455	.0465	.0475	.0485	.0495	.0505	.0516	.0526	.0537	.0548	−1.6
.0559	.0571	.0582	.0594	.0606	.0618	.0630	.0643	.0655	.0668	−1.5
.0681	.0694	.0708	.0721	.0735	.0749	.0764	.0778	.0793	.0808	−1.4
.0823	.0838	.0853	.0869	.0885	.0901	.0918	.0934	.0951	.0968	−1.3
.0985	.1003	.1020	.1038	.1056	.1075	.1093	.1112	.1131	.1151	−1.2
.1170	.1190	.1210	.1230	.1251	.1271	.1292	.1314	.1335	.1357	−1.1
.1379	.1401	.1423	.1446	.1469	.1492	.1515	.1539	.1562	.1587	−1.0
.1611	.1635	.1660	.1685	.1711	.1736	.1762	.1788	.1814	.1841	−0.9
.1867	.1894	.1922	.1949	.1977	.2005	.2033	.2061	.2090	.2119	−0.8
.2148	.2177	.2206	.2236	.2266	.2296	.2327	.2358	.2389	.2420	−0.7
.2451	.2483	.2514	.2546	.2578	.2611	.2643	.2676	.2709	.2743	−0.6
.2776	.2810	.2843	.2877	.2912	.2946	.2981	.3015	.3050	.3085	−0.5
.3121	.3156	.3192	.3228	.3264	.3300	.3336	.3372	.3409	.3446	−0.4
.3483	.3520	.3557	.3594	.3632	.3669	.3707	.3745	.3783	.3821	−0.3
.3859	.3897	.3936	.3974	.4013	.4052	.4090	.4129	.4168	.4207	−0.2
.4247	.4286	.4325	.4364	.4404	.4443	.4483	.4522	.4562	.4602	−0.1
.4641	.4681	.4721	.4761	.4801	.4840	.4880	.4920	.4960	.5000	−0.0

Cumulative normal probabilities $p(Z \leq z)$ for $z \geq 0$

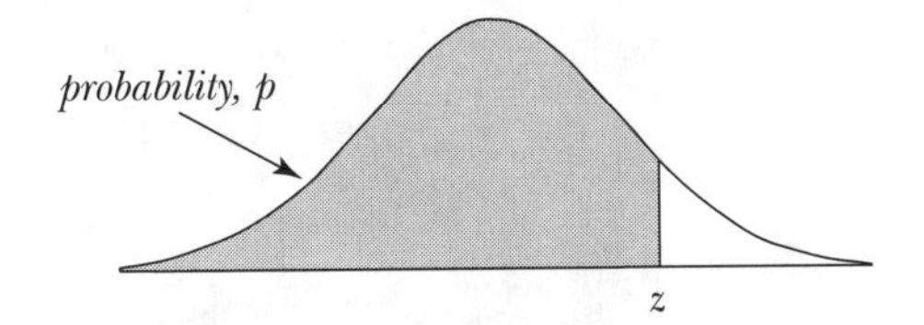

z	.00	.01	.02	.03	.04	.05	.06	.07	.08	.09
0.0	.5000	.5040	.5080	.5120	.5160	.5199	.5239	.5279	.5319	.5359
0.1	.5398	.5438	.5478	.5517	.5557	.5596	.5636	.5675	.5714	.5753
0.2	.5793	.5832	.5871	.5910	.5948	.5987	.6026	.6064	.6103	.6141
0.3	.6179	.6217	.6255	.6293	.6331	.6368	.6406	.6443	.6480	.6517
0.4	.6554	.6591	.6628	.6664	.6700	.6736	.6772	.6808	.6844	.6879
0.5	.6915	.6950	.6985	.7019	.7054	.7088	.7123	.7157	.7190	.7224
0.6	.7257	.7291	.7324	.7357	.7389	.7422	.7454	.7486	.7517	.7549
0.7	.7580	.7611	.7642	.7673	.7704	.7734	.7764	.7794	.7823	.7852
0.8	.7881	.7910	.7939	.7967	.7995	.8023	.8051	.8078	.8106	.8133
0.9	.8159	.8186	.8212	.8238	.8264	.8289	.8315	.8340	.8365	.8389
1.0	.8413	.8438	.8461	.8485	.8508	.8531	.8554	.8577	.8599	.8621
1.1	.8643	.8665	.8686	.8708	.8729	.8749	.8770	.8790	.8810	.8830
1.2	.8849	.8869	.8888	.8907	.8925	.8944	.8962	.8980	.8997	.9015
1.3	.9032	.9049	.9066	.9082	.9099	.9115	.9131	.9147	.9162	.9177
1.4	.9192	.9207	.9222	.9236	.9251	.9265	.9279	.9292	.9306	.9319
1.5	.9332	.9345	.9357	.9370	.9382	.9394	.9406	.9418	.9429	.9441
1.6	.9452	.9463	.9474	.9484	.9495	.9505	.9515	.9525	.9535	.9545
1.7	.9554	.9564	.9573	.9582	.9591	.9599	.9608	.9616	.9625	.9633
1.8	.9641	.9649	.9656	.9664	.9671	.9678	.9686	.9693	.9699	.9706
1.9	.9713	.9719	.9726	.9732	.9738	.9744	.9750	.9756	.9761	.9767
2.0	.9772	.9778	.9783	.9788	.9793	.9798	.9803	.9808	.9812	.9817
2.1	.9821	.9826	.9830	.9834	.9838	.9842	.9846	.9850	.9854	.9857
2.2	.9861	.9864	.9868	.9871	.9875	.9878	.9881	.9884	.9887	.9890
2.3	.9893	.9896	.9898	.9901	.9904	.9906	.9909	.9911	.9913	.9916
2.4	.9918	.9920	.9922	.9925	.9927	.9929	.9931	.9932	.9934	.9936
2.5	.9938	.9940	.9941	.9943	.9945	.9946	.9948	.9949	.9951	.9952
2.6	.9953	.9955	.9956	.9957	.9959	.9960	.9961	.9962	.9963	.9964
2.7	.9965	.9966	.9967	.9968	.9969	.9970	.9971	.9972	.9973	.9974
2.8	.9974	.9975	.9976	.9977	.9977	.9978	.9979	.9979	.9980	.9981
2.9	.9981	.9982	.9982	.9983	.9984	.9984	.9985	.9985	.9986	.9986
3.0	.9987	.9987	.9987	.9988	.9988	.9989	.9989	.9989	.9990	.9990
3.1	.9990	.9991	.9991	.9991	.9992	.9992	.9992	.9992	.9993	.9993
3.2	.9993	.9993	.9994	.9994	.9994	.9994	.9994	.9995	.9995	.9995
3.3	.9995	.9995	.9995	.9996	.9996	.9996	.9996	.9996	.9996	.9997
3.4	.9997	.9997	.9997	.9997	.9997	.9997	.9997	.9997	.9997	.9998
3.5	.9998									

INDEX